Potential for Nature-Based Solutions in Urban Green Infrastructure

Potential for Nature-Based Solutions in Urban Green Infrastructure

Guest Editor

Petra Schneider

MDPI

Basel • Beijing • Wuhan • Barcelona • Belgrade • Novi Sad • Cluj • Manchester

Guest Editor
Petra Schneider
Department of Water, Environment,
Civil Engineering, and Safety Magdeburg-Stendal
University of Applied Sciences
Magdeburg
Germany

Editorial Office
MDPI AG
Grosspeteranlage 5
4052 Basel, Switzerland

This is a reprint of the Special Issue, published open access by the journal *Land* (ISSN 2073-445X), freely accessible at: https://www.mdpi.com/journal/land/special_issues/1W92T3679B.

For citation purposes, cite each article independently as indicated on the article page online and as indicated below:

Lastname, A.A.; Lastname, B.B. Article Title. *Journal Name* **Year**, *Volume Number*, Page Range.

ISBN 978-3-7258-7288-6 (Hbk)
ISBN 978-3-7258-7289-3 (PDF)
https://doi.org/10.3390/books978-3-7258-7289-3

Cover image courtesy of Petra Schneider

Contents

About the Editor

Petra Schneider

Petra Schneider holds a PhD in hydrogeology from the Technical University Mining Academy in Freiberg, Germany, and worked about 20 years in the private sector in the field of mining landscape rehabilitation. In 2015, she was appointed to the professorship of International Water Management at Magdeburg-Stendal University of Applied Sciences. Moreover, she heads the master program and the research group Ecological Engineering. Based on her passion for urban green infrastructure, she focuses on integrating natural systems into cities to improve climate resilience, health, and social equity.

land

Editorial

Potential for Nature-Based Solutions in Urban Green Infrastructure: An Editorial Overview

Petra Schneider

Department of Water, Environment, Civil Engineering, and Safety, Magdeburg-Stendal University of Applied Sciences, Breitscheidstraße 2, D-39114 Magdeburg, Germany; petra.schneider@h2.de

1. Introduction

The Special Issue "Potential for Nature-Based Solutions in Urban Green Infrastructure" is published within the section Urban Contexts and Urban–Rural Interactions and focuses on the increasing importance of ecological engineering (EE) and nature-based solutions (NbS) for shaping sustainable urban and peri-urban landscapes. EE, understood as the integration of ecological principles into engineering design and planning [1–5], has become increasingly relevant in addressing the complex challenges associated with urbanization, climate change, and land-use transformation [6,7]. At its core, it focuses on the design, restoration, and management of ecosystems that are strongly influenced by human activity, with the aim of ensuring their long-term functionality, resilience, and societal benefit [5,8]. Within this framework, NbS provide a key operational and conceptual foundation for developing adaptive and multifunctional green infrastructure systems that simultaneously support ecological processes and human well-being [8].

This perspective builds on ecological and ecosystem science, including principles of resource-efficient system design, reduced material and energy inputs, and optimized nutrient and water cycling [3,9,10]. Equally important is the development of spatially integrated approaches that emphasize multifunctionality, cross-scale connectivity, and the establishment of coherent green and blue infrastructure networks [4,10,11]. These systems rely on decentralized and redundant structures that enhance robustness and adaptability while applying ecologically sound construction and management principles [10,11].

According to [11], ecological engineering provides an umbrella for a range of established and emerging concepts, including nature-based solutions, constructed wetlands and vegetated soil filters, greening of buildings (including roofs, facades, cooling systems, indoor climate, and energy supply), the biosphere, circular systems, and water reuse mechanisms, among others. The emphasis on integrating ecological principles and functions into adaptive engineering design, with the aim of fostering the development of robust, sustainable systems, distinguishes ecological engineering from other disciplines.

The overarching objective of ecological engineering is to ensure sustainability, thereby enabling the designed ecosystem to function effectively and enduringly for the benefit of both society and the environment. This entails the application of insights derived from ecology and ecosystem research, specifically, a holistic approach, the management of internal nutrient cycles with minimal material loss, the minimization of technical energy inputs, and the efficient utilization of energy and materials. Furthermore, the aim is to achieve multifunctional land use, the integration of various spatial scales and the creation of networks, and the establishment of decentralized and redundant decision-making structures, all while simultaneously employing the most ecologically sound construction methods and treatment processes available. Central to this endeavor is the sustainable integration of human-induced modifications into the surrounding ecosystems. Ecological engineering design is thus conceived and executed as an integral component of an ecosystem.

Against this background, EE applies NbS to the planning and development of coherent networks of green and blue infrastructure in both rural and urban contexts. These interconnected systems aim not only to strengthen ecological functions but also to enhance the livability, resilience, and sustainability of human settlements. Building on these considerations, this Special Issue brings together contributions that explore the potential of NbS in urban green infrastructure across different spatial scales and thematic dimensions. Key topics include sustainable settlement ecosystems and urban green infrastructure, multifunctional structural elements implemented as NbS, strategies to promote urban biodiversity, sustainable landscape development in rural areas, circular green and blue systems, and approaches to land recycling and land conversion.

2. Materials and Methods

The contributions included in this Special Issue were selected through a standard double-blind peer-review process in accordance with the editorial policies of the journal Land (MDPI). All submitted manuscripts were first subject to an initial editorial assessment to ensure relevance to the scope of the Special Issue, followed by external peer review by independent experts in the respective fields. Based on reviewers' recommendations and editorial judgment, manuscripts were either accepted, revised, or rejected.

The editorial process aimed to ensure scientific quality, methodological rigor, and thematic coherence with the overarching focus on NbS in urban green infrastructure. Particular attention was given to the relevance of each contribution to ecological engineering approaches, multifunctional landscape design, and urban–rural sustainability transitions.

For the purpose of this Editorial, the published contributions were further analyzed and structured into thematic groups reflecting the key dimensions of the Special Issue. This synthesis approach allows for an integrative overview of the diverse methodological perspectives, spatial scales, and application domains addressed by the individual papers.

Rather than providing a simple summary of each contribution, the Results section synthesizes the findings across papers in order to identify cross-cutting themes, shared methodological approaches, and emerging research directions in the field of NbS and urban green infrastructure. The synthesis focuses on identifying recurring patterns related to (i) social and environmental equity, (ii) governance and participatory processes, (iii) technical design and performance of NbS, (iv) microclimatic and morphological influences, (v) policy implementation and institutional frameworks, (vi) experiential and perceptual dimensions of urban nature, (vii) geotechnical and landscape-scale stabilization processes, and (viii) economic valuation of urban green infrastructure.

3. Overview on NbS

A widely accepted definition of NbS is provided by the International Union for Conservation of Nature (IUCN), which describes NbS as "actions to protect, sustainably manage, and restore natural or modified ecosystems that address societal challenges effectively and adaptively, while simultaneously providing benefits for human well-being and biodiversity" [8]. In particular, NbS are increasingly recognized for their potential to address pressing global challenges, including climate change, food and water security, disaster risk reduction, and risks to human health and socio-economic development [8]. According to [8], the following interdisciplinary measures contribute to NbS: Ecological Restoration, Ecological Engineering, Forest Landscape Restoration, Green Infrastructure, Natural Infrastructure, Ecosystem-based Management, Ecosystem-based Adaptation, Ecosystem-based Mitigation, Ecosystem-based Disaster Risk Reduction, Climate Adaptation Services, and Area-based

https://doi.org/10.3390/land15060971

Conservation. In this regard, these measures are means for sustainable urbanization as all of them contribute to the provision of ecosystem services. However, in the authors' view, EE and green infrastructure are also design tools that support the development of resilient cities (Figure 1).

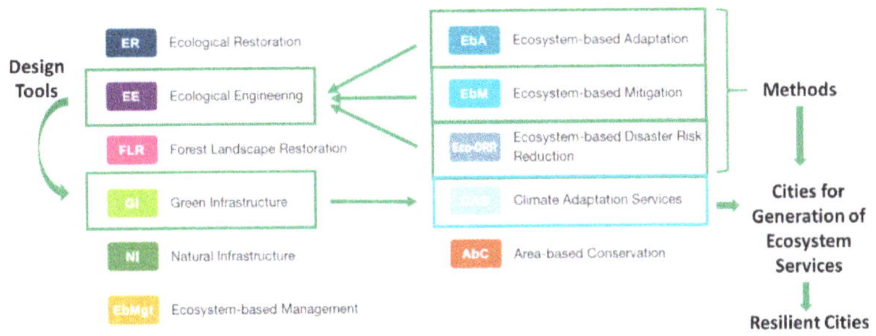

Figure 1. NbS, according to [8], interpreted as design tools for sustainable urbanization (figure modified after [8]). Green squares left—design tools, green squares right—ecosystem-based approaches for ecological engineering, blue square right—effect of green infrastructure for climate adaptation.

Each NbS in the selected papers in the Special Issue represents at least one element mentioned in Figure 1, and moreover refers to the five functional classes for eco-conceptual planning identified by Mitsch & Jorgensen [12]:

- Ecosystems designed to reduce pollution problems. Example: phytoremediation, wastewater wetlands, and bioretention of stormwater to filter out excess nutrients and metal contaminants.
- Mimicked or replicated ecosystems designed to address resource issues. Example: forest restoration, wetland replacement, and the installation of rain gardens in urban areas to enhance shading and optimize cooling in residential zones and cities.
- Restored ecosystems. Example: the restoration of former mining sites, lakes, and watercourses featuring ecologically valuable riparian corridors.
- Ecologically modified ecosystems. Example: selective timber harvesting, biomanipulation, and the introduction of predatory fish to reduce planktivorous fish populations, increase zooplankton levels, consume algae or phytoplankton, and purify water.
- Ecosystems utilized for specific purposes without disrupting ecological balance. Example: sustainable agroecosystems, multi-species aquaculture, and the integration of agroforestry plots into residential properties to generate primary production across multiple vertical strata.

4. Results of the Special Issue (Synthesis of Contributions)

The contributions included in this Special Issue address NbS in urban green infrastructure from a wide range of disciplinary and methodological perspectives, spanning social, ecological, governance, climatic, technical, morphological, policy implementation, experiential, geotechnical, and economic dimensions. Across the papers, several interlinked thematic strands emerge, including environmental justice and social vulnerability, participatory governance and planning processes, the design and performance of urban blue-green infrastructure systems, enabling and constraining implementation conditions, the influence of urban form on microclimatic and human thermal experience, policy translation into practice, perceptual and experiential relationships with urban nature, geotechnical stabilization and landscape restoration, and the economic valuation of urban green infrastructure within land and property markets.

https://doi.org/10.3390/land15060971

A first group of studies emphasizes the social and perceptual dimensions of urban green infrastructure and highlights the importance of equity-oriented planning approaches. In this context, Walter et al. (2025) (Contribution 1) examined housing conditions, perceptions of green space, and climate-related stress in Polish cities using a cluster analysis based on a survey of 963 respondents. Their findings reveal distinct population groups with varying sensitivities to heat waves, levels of environmental awareness, and access to urban green spaces. Notably, one group exhibits high tolerance to heat stress despite unfavorable housing conditions, illustrating the complex and non-linear relationship between vulnerability and lived experience. The study further demonstrated that the use and appreciation of green spaces are strongly influenced by their spatial accessibility and distribution within neighborhoods. This underscores that NbS in urban contexts cannot be understood solely as ecological interventions, but must also address environmental justice, housing inequality, and differentiated exposure to climate risks.

Complementing this perspective, Plüschke-Altof et al. (2025) (Contribution 2) investigated the role of public participation in NbS implementation processes in Northern European cities, focusing on experiences from seven urban planning cases. Their qualitative analysis identified several key tensions that shape participatory NbS planning, including trade-offs between open-ended environmental and participatory objectives, constraints arising from the time-intensive nature of participatory processes within project-based planning frameworks, and differing roles and expectations regarding expert and lay knowledge. These tensions can complicate both the implementation of NbS projects and the achievement of broader sustainability goals, while also pointing toward more adaptive and inclusive governance approaches.

Expanding into the technical dimension of urban NbS, Moeller et al. (2025) (Contribution 3) presented a four-year monitoring study of tree infiltration trenches in Leipzig as a form of decentralized stormwater management. The study evaluated three differently designed systems, highlighting how design variations in retention layers and construction materials influence hydrological performance and tree vitality. The results showed that systems with clay-based retention layers achieve superior water storage capacity and improved vegetation performance, while design refinements in inlet structures are identified as key for future optimization. These findings underlined the importance of integrating long-term monitoring into NbS planning to ensure functional reliability and adaptive improvement.

At a complementary spatial scale, Breulmann et al. (2025) (Contribution 4) investigated the performance of green roofs under extreme rainfall conditions, focusing on stormwater retention and evapotranspiration processes across different substrate configurations. Their lysimeter-based experiments demonstrated that retention performance is highly dynamic and strongly dependent on the saturation state of the system. While fully saturated systems behave similarly to conventional roofs with limited retention capacity, systems with available storage can retain up to 99% of rainfall. The study further highlighted trade-offs between substrate depth, evapotranspiration, and water storage, showing that deeper substrates may reduce immediate evaporation rates but enhance long-term water availability for plant transpiration and overall system functioning. These results emphasized the importance of adaptive design strategies for green roofs to increase urban resilience under increasingly extreme precipitation regimes.

Moving from individual technologies to broader implementation frameworks, Zarei & Shahab (2025) (Contribution 5) provided a systematic review and bibliometric analysis of 90 peer-reviewed studies on NbS in urban green infrastructure. Their work identified key success factors and persistent barriers to implementation, including financial, technical, social, and political constraints. The analysis highlighted the importance of enabling conditions such as spatial justice, governance integration, financial viability, and institutional

https://doi.org/10.3390/land15060971

capacity. It also emphasized the roles of diverse stakeholders, including local governments, private actors, and communities, in the planning, execution, and maintenance of NbS projects. Overall, the study revealed both conceptual convergence and contextual variability in NbS implementation, underscoring the need for context-sensitive and integrated approaches to urban green infrastructure development.

Extending the focus to the interface between urban form, microclimate, and human experience, Qian et al. (2025) (Contribution 6) investigated how multidimensional urban morphology influences thermal sensation in Shanghai. Using a combined framework of thermal sensing feedback and land surface temperature, the study integrates both objective climatic indicators and subjective human thermal perception data. The results demonstrated that key morphological variables, including vegetation density, building density, floor area ratio, impervious surface characteristics, and sky view factor, significantly influence perceived thermal conditions, with seasonal variations revealing contrasting effects between summer and winter. Importantly, the study highlighted that certain urban form parameters exert differing influences on thermal sensation compared to land surface temperature, reinforcing the need for human-centered approaches in urban climate research. These findings contribute to a more nuanced understanding of how urban morphology shapes both physical and perceived thermal environments, with direct implications for the design of climate-sensitive and thermally comfortable urban spaces.

Positioned at the interface between national policy implementation and local urban forestry practice, Magliocco & Sabbion (2025) (Contribution 7) examined the advancement of urban and extra-urban afforestation within the framework of the Italian National Urban Forestry Plan, using the Metropolitan City of Genoa as a case study. Against the backdrop of Italy's National Ecological Transition Plan, the study explored how afforestation initiatives funded through the National Recovery and Resilience Plan contribute to climate mitigation, biodiversity restoration, ecological connectivity, and the broader promotion of NbS in urban regions. The authors identified three central dimensions shaping the effectiveness of urban forestry implementation: ecological planning and site selection, citizen engagement and communication strategies, and the monitoring of environmental impacts. By combining spatial and policy-oriented analysis with quantitative assessments of CO_2 sequestration potential, the study demonstrated both the opportunities and contextual limitations associated with large-scale urban greening programs in densely structured metropolitan environments. The findings further highlighted that the long-term success of afforestation strategies depends not only on planting efforts themselves, but equally on adaptive governance structures, public participation, and transparent evaluation frameworks capable of assessing ecological and social co-benefits over time. In this way, the study contributes to a deeper understanding of how national NbS policies can be operationalized at the metropolitan scale while addressing the complexities of urban ecological transformation.

The experiential dimension of urban nature is further expanded by Liu & Green (2025) (Contribution 8), who explored children's perceptions of nature within urban landscapes in Beijing. Through interviews with children aged 8 to 12 and a photo-based Q-sort methodology, the study identifies three primary ways in which urban nature is conceptualized: ecological, emotional, and visual. The findings highlighted that children's understanding of nature is shaped by a combination of personal experience, education, media exposure, and innate perceptual tendencies. Importantly, the results suggested that urban nature is not perceived uniformly, but is instead interpreted through multiple cognitive and affective lenses. These insights provide valuable implications for urban design and environmental education, emphasizing the need to create urban environments that are both accessible

https://doi.org/10.3390/land15060971

and meaningful to younger populations, thereby fostering early connections with nature in cities.

Extending the scope into geotechnical and large-scale landscape systems, Zedek et al. (2025) (Contribution 9) demonstrated how NbS can be incorporated into slope stability assessment in post-mining landscapes. Using the KurZeS method, the study integrated vegetation-based root reinforcement into spatially explicit stability modeling across large areas. The approach was validated against established geotechnical simulations and empirical land-use data from a former open-cast mining region in Lusatia, Germany. The results showed that vegetation significantly contributes to slope stability, highlighting the potential of NbS to function as measurable stabilizing agents in degraded and reconfigured landscapes. This expands the application domain of NbS from urban infrastructure into large-scale land restoration and hazard mitigation contexts.

Finally, Fauk & Schneider (2025) (Contribution 10) investigated the relationship between urban green infrastructure and land and property values in Magdeburg, Germany. Their econometric analysis identified significant correlations between green infrastructure characteristics, such as tree density and proximity to allotment gardens, and standard land reference values. The results suggested that green infrastructure can exert both positive and context-dependent effects on property valuation, although broader market-driven factors remain dominant. The study highlighted the need for more consistent valuation frameworks that better integrate ecological and spatial planning considerations into land market assessments, as well as improved training and methodological alignment in ecologically informed land valuation practices.

Taken together, these studies illustrate that the successful implementation of NbS in urban green infrastructure and related landscape systems depends not only on ecological design and spatial planning, but equally on social acceptance, governance frameworks, technical performance, enabling institutional conditions, policy translation, urban form dynamics, perceptual and experiential relationships with urban nature, geotechnical stabilization processes, and economic valuation mechanisms across multiple spatial, temporal, and systemic scales.

5. Conclusions

This Special Issue highlights the broad and multidimensional nature of NbS in urban green infrastructure and related landscape systems. Rather than functioning solely as ecological design interventions, NbS emerge as integrated socio-ecological and technical systems shaped by governance structures, planning practices, social perceptions, institutional frameworks, and economic valuation mechanisms. The contributions assembled here demonstrate their applicability across multiple spatial scales, ranging from building- and neighborhood-level interventions to urban-regional systems and large-scale post-mining landscapes.

A central insight is that the effectiveness of NbS depends on the interaction of multiple interdependent dimensions. Social equity, environmental perception, and lived experience influence access to and valuation of urban green infrastructure, while governance and participatory planning frameworks shape implementation processes and long-term management. In parallel, technical design parameters determine the functional performance of NbS, particularly in relation to stormwater regulation, climate adaptation, and ecosystem service provision. These dimensions are further embedded in institutional and policy contexts, as well as in economic valuation systems that increasingly guide land-use decisions. Importantly, the contributions underscore the need to consider the human–environment interface more explicitly. Urban morphology and microclimatic conditions shape both physical environmental conditions and subjective thermal perception, while

https://doi.org/10.3390/land15060971

early-life interactions with urban nature influence long-term environmental awareness and engagement. Extending beyond urban contexts, the application of NbS in geotechnical stabilization and landscape restoration demonstrates their relevance for addressing large-scale environmental degradation and risk mitigation.

Despite the diversity of approaches, several cross-cutting challenges emerge. These include the need for stronger interdisciplinary integration, persistent gaps between planning concepts and implementation practice, and limited comparability in the assessment and valuation of co-benefits. Addressing these challenges requires improved methodological alignment, enhanced institutional coordination, and more adaptive, context-sensitive governance approaches. Future research should therefore focus on systemic integration of NbS across sectors and scales, the development of long-term monitoring frameworks, and the refinement of participatory planning and governance models. Equally important is the advancement of robust approaches to quantify and communicate ecological, social, climatic, and economic co-benefits in support of evidence-based decision-making.

Overall, this Special Issue demonstrates that NbS represent a transformative planning and design paradigm for urban and landscape development. Realizing their full potential requires sustained interdisciplinary collaboration and stronger alignment between research, planning practice, and policy implementation frameworks.

Funding: This research received no external funding.

Conflicts of Interest: The author declares no conflicts of interest.

List of Contributions:

1. Walter, A.; Wanner, A.; Pröbstl-Haider, U. Clustering Comfort: A Cluster Analysis on Housing Conditions and Nature-Based Solutions in Polish Cities. *Land* **2025**, *14*, 1884. https://doi.org/10.3390/land14091884.
2. Plüschke-Altof, B.; Loewen, B.; Calderon, C.; Chebotareva, M.; Tuula-Fjodorov, R.; Gäckle, J. Nature-Based Solutions and Public Participation: Unpacking Tensions in Sustainable City Development in Northern Europe. *Land* **2025**, *14*, 1649. https://doi.org/10.3390/land14081649.
3. Moeller, L.; Bernhard, K.; Kruckow, S.; Wolf, S.; Georgi, A.; Friesen, J.; Mackenzie, K.; Müller, R.A. Tree Infiltration Trenches in the City of Leipzig—Experiences from Four Years of Operation. *Land* **2025**, *14*, 1315. https://doi.org/10.3390/land14071315.
4. Breulmann, M.; Merbach, A.; Bernhard, K.; Moeller, L. Enhancing Urban Resilience: Stormwater Retention and Evapotranspiration Performance of Green Roofs Under Extreme Rainfall Events. *Land* **2025**, *14*, 977. https://doi.org/10.3390/land14050977.
5. Zarei, M.; Shahab, S. Nature-Based Solutions in Urban Green Infrastructure: A Systematic Review of Success Factors and Implementation Challenges. *Land* **2025**, *14*, 818. https://doi.org/10.3390/land14040818.
6. Qian, H.; Wang, M.; Zheng, S.; Qiu, B.; Zhang, F. Does Multidimensional Urban Morphology Affect Thermal Sensation? Evidence from Shanghai. *Land* **2025**, *14*, 769. https://doi.org/10.3390/land14040769.
7. Magliocco, A.; Sabbion, P. Advancing Urban and Extra-Urban Afforestation: A Case Study of the Italian National Urban Forestry Plan in the Metropolitan City of Genoa. *Land* **2025**, *14*, 695. https://doi.org/10.3390/land14040695.
8. Liu, J.; Green, R.J. Nature Through Young Eyes: Exploring Children's Understanding of Nature in Urban Landscapes in Beijing, China. *Land* **2025**, *14*, 624. https://doi.org/10.3390/land14030624.
9. Zedek, L.; Šembera, J.; Kurka, J. Inclusion of Nature-Based Solution in the Evaluation of Slope Stability in Large Areas. *Land* **2024**, *13*, 372. https://doi.org/10.3390/land13030372.
10. Fauk, T.; Schneider, P. Does Urban Green Infrastructure Increase the Property Value? The Example of Magdeburg, Germany. *Land* **2023**, *12*, 1725. https://doi.org/10.3390/land12091725.

https://doi.org/10.3390/land15060971

References

1. Mitsch, W.J.; Jørgensen, S.E. *Ecological Engineering: An Introduction to Ecotechnology*; Wiley: New York, NY, USA, 1989; ISBN 0-471-62559-0.
2. Odum, H.T.; Slier, W.L.; Beyers, R.J.; Armstrong, N. *Experiments with Engineering of Marine Ecosystems*; Institute of Marine Science of the University of Texas: Port Aransas, TX, USA, 1963; pp. 374–403.
3. Bergen, S.D.; Bolton, S.M.; Fridley, J.L. Design principles for ecological engineering. *Ecol. Eng.* **2001**, *18*, 201–210. [CrossRef]
4. Barrett, K.R. Ecological Engineering in Water Resources: The Benefits of Collaborating with Nature. *Water Int.* **1999**, *24*, 182–188. [CrossRef]
5. Schönborn, A.; Junge, R. Redefining Ecological Engineering in the Context of Circular Economy and Sustainable Development. *Circ. Econ. Sustain.* **2021**, *1*, 375–394. [CrossRef]
6. Wu, H.; Yang, C.; Xie, C.; Man, Z.; He, S.; Qin, Y.; Che, S. Quantification of contribution of climate change and land use change on urban ecosystem service using multi-scale approach. *Ecol. Indic.* **2024**, *167*, 112619. [CrossRef]
7. Ghazian, N.; Lortie, C.J. A Review of the Roots of Ecological Engineering and its Principles. *J. Ecol. Eng.* **2024**, *25*, 345–357. [CrossRef]
8. Cohen-Shacham, E.; Andrade, A.; Dalton, J.; Dudley, N.; Jones, M.; Kumar, C.; Maginnis, S.; Maynard, S.; Nelson, C.R.; Renaud, F.G.; et al. Core principles for successfully implementing and upscaling Nature-based Solutions. *Environ. Sci. Policy* **2019**, *98*, 20–29. [CrossRef]
9. Mitsch, W.J.; Jørgensen, S.E. Ecological engineering: A field whose time has come. *Ecol. Eng.* **2003**, *20*, 363–377. [CrossRef]
10. Dale, G.; Dotro, G.; Srivastava, P.; Austin, D.; Hutchinson, S.; Head, P.; Goonetilleke, A.; Stefanakis, A.; Junge, R.; L., J.A.F.; et al. Education in Ecological Engineering—A Need Whose Time Has Come. *Circ. Econ. Sustain.* **2021**, *1*, 333–373. [CrossRef]
11. Schneider, P.; Hack, J. Ökologisch orientierte Planung in der Wasserwirtschaft. *Wasser Abfall* **2021**, *23*, 14–18. [CrossRef]
12. Mitsch, W.J.; Jørgensen, S.E. *Ecological Engineering and Ecosystem Restoration*; John Wiley & Sons: New York, NY, USA, 2004; ISBN 978-0-471-33264-0.

https://doi.org/10.3390/land15060971

land

MDPI

Article

Does Urban Green Infrastructure Increase the Property Value? The Example of Magdeburg, Germany

Tino Fauk and Petra Schneider *

Department Water, Environment, Civil Engineering and Safety, Magdeburg-Stendal University of Applied Sciences, Breitscheidstr. 2, D-39114 Magdeburg, Germany; tino.fauk@h2.de
* Correspondence: petra.schneider@h2.de

Abstract: Are there any correlations between land use and the associated prices charged for the soil? What is the significance of green infrastructure and what is the significance of public facilities and transport? For the analysis of the data, various methods of factor reduction and analysis were used to identify a multiple regression model that explained the price building. An effect was found between the pricing of the standard land reference value (SLRV), number of trees and distance to allotments. Summarizing the results, less than 231 trees in an SLRV zone causes an SLRV increase, the opposite is the case with a larger number of trees. The more accessible an allotment garden is (in terms of distance <421 m), the lower the SLV in the adjacent area. If the distance that must be covered to the allotment garden increases, the SLRV of the area increases. However, a more significant influence on the SLRV was concluded by the market economy variables. In summary, the present study indicates that (a) a uniform evaluation matrix for the SLRV should be created, and (b) the present subjective land assessments by the relevant experts should be complemented through targeting further training in the ecologically oriented planning context.

Keywords: urban green infrastructure; property value; ecosystem services; multiple regression; standard land reference value

1. Introduction

Land management represents an interdisciplinary activity field being in the competitive area between land use planning, economic development, marketing and financing. Land management is defined in Germany as "...an instrument at municipal or regional level that makes it possible to deal with land and land efficiently" [1,2]. It thus serves to optimize land use and the provision of construction land. However, this definition does not consider the market situation, which naturally largely determines the marketability of properties. Having in view the still increasing land consumption, it can be assumed that the value of land and property will continue to increase steadily, from an economic point of view. Even polluted land, for instance, brownfields, is increasingly recovered for land recycling purposes. Land recycling was defined by the Engineering Association of Contaminated Sites Management and Land Recycling e.V. (ITVA) as the "use-related reintegration of such properties into the economic and natural cycle that have lost their previous function and use-such as abandoned industrial and commercial areas, military properties, traffic areas, etc., by means of planning, environmental and economic policy measures" [3]. The interesting point in this definition is the reintegration into the natural cycle that supports the use of fallow land in cities as well for green infrastructure. In an urban society, green spaces are to be regarded as social and economic goods [4]. However, the competition between green and building areas is fierce [5,6], and the number of publications is still comprehensible.

Debus (2015) [7] summarized that the location quality of a property results from numerous factors such as infrastructure, local public transport and, above all, from the proximity to parks, green spaces, privately navigable canals and publicly owned water areas as non-privatizable common goods. The determination and monetization of an influence of green spaces on the (subsequent) market value according to the construction law is primarily based on the standard land reference value (SLRV) [5]. On the basis of the purchase price collection, average location values for the land are to be determined taking into account the development status. The value-influencing features of the land value property are to be determined. Finally, the question is also whether green spaces and other elements of green-blue infrastructure primarily represent a cost or a beneficial factor. Admittedly, the SLRV's are not binding, nor can any claims be derived from them against the bodies responsible for land use planning and land reorganization [8]. The basis for determining the SLRV as average location values in euros/m^2 are the respective purchase price collections of the expert committees. With the help of conversion coefficients and value factors, properties with different buildings and uses can be adapted to a selected sample property [5]. In the urban area, there are a number of examples that demonstrate the upgrading of the property quality and, thus, also the increase in the property value (via the SLRV) after the construction of green infrastructure. To date, however, the published scientific literature on this topic is scarce. The research questions related to urban greenery and land value are comparable on a global scale: Who benefits from the existing impact of green space on real estate market values? Who benefits from the usually positive influence: private or public property owners, e.g., the green space authorities? Does the community generally pay for the initial setup, care and maintenance of green spaces, and due to this favorable location does the private owner have better chances of renting at a higher rent and selling at a higher market price? How do SLRV develop in green areas compared to ecologically poorly equipped areas?

For the American Society of Landscape Architects (2019) [9], green infrastructure can be at the heart of smart regional and urban planning, guaranteeing communities a livable environment with clean air and water for future generations. According to Kardan et al. (2015) [10] a positive health effect was correlated to the tree density for Toronto. For Example, an increase of 408 cm^2 per m^2 leaf area index caused a gain of 1.4 life years in the context of cardio-metabolic conditions [10]. GI planning also includes the restoration of degraded ecosystems [6]. Ecosystem restoration needs to be supported by predicting habitat evolution during restoration activities. Typically, such a type of model takes into account two steps: (a) the reference state and (b) the ecological direction, i.e., the diversity of life forms present, the structure of the species to be created and the role that these species play in the ecological system [11,12]. Well-functioning ecosystems can provide a constant flow of ecosystem services that are central to human well-being [10]. A comprehensive overall assessment of ecosystem services and their costs and benefits is indicated in planning decision-making processes, so that ecosystems are protected and harmful interventions prevented by quantifying the ecological value [13].

For Germany, there are only a few published studies on property value and urban greenery. Wittowsky et al. (2020) [14], based on [15–21], summarized the neighborhood related aspects, namely, surrounding walkability, crime level, school quality, racial or ethnic population composition, green spaces and amenity investment effects, as well as presence or absence and quality of sidewalks. Moreover, the applicable price models are hedonic price modelling [22] and accessibility [21]. The authors of Ref. [14] applied the accessibility approach to Dortmund, a former industrial city located in the German Ruhr area. While the inner city is dominated by apartment buildings, the majority of owner-occupied houses are located in the outskirts. House prices in Dortmund have increased between 2012 and 2016 for detached houses (+25%), semidetached new-builds (+19%) and older semidetached houses (+25%). Moreover, Hanhörster and Lobato (2021) [23] identified shortages of low-rent housing in many cities of Germany as open door for discrimination, particularly to immigrants. Galvin (2023) [24] focused on western Germany's post-war apartments and the question of whether rental and sales markets incentivize energy-efficient retrofitting. These buildings are a notoriously energy-inefficient building cohort. The authors of Ref. [24]

found that the sales market premium for energy efficiency will only support energy efficiency retrofitting from the standpoint of the vendor if was received standard federal subsidies for energy efficiency upgrading, but not the purchaser or property owner, who is likely to recoup less than one-third of the purchase premium through energy savings over the lifetime of the retrofit measures. Besides greenery, energy efficiency is one of the main motivations to accept increased property values.

In the internal investigation of Köhler (2022) [25], SLRV were evaluated for the Goethe Street green belt in Magdeburg in the period from 2006 to 2020 and compared with green spaces in two other cities (Halle City and the town Aschersleben). While the Goethe green belt is a greenery-upgraded 825-m street in Magdeburg, the upgrade in Aschersleben concerned a Park of the State Garden Show in 2010 leading to increased property values. In addition, four areas in Halle were considered, (a) in the vicinity of the Saale river banks and the botanical garden, (b) a green residential area, (c) three more heavily built-up areas, and (d) a residential area directly on the Saale river with a lower greenery share [25]. The increase property value in Magdeburg of 68% was, essentially, observed since 2016 [25], compared to 5.2% in the mixed-use area since 2006. Overall, in the 2018–2020 period was observed a major property value growth ranging from 50% to 70%. Moreover, the results showed a higher increase in value per m^2 for the greenery areas compared to other surrounding mixed areas (Halle 180%, Aschersleben 5%). Also in Halle City, the property value increased since 2016. In Aschersleben, the values are generally lower than in the two major cities, which reflects the general situation in a structurally weak region. The State Garden Show 2010 in Aschersleben, which brought about structural and horticultural changes in the city, did not immediately lead to an increase in value, which can probably be attributed to the effects of the financial crisis of 2008/2009. Further investigations for Germany were performed by Ali et al. (2020) [26]. They investigated gentrification through green regeneration in the Lene-Voigt-Park in Leipzig, Eastern Germany. Bokarjova et al. (2020) [27] and Rigolon and Németh (2018) [28] confirmed this subject as well. On the international level, I-Chun Tsai et al. (2023) [29] focused the influence of migration policy risk on market segmentation of housing and rental markets in the Euro Area. Moreover, compared with housing market connectedness, the rental market connectedness is less influenced by migration policy risk and migration fear in the UK than in Germany [29]. The findings of Mell et al. (2016) [30] and Dell'Anna et al. (2022) [31] support the wider literature evaluating the economic value of GI which argues that investment in urban greenspace can have a significant impact on local housing and commercial markets where it produces more attractive and functional landscapes. The objective of this study was to investigate the relationship between financial, socioeconomic and urban ecosystems providing ecosystem services, on SLRV. For this purpose, a principal component analysis was used to assign the selected indicators to the corresponding components. A multiple regression was used to discuss the impacts on SLRV. For this research, a case study was conducted for the city of Magdeburg, Saxony-Anhalt, Germany.

2. Materials and Methods

For the present analysis, a data set was created to identify the relationships between monetary, biotic and socio-economic aspects of urban land use. This analysis was focused on the state capital Magdeburg of the federal state of Saxony-Anhalt. The selected areas are in the districts of Altstadt (first documented mention in year 805), Stadtfeld Ost (first development in 1920s) and Stadtfeld West (first development in 1920s).

The SLRV level was determined on the basis of comparable areas. Accordingly, for reasons of consistency in the data set, a selective consideration of purely residential areas was undertaken. This was intended to exclude the effects of increased economic use. The observed area of Altstadt contains two spaces defined as residential area (RA), the rest was a mixed area (industrial and commercial side by side with residential area). Table 1 shows the total areas of each residential area with corresponding districts and their green spaces.

Table 1. The district, with the corresponding residential areas with the entire areal area and green areas based on the normalized difference vegetation index NDVI.

District	Residential Area j	Total Area in ha A_{SLVR}	Area of Green Spaces in ha A_{GI}
Stadtfeld Ost	RA 1	12.5	4.0
Stadtfeld Ost	RA 2	80.0	28.6
Stadtfeld Ost	RA 3	12.1	6.9
Stadtfeld Ost	RA 4	102.2	30.0
Stadtfeld West	RA 5	98.6	47.8
Stadtfeld West	RA 6	18.6	9.4
Stadtfeld West	RA 7	9.7	5.5
Stadtfeld West	RA 8	32.1	19.5
Stadtfeld West	RA 9	8.7	4.6
Stadtfeld West	RA 10	16.1	9.8
Altstadt	RA 11	38.7	15.3
Altstadt	RA 12	4.1	1.8

Twenty indicators were used, which might be related to land use. The 20 indicators whose correlations were examined in the following are: standard land reference value (SLRV), offered cold rents (OCR), interest on savings deposits (ISD), economic income (INC), inflation rate (IR), population density (INH), population density with migration background (IMIG), relative leaf area for trees on public green spaces and beside roads (LAI), number of tree individuals (Ntree), Evenness (Es), proportion of green spaces including private gardens per inhabitants (GS), biodiversity index according to Gloor [32] (BDI), weighted average distance to the nearest allotment garden (AlG), as well as to the nearest large green area (min_GI), the nearest water surface (Water), to schools (school), to medical treatment (min_med), to public transport (tra), the nearest retailer (ret), and to the nearest restaurant (rest_pkt).

The respective prices of the SLRVs are taken for the corresponding points in time from [33] and entered manually in a spreadsheet document. The stored time series amounted to between 2006 and 2020. The extents of the areas under consideration were manually accounted for in Quantum GIS (QGIS). A corresponding layer has been created for this purpose. There are a total of twelve residential areas, which were specifically considered as described below. An illustration can be seen in Figure 1.

The OCR was provided as an average per block [34]. For the year 2020, the 1st quarter of 2021 was scheduled. The different OCRs were calculated for the respective SLRV ranges using a weighted average. For this purpose, the mean OCR per block was summed over the number of blocks in the SLVR and then divided by the number of blocks in the SLVR. The OCR could not be generated completely until 2006. Only time series from 2011 to 2022 were available [35]. Furthermore, a linear growth of OCR was assumed. For this purpose, a model equation in the form y = mx + n was used for the respective rental apartment sizes (30 m^2, 60 m^2 and 100 m^2). Here, y is the searched OCR at time x (year; Y). The following Equations (1)–(5) are used for OCR:

$$OCR_{30,i} = 0.1236 \times Y - 243.0481 \pm 0.20\ EUR/m^2, \quad (1)$$

$$OCR_{60,i} = 0.1301 \times Y - 256.8140 \pm 0.09\ EUR/m^2, \quad (2)$$

$$OCR_{100,i} = 0.1825 \times Y - 361.7568 \pm 0.29\ EUR/m^2, \quad (3)$$

$$meanOCR_i = (OCR_{30,i} + OCR_{60,i} + OCR_{100,i})/3 \quad (4)$$

$$OCR_{m,i} = 0.1454 \times Y - 287.2063 \pm 0.17\ EUR/m^2. \quad (5)$$

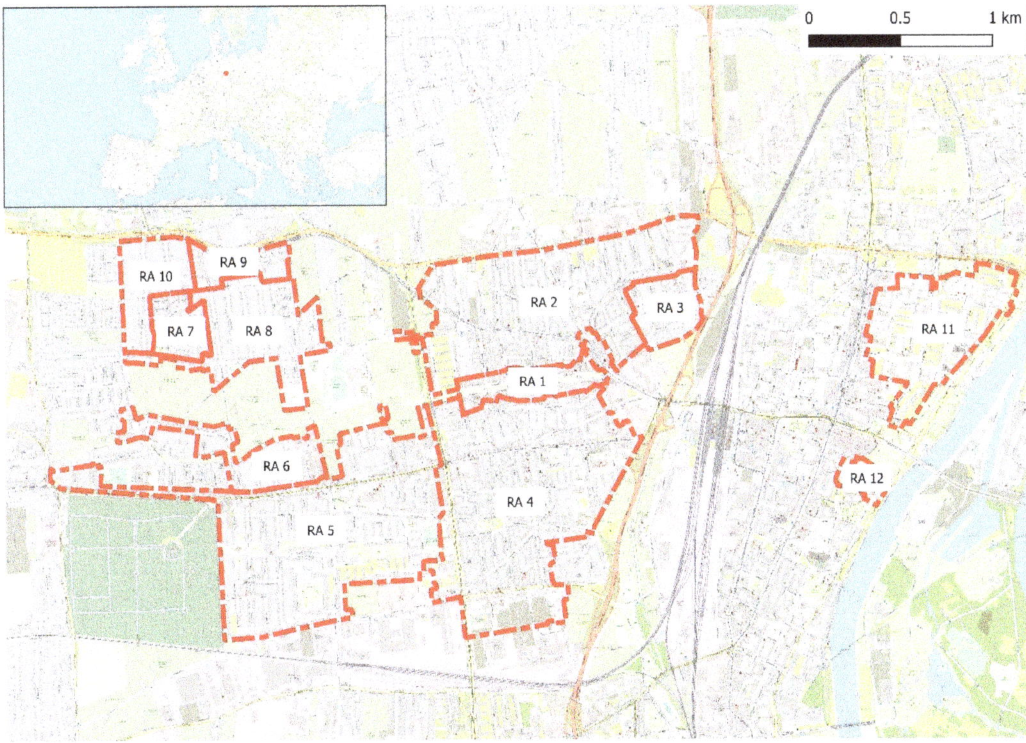

Figure 1. Areas of interest, City of Magdeburg. © https://www.openstreetmap.org and contributors.

OCR_m is the estimation function for the mean values with equal distribution of the different apartment sizes, which is rarely the case. Accordingly, a correction factor was introduced for each of the different SLRVs, which considers the inhomogeneous distribution of housing supply. Equation (6) shows the determination of the correction factor.

$$f_{c,j} = OCR_{j,2020}/OCR_{m,2020} \quad (6)$$

This assumes that the correction invoice does not undergo any significant change over time. This was based on the ratio of the forecast value OCRm, 2020 (in 2020) with the actual OCR in the respective SLRV j (j = RA1 to RA12) in 2020. The estimation of the mean OCR for the area to be considered is performed using Equation (7).

$$OCR_{m,j,i} = f_{c,j} \times OCR_{m,i} \quad (7)$$

The next chosen indicator was interest rates on savings deposits. The underlying consideration is the relocation of financial values to real estate due to better returns. The evaluation "Development of the average interest rate for savings deposits in Germany in the years from 1975 to 2022" [36] was used as the data basis.

Next, the assets are considered. These are taken as income values for society. No regionalization was carried out here, as investments in real estate are usually not made by residents. This is especially true for properties that are used for rent. The underlying database consists of the one hand of the number of employees within Germany (E) [37] and on the other hand the mean income of the employees (INCa) [38], these two values are given to the Income Indicator (INC) by Equation (8):

$$INC_j = E_j \times INCa_j \quad (8)$$

The inflation rate was taken into account based on Deutsche Bank's records [39].

The density of people was provided for the SLRV areas examined by the City of Magdeburg, City Planning Office. An address-related distribution of persons is not possible for data protection reasons. In order to be able to incorporate the distribution of people for the next steps, the inhabitants were assigned to the buildings. The buildings were made available online by the State Office for Surveying and Geoinformation Saxony-Anhalt [40]. The buildings were first tailored to the SLRV areas studied and then the approximate people were distributed. For the three districts that are home to the SLRV areas studied, corresponding migration movements are known. Ziegler's [41] explanations provide the database for the respective districts with the corresponding years as an assignment variable. For the calculation of the initial variable, the current persons were divided by the persons in the corresponding district. The relative person now obtained (persons of the SLRV area per person of the district) was multiplied by the annual person sizes of the districts. Persons with a migration background are also included in the statistics and taken into account as a separate influencing factor. The persons per SLRV area were divided by the corresponding area (A_{SLVR}, see Table 1). Furthermore, the density of persons and persons with migration background (P) are obtained.

$$INH_i \text{ or } IMIG_i = P_{i,j,k} / A_{SLVR,j} \quad (9)$$

The next indicator was the leaf area index (LAI) at the respective time i in the respective SLRV area j. For this purpose, the tree cadastre of the state capital Magdeburg was used [42]. Following Kardan [10] and Kalusche [43], we used a maximum unilateral leaf area of 500 m^2 per tree individual. In general, leaf area differs among tree species. We reduced these individual allometric growth conditions to a general approach. The tree individuals were determined for each area studied. After that, the crown diameter (cd) and tree height (h) were chosen to approach the leaf area per tree individual (indices t). According to Fauk and Schneider [44], the individual growth of trees has many influencing factors. These complex interactions between biotic and abiotic environment cannot be considered in this study. However, we used the knowledge about the general growth conditions for urban trees in the city of Magdeburg to introduce a simple model for crown expansion and height growth. We assumed an annual crown diameter growth of 0.5 cm and a height increase of about 1 cm. These values are to be considered conventional. Equation (10) shows the approach used.

$$LA_{i,j,t} = [500 \times (1 - \exp(-((cd - (0.005 \times \Delta Y)) \times ((h - (0.01 \times \Delta Y))/3)/100))) + 1] \quad (10)$$

To finally establish a relationship between the individual components or indicators, the individual leaf area had to be summed up first for the corresponding year i and the residential area j (see Equation (11)).

$$LA_{i,j} = \sum LA_{i,j,t} \quad (11)$$

For the summed leaf area to ultimately be converted to functional and comparable unit, an area reference to the respective residential areas had to be established. This was performed using Equation (12).

$$LAI_{i,j} = LA_{i,j} / A_{SLVR,i,j} \quad (12)$$

Next, the biodiversity of the trees was determined. For this purpose, the Shannon Index was used first (see Equation (13)). The individuals per tree species was divided by the total number of tree individuals; the proportion of the respective tree species obtained in this way was multiplied by $p_{j,i}$. Ultimately, the Shannon Index was calculated as a measure of plant diversity as follows:

$$HS_{j,i} = -\sum p_{j,i} \times \ln(p_{j,i}) \quad (13)$$

This index only gives information about the number of species and their distribution, i.e., the diversity. However, it was important for us to consider the balance of species in the respective residential area. For this purpose, we used the species equitability (Es). For the calculation, the number of species (S) of all studied residential areas was used. The species equitability is also called evenness and was determined with Equation (14).

$$Esj,i = HS_{j,i}/\ln(S) \quad (14)$$

The green spaces (GS) for the respective residential areas were measured using NDVI (normalized difference vegetation index) [45]. The NDVI was determined using orthophotos (RGBI). These were provided by the State Office for Surveying and Geoinformation [40]. The NDVI gives values between −1 and 1. In order to determine GS, a reclassification had to be made. Based on comparative analyses, the green areas for the respective time point were matched with the NDVI values and a threshold of 0.15 was defined. These values can differentiate depending on the time of recording and must be determined manually or with machine learning algorithm. All values of the NDVI map that were greater than 0.15 were marked with 1 and the rest with 0. Subsequently, all pixels (0.2 × 0.2 m) within the respective residential areas were summed and multiplied by 0.04 m^2. For this purpose, the raster calculator and the zone statistics of the QGIS software were used and the data was transferred to the residential areas. Finally, the calculation of the GS proportion was carried out per inhabitants for the respective residential area and is shown in Equation (15).

$$GS_j = (A_{GI,j}/\sum P_{j,k}) \quad (15)$$

The next indicator is the Biodiversity Index (BDI) according to Gloor [32]. The BDI contains the faunistic biodiversity based on indicators species (birds, butterflies, bees, cicadas) for typical urban trees such as oak (*Quercus robur*, L.) or tree of heaven (*Ailanthus altissima*, (MILL.) SWINGLE). In general, neophytes show a smaller contribution to the biodiversity of native fauna than their native equivalent. The previously selected tree species and their number were multiplied by the corresponding BDI according to Gloor for the respective residential area and divided by the number of tree individuals in the respective residential area (weighted mean).

The indicators and their selection are briefly explained. We have generally opted for distance as the value for accessibility. At this point, we simplified accordingly and only considered the distance as the crow flies instead of the actual route.

The first indicator to be explained is the allotment gardens (AlG). Schneider and Fauk (2022) [46] pointed out that allotment gardens play an important role in the future production of regional food, while at the same time they form a potential to link natural areas as well as to preserve genetic diversity. This is underlined by our own observations in the allotment garden association Schnarsleben; the vegetation studies of different fallow areas indicate a settlement and establishment of protected (*Aquilegia vulgaris*, L.) and or rare (*Dictamnus albus*, L.) plant species. In view of the monocultural landscape character in the area concerned, a previous deliberate establishment of the plant species through horticultural activity is to be assumed. Accordingly, the cultural value of allotment gardens is to be estimated.

The nearest large green area (min_GI) where recreation and relaxation could be found, and, thus, an important contributor to cultural and regulating ecosystem services, were also determined as the crow flies. In this indicator we have considered cemeteries, grasslands, woodlands and nature reserves. We did not consider individual preferences of the population, but only the shortest distance.

The distance to the nearest water area (Water), which was also determined as the crow flies, should take into account both the function as a ventilation pathway in the terms of urban climatology and the importance of aesthetics. We have taken large as well as smaller standing and flowing waters as a basis.

The distance to schools (school) indicator was determined independently of the age structure of the population. A correspondingly in-depth database was not available. Furthermore, the distance was determined as the crow flies from the respective buildings and averaged over the residential areas. The shortest distance to medical treatment (min_med), the distance to public transport (tra), the distance to the nearest retailer (ret) and the distance to the nearest restaurant (rest_pkt) also determined as a crow flies. These indicators are intended to represent care and connectivity.

All distance indicators were measured using the distance tool in QGIS version 3.18.3. The linear distance was used, which takes into account the distance from the respective building in the corresponding residential area to the center of the indicator under consideration. These distances were weighted according to the buildings and estimated persons. The geodata were loaded into QGIS by the open street map extension and then used for the evaluation.

First, the connection, as shown in Figure 2, was assumed.

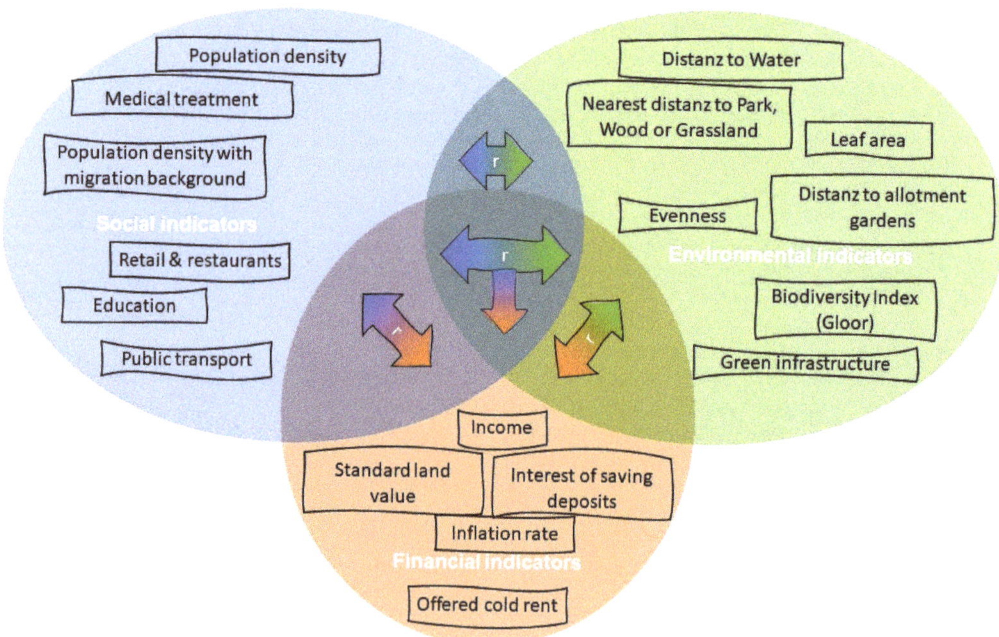

Figure 2. The relationship examined between the respective land use indicators.

For further analysis, the software RStudio was used. First, the data set was standardized and centered. The data matrix has the dimension (156 $\times$ 20) and is denoted as Z.

First, the data set was tested for the hypothesis that:

$\mathbf{H_0}$: *The variables in the survey population are uncorrelated.*

$\mathbf{H_1}$: *The variables in the survey population are correlated.*

For this purpose, Bartlett's test of sphericity [47] was used. The test value is strongly influenced by the sample size [48]. Chi-square (X^2) was used as a test variable for assessment. The significance level was determined from the sample size (N), the number of variables (v) and the determinants of the correlation matrix (det(R)) using the degree of freedom (df = (v × (v − 1))/2).

$$X^2 = -\ln(\det(R)) \times (N - 1 - (2 \times v + 5)/6) \quad (16)$$

At this point, the different correlation methods should be pointed out. These will play a significant role in the following assessments. Basically, a distinction is made between rank correlation and correlation coefficients. The rank correlation coefficient is determined according to the Kendall Equation (17) and the Spearman Equation (18). As the name "rank" suggests, only the place in an ordered list is considered. For Kendall's rank regression, the concordant (C) and discordant (D) data pairs are used. For the subsequent normalization, the ties between variable Y (Ty) and variable X (Tx) are taken into account.

$$\tau = (C - D)/[(C + D + Tx) \times (C + D + Ty)]^{0.5} \quad (17)$$

For the Spearman rank regression, the respective ranks (R) and their mean values (R_m) are considered.

$$r_s = \sum_i (R(x_i) - R_{m,x}) \times (R(y_i) - R_{m,y})/[(\sum_i (R(x_i) - R_{m,x})^2)^{0.5} \times (\sum_i (R(y_i) - R_{m,y})^2)^{0.5}] \quad (18)$$

The Pearson correlation coefficient is determined using the data values (X, Y).

$$r_p = \sum_i (X_i - X_m) \times (Y_i - Y_m)/[(\sum_i (X_i - X_m)^2)^{0.5} \times (\sum_i (Y_i - Y_m)^2)^{0.5}] \quad (19)$$

Subsequently, the suitability of the data set was tested using the Kaiser–Meyer–Olkin (KMO) and the MSA criterion (package: EFAtools [49] and factoextra [50]). The Kendall, Pearson and Spearman regression methods were used. The KMO criterion considers all variables of the data set [48] and was calculated with the equation.

$$KMO = \sum_i \sum_{i'} (r_{xixi'})^2/[\sum_i \sum_{i'} (r_{xixi'})^2 + \sum_i \sum_{i'} partial\_(r_{xixi'})^2] \quad (20)$$

The measure of sampling adequacy (MSA) tested the single variable for fitness [48]. For this purpose, we have used the equation.

$$MSA = \sum_{i'} (r_{xixi'})^2/[\sum_{i'} (r_{xixi'})^2 + \sum_{i'} partial\_(r_{xixi'})^2] \quad (21)$$

A partial correlation (partial_($r_{xixi'}$)) is defined as a correlation between two variables with no further influence from the other variables [48].

The next step was a principal component analysis. This was used for the initial interpretation of the data. The unknown matrices S, which contains the factor values and L, which contains the factor charges, reproduce the standardized initial matrix Z. This is generally expressed by Equation (22):

$$Z = S \times L \quad (22)$$

By transforming and substituting the correlation matrix (R = 1/(N − 1) × Z′ × Z), assuming the uncorrelated nature of the data for principal component analysis, it follows (Backhaus et al.) [48]:

$$R = L \times L' \quad (23)$$

The PCA was performed without and with rotation of axes.

Figure 3 shows that the raw results from the PCA have been further processed. The scores of the indicators on the newly created axes can be interpreted differently depending on the application. This individuality offers a correspondingly large space for interpretation of the results and corresponding discussions. For this purpose, reference will be made to the relationship between financial, socio-economic and urban ecosystems (which can provide ecosystem services) on standard land values. For our investigation, it was not important whether an indicator loads on one of the new components, but rather the multiple loading of an indicator. The goal was to identify as large an explanation of the variance as possible, especially for the SLRV. In accordance with good practice, the results of the new dimensions (scores) were considered with the communality ($h_j^2 = \sum_{q=1}^{Q}(a_{jq})^2$) per variable. We did not use the normal criteria to remove variables from the data set for further principial component analysis. The first step was to find the dimension where the score for the SLRV was greater than or equal to 0.5. Subsequently, the individual variables were considered in the respective dimension. If a score was smaller than 0.5, we first looked for other dimensions in which SLRV and the respective variable loaded above 0.5. If no further dimensions could be found, it was checked whether the variables were smaller than 0.3, and analogous to the previous procedure, a further step was introduced. Now the explanatory variance of the respective dimensions was tested, as well as the significance of the variables based on the communality after exclusion of the unused dimensions.

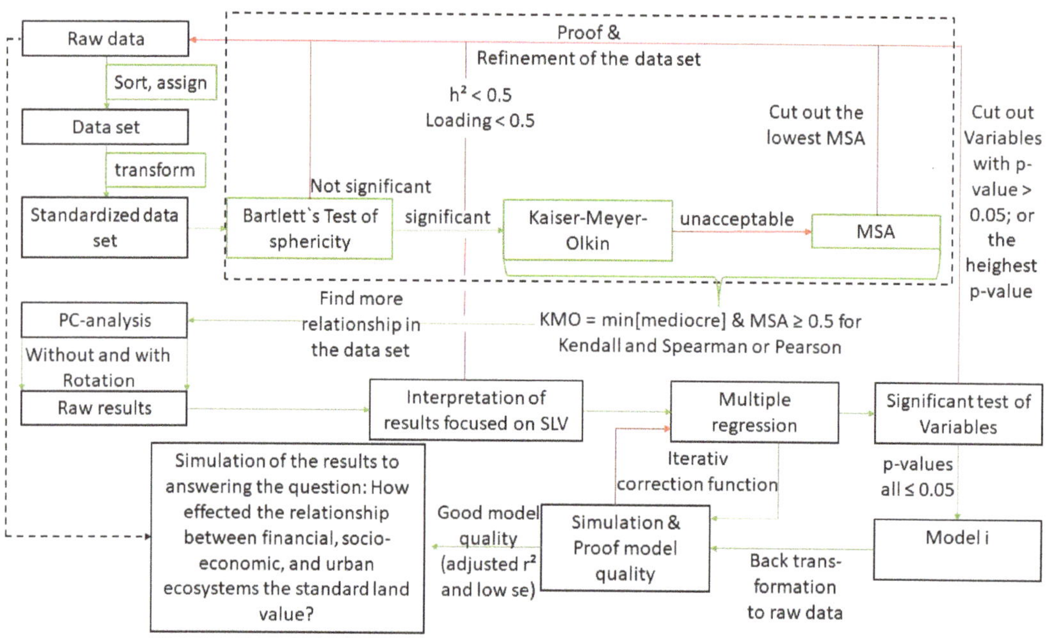

Figure 3. Methodological approach for the identification of linear relationships between the indicators under study; Note: the Kaiser–Meyer–Olkin criterion must have at least "mediocre" in min one of the regression methods (Spearman, Pearson, Kendall). Better scores (e.g., middling, meritorious) are explicitly desirable; the MSA must be greater or equal 0.5; the green arrows indicate the steps towards the result, and the red arrows indicate preparatory/corrective steps; the linkage of the initial data set with the regression results (to illustrate the effects) is symbolized with the black dashed arrow.

The final step was to interpret the PCA results and build a multiple regression model to predict SLRV based on the identified indicators or variables (I). For this purpose, we established a linear relationship to the indicators or variables. The general form of the equation can be seen in the equation. For the first model, the coefficients of each variable were determined using the standardized data matrix. Since the manipulation of data sets in terms of standardization or logarithmization may have a bias in the back transformation, two fitting models were implemented.

$$SLV = c_1 I_i + c_2 I_{i+1} + \ldots + c_n I_{i+n} \tag{24}$$

The simple linear fitting model 1 Equation (25).

$$SLV_1 = b + m \times SLV \tag{25}$$

The simple quadratic fitting model 2 Equation (26).

$$SLV_2 = b + m \times SLV + n \times SLV^2 \tag{26}$$

3. Results

3.1. Results of Data Matrix Test and Data Refinement Procedure

The Bartlett's test for sphericity was significant at an alpha level of 0.05 for all iteration steps (Pearson: start ($X^2(190) = 6667.64$, $p < 0.001$), iteration 8 ($X^2(66) = 2099.27$, $p < 0.001$)). These data are probably suitable for factor analysis. A singular matrix was reported for the rank correlation (Kendall, Spearman). This indicates correlated values in the initial matrix. Accordingly, the further procedure (KMO and MSA) was based on Pearson's method, while Kendall's and Spearman's methods were not considered further. Since Bartlett's test only shows the principal suitability of the data set, the Kaiser–Meyer–Olkin criterion with the respective MSAs served to fit the data set. Each exclusion of a variable can have a large effect on KMO and MSA, only single variables (indicators) were removed. The initial data set had a KMO value of 0.374 and was classified as unacceptable. Based on the MSA criterion, min_gi (0.160) was removed. For iteration 1, there was an increase in KMO to 0.407 (unacceptable). The indicator BDI (0.145) was excluded. Iterations 2 (KMO = 0.395 and MSA excl. = ret (0.163)), 3 (KMO = 0.368 and MSA excl. = tra (0.203)) and 4 (KMO = 0.466 and MSA excl. = LAI (0.284)) were also unacceptable. The first iteration to show a miserable result for the KMO value (0.562) was number 5. Evenness was excluded based on the MSA of 0.301. Analogous to iteration 5, iteration 6 also showed a miserable KMO score (0.595). Based on the lowest MSA, Water (0.382) was extracted. Iteration 7 and 8 were both mediocre (Iter. 7 = 0.641 and Iter. 8 = 0.661). After iteration 7, IMIG (0.483) was removed. Iteration 8 indicated SLRV (0.495) as an indicator to be extracted, so the final data set was reached at this point. The full results are given in Table 2.

Table 2. The test results for the MSA test with regression method Pearson for each indicator and iteration. Red = deleted variable, wavy undersigned = Indicator of Interest (<0.5); Italic: KMO value of the respective iteration.

Design.	Start	Iter. 1	Iter. 2	Iter. 3	Iter. 4	Iter. 5	Iter. 6	Iter. 7	Iter. 8
SLRV	0.577	0.632	0.568	0.533	0.464	0.557	0.564	0.566	0.495
INH	0.518	0.546	0.382	0.362	0.463	0.639	0.633	0.670	0.675
IMIG	0.655	0.584	0.505	0.407	0.321	0.507	0.445	0.483	
OCR	0.677	0.507	0.391	0.299	0.399	0.521	0.740	0.754	0.823
INC	0.704	0.652	0.670	0.642	0.607	0.560	0.602	0.597	0.567
ISD	0.659	0.545	0.626	0.600	0.679	0.757	0.746	0.742	0.733
IR	0.658	0.570	0.541	0.478	0.566	0.707	0.673	0.663	0.650
Es	0.216	0.368	0.298	0.565	0.330	0.301			
LAI	0.242	0.406	0.298	0.238	0.284				
Ntree	0.237	0.327	0.455	0.320	0.532	0.427	0.491	0.629	0.663
GS	0.578	0.461	0.440	0.774	0.595	0.608	0.580	0.791	0.733
BDI	0.199	0.145							
AlG	0.323	0.528	0.451	0.356	0.489	0.680	0.648	0.584	0.617
min_gi	0.160								
Water	0.209	0.339	0.324	0.241	0.403	0.433	0.382		

Table 2. *Cont.*

Design.	Start	Iter. 1	Iter. 2	Iter. 3	Iter. 4	Iter. 5	Iter. 6	Iter. 7	Iter. 8
school	0.265	0.274	0.291	0.261	0.583	0.568	0.538	0.563	0.561
min_med	0.471	0.328	0.329	0.292	0.458	0.653	0.642	0.575	0.711
tra	0.267	0.241	0.241	0.203					
ret	0.365	0.211	0.163						
rest_pkt	0.367	0.344	0.407	0.326	0.417	0.531	0.654	0.748	0.803
KMO	*0.374*	*0.407*	*0.395*	*0.368*	*0.466*	*0.562*	*0.595*	*0.641*	*0.661*

3.2. Evaluation of the PC Analysis

Due to the good test results in the Bartlett test and the KMO test, the principal component analysis was performed. The correlation matrix provides 12 dimensions (principal components), 8 of which explain ≈ 98% of the variance in the data set. Before the important dimensions are considered more decidedly, they should first be derived optically or on the basis of the explained variance. Figure 4 shows a first clear bend on the third dimension. However, a roughly equal explained variance could be found for dimension four (explained variance > 5%). The summed explained variance in the data set was 79.4% for the third dimension (principial components) and86.3% for dimension four.

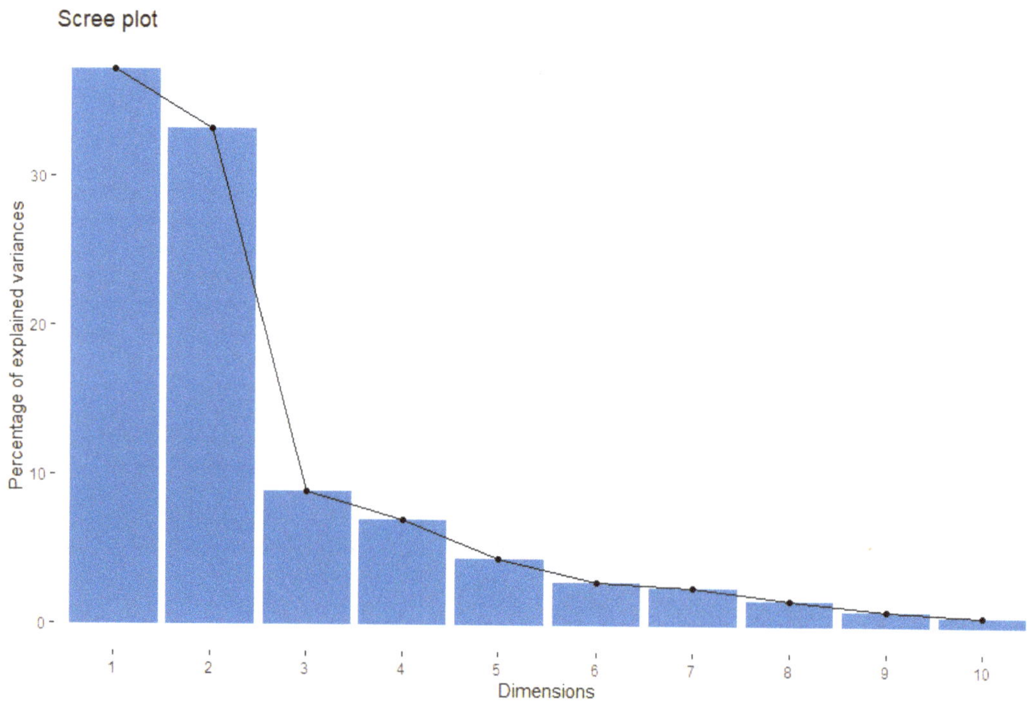

Figure 4. Scree Plot of the PCA and FA result based on eigenvalue and factor or component number.

The first two dimensions represent 70.46% of the variance in the data set and can be seen in the Figure 5. Dimension one correlates most strongly with GS (r = 0.7615) and INH (r = −0.7719). Other positively influenced indicators by dimension 1 are: rest_pkt (r = 0.7248), min_med (r = 0.7405) and school (r = 0.5361). On the other hand, there are the negatively influenced indicators, such as AIG (r = −0.6744) and SLRV (r = −0.6730). Overall, 37.20% of the variance was described in this dimension. Dimension two correlates most strongly with ISD (r = 0.8346). Negatively correlated are INC (r = −0.8631) and OCR (r = −0.8388) and IR (r = −0.8500). This dimension describes 33.26% of the variance in the data set.

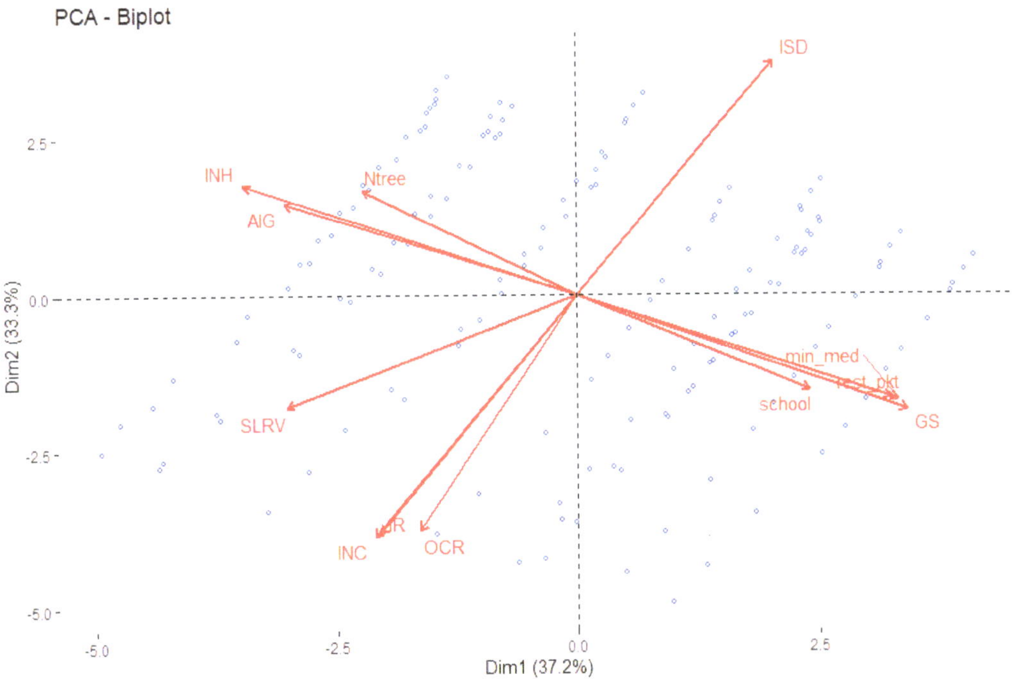

Figure 5. Variable or indicator as Biplot for dimension 1 and 2.

Dimensions three and four describe 15.88% of the variance in the data set and are shown in the Figure 6. Dimension 3 correlates most strongly with the indicator Ntree (r = −0.6464). This dimension describes about 8.91% of the variance. The fourth dimension correlates with school (r = 0.5132) and describes 6.97% of the variance.

Now the PCA analysis was used to check whether further indicators can be reduced. For this purpose, the dimensions were first reduced without lowering the explanatory variance of the indicator below 0.3. The PCA result was reduced to three dimensions. The explanatory variance of the data set was thus reduced to 79.37%. Moving on, the focus was now on SLRV, for which h^2 was used after reducing the dimensions. The initial h^2 for SLRV was 1 (normal for PCA analysis). After reducing the dimensions, the commonality was 0.8220. h^2 for all other indicators was greater than 0.5 and the absolute minimal score in one of the three dimensions was greater than 0.5. All in all, no further indicator exclusion was performed by the PCA analysis. In the case of a PCA, the indicators shown in green would each have to be assigned to the dimensions individually. Dimension 1 showed that the standard land value could apparently only be described by social factors and the green infrastructure. However, the communality for SLRV was then only 0.4530 (45.30% explained variance of the variable SLRV). In the second dimension, only loadings of the financial indicators greater than 0.5 were evident, and an absolute loading greater than 0.3 was still recorded for SLRV. We do not exclude SLRV and with the dimension 2, we obtain a communality of 0.6124. This 61.24% explained variance was not enough for us either. Furthermore, the number of trees would have no influence at this point and this variable would be described incompletely (h^2 = 0.3848). With the dimension 3, on the one hand, a significant part of the communality for SLRV (h^2 = 0.8220) could be described; on the other hand, the contribution of the city trees was identified by an absolute loading >0.5. The SLRV charge was close to 0.5, which was why a connection had to be assumed. The fourth dimension can be largely ruled out, (a) no significant loading of the variable SLRV, (b) double loading of School, which is why it was already described in the previous dimension. The PCA analysis showed that initially all 12 variables should be used for further multiple regression. The PCA scores are shown in the Table 3 for three Dimensions. The full PCA result can be seen in the Table 4.

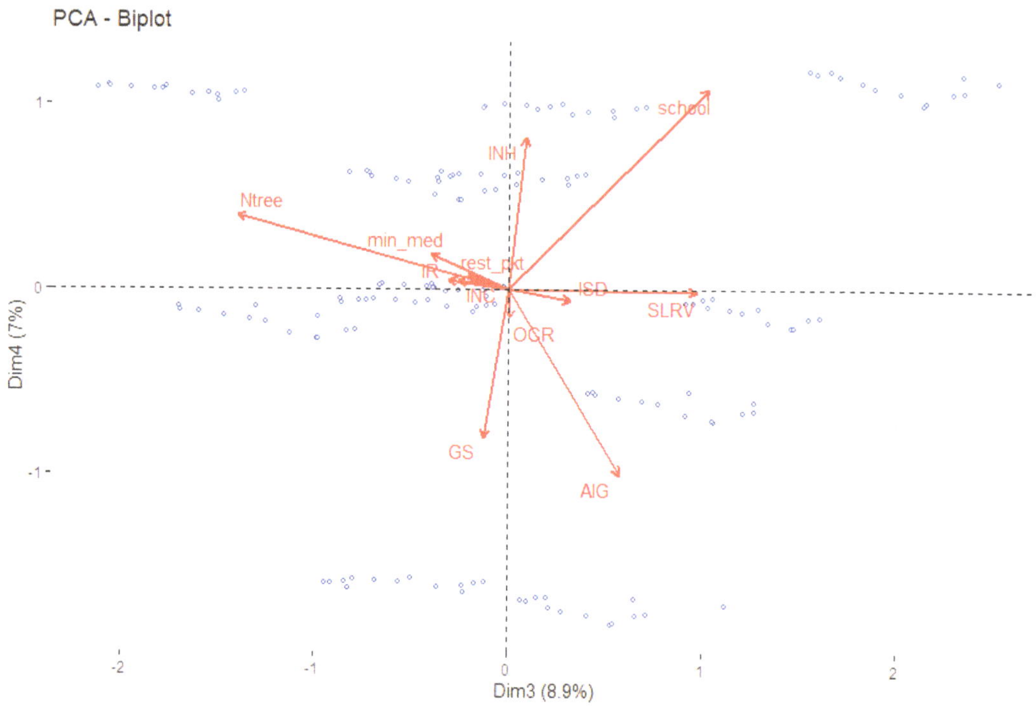

Figure 6. Variable or indicator as Biplot for dimension 3 and 4.

Table 3. PCA loadings with communality (h^2), eigenvalue (EV), explained Variance (Var) and cumulative Variance (cum Var) after dimension reduction. Green highlight absolute charge > 0.5 & orange highlight absolute charge > 0.3.

Indicator \ Dimension	Dim.1	Dim.2	Dim.3	h^2
SLRV	−0.6730	−0.3993	0.4578	0.8220
INH	−0.7719	0.3924	0.0403	0.7515
OCR	−0.3679	−0.8388	0.0022	0.8390
INC	−0.4715	−0.8631	−0.1230	0.9824
ISD	0.4523	0.8346	0.1495	0.9235
IR	−0.4588	−0.8500	−0.1482	0.9550
Ntree	−0.4972	0.3709	−0.6629	0.8242
GS	0.7615	−0.4143	−0.0620	0.7554
AlG	−0.6744	0.3258	0.2712	0.6345
school	0.5361	−0.3397	0.4851	0.6380
min_med	0.7405	−0.3764	−0.1894	0.7258
rest_pkt	0.7248	−0.3704	−0.1007	0.6727
EV	4.46	3.99	1.07	
Var	37.20%	33.26%	8.91%	
cum Var	37.20%	70.46%	79.37%	

Table 4. PCA loadings with communality (h^2), eigenvalue (EV), explained Variance (Var) and cumulative Variance (cum Var). Green highlight absolute charge > 0.5 & orange highlight absolute charge > 0.3.

Indicator \ Dimension	Dim.1	Dim.2	Dim.3	Dim.4	Dim.5	Dim.6	Dim.7	Dim.8	Dim.9	Dim.10	Dim.11	Dim.12	h^2
SLRV	−0.6730	−0.3993	0.4578	−0.0058	−0.2939	0.0080	0.0352	−0.2983	0.0246	0.0036	−0.0211	0.0157	1.0000
INH	−0.7719	0.3924	0.0403	0.3873	0.1722	−0.1718	−0.0620	−0.0528	−0.0481	0.1069	0.1378	−0.0059	1.0000
OCR	−0.3679	−0.8388	0.0022	−0.0732	0.2674	−0.0219	−0.1471	−0.0268	−0.2309	0.0186	−0.0875	0.0007	1.0000
INC	−0.4715	−0.8631	−0.1230	0.0240	−0.0489	0.0097	0.0439	−0.0024	0.0570	−0.0546	0.0299	−0.0742	1.0000
ISD	0.4523	0.8346	0.1495	−0.0305	0.0161	−0.0024	−0.0448	−0.1286	−0.1457	−0.1801	0.0520	−0.0198	1.0000
IR	−0.4588	−0.8500	−0.1482	0.0245	−0.0079	0.0064	0.0303	0.0911	−0.0039	−0.1435	0.1113	0.0455	1.0000
Es	−0.4972	0.3709	−0.6629	0.1879	−0.2411	0.1831	0.1748	−0.0631	−0.1144	0.0191	−0.0278	0.0030	1.0000
LAI	0.7615	−0.4143	−0.0620	−0.3806	−0.0274	0.2441	−0.0327	−0.0894	−0.0535	0.1070	0.1266	−0.0026	1.0000
Ntree	−0.6744	0.3258	0.2712	−0.4759	−0.0041	−0.0989	0.3238	0.1273	−0.0842	0.0253	0.0186	−0.0055	1.0000
GI	0.5361	−0.3397	0.4851	0.5132	−0.0539	0.2036	0.1782	0.1240	−0.0828	0.0154	0.0016	−0.0043	1.0000
AIG	0.7405	−0.3764	−0.1894	0.0901	0.3126	−0.1763	0.3209	−0.1821	0.0318	−0.0054	−0.0086	0.0064	1.0000
school	0.7248	−0.3704	−0.1007	0.0345	−0.4206	−0.3681	−0.0364	0.0530	−0.0917	0.0311	0.0130	−0.0022	1.0000
EV	4.46	3.99	1.07	0.84	0.53	0.34	0.30	0.20	0.12	0.08	0.06	0.01	
Var	37.20%	33.26%	8.91%	6.97%	4.39%	2.84%	2.53%	1.64%	1.00%	0.68%	0.50%	0.07%	
cum Var	37.20%	70.46%	79.37%	86.34%	90.73%	93.57%	96.11%	97.75%	98.75%	99.43%	99.93%	100.00%	

3.3. Results of the Multiple Regression

The multiple regression for the 11 indicators (after rearranging the indicators for multiple regression) shows no significance for the variable OCR (p = 0.7449) in the first iteration step. We have removed the variable before the next iteration. The second iteration showed no significance for the indicator GS (p = 0.2640). The third iteration showed all variables to be significant. For further information see the Table 5.

Table 5. Results for multiple regression Model 1 (SLRV model basic; Equation (24)) r = 0.9192, adj. r^2 = 0.8298, se = 0.4043.

Variable	Coefficienten	Standard Error	t-Statistic	*p*-Value	Lower 95%	Upper 95%
INH	0.2077	0.0506	4.1019	0.0001	0.1077	0.3078
INC	2.5286	0.1950	12.9670	0.0000	2.1432	2.9139
ISD	0.6203	0.1054	5.8850	0.0000	0.4120	0.8286
IR	−1.3472	0.1566	−8.6040	0.0000	−1.6566	−1.0377
Ntree	−0.1871	0.0430	−4.3529	0.0000	−0.2721	−0.1022
AlG	0.3255	0.0485	6.7130	0.0000	0.2297	0.4213
min_med	−0.3928	0.0482	−8.1507	0.0000	−0.4881	−0.2976
rest_pkt	0.1095	0.0511	2.1439	0.0337	0.0086	0.2104
school	0.2610	0.0442	5.9002	0.0000	0.1736	0.3484

The basic model for determining the individual effect sizes of the variables examined was based on a standardized data set (mean = 0 and standard deviation = 1). A corresponding comparability of the baseline situation was given. Before the effects are quantified in the following chapter, a basic evaluation of the charges on the variables was carried out at this point. The largest charges or effects are given by the variables INC (c = 2.5286, $p < 0.05$), IR (c = −1.3472, $p < 0.05$) and ISD (c = 0.6203, $p < 0.05$). It became clear that the financial sector, with the variables' Germany-wide labor market situation with corresponding income, as well as the inflation rate had the greatest influence on the standard land values. A variable from the financial sector (ISD) was also identified as having less influence but was the third largest. Distance to the nearest medical facility (min_med c = −0.3928, $p < 0.05$) and to the nearest allotment garden (AlG c = 0.3255, $p < 0.05$) were the next largest influencing variables. The last variable to take on a more profound significance in our research question was the number of trees with c = −0.1871 ($p < 0.05$). For a complete overview of the variables and the corresponding statistics, please refer to Table 5.

Since standardized values were used for the initial model (see Figure 7), the mean value of the SLRV could be used to back-calculate to the functional unit. The back calculation had this form SLRV = SLRV mean + SLRV result × SLRV mean. Due to standardization, SLV result = 0 means that SLRV must take the value of SLRV mean. Since a drift of the forecast for a residential area was evident due to the standardization (see Figure 8). The back-calculated model 1 has a RMSE of EUR 88.65 per m^2. Now, another improvement of the model has been made. For this purpose, linear and quadratic functions were used. The linear function could not improve the multiple correlation coefficient (r = 0.9192), but the model quality improved from adj. r^2 = 0.8297 to adj. r^2 = 8440. The RMSE was EUR 30.12 per m^2. The quadratic equation for model fitting yielded an improvement in the multiple correlation coefficient (r = 0.9504) and that of the model goodness (r^2 = 0.9019). The RMSE was EUR 23.80 per m^2. The coefficients, as well as the parameter statistics, can be found in the Tables 6 and 7. The graphical Illustration can be seen in the Figures 9 and 10. The comparison of the individual models was realized in the Figure 11.

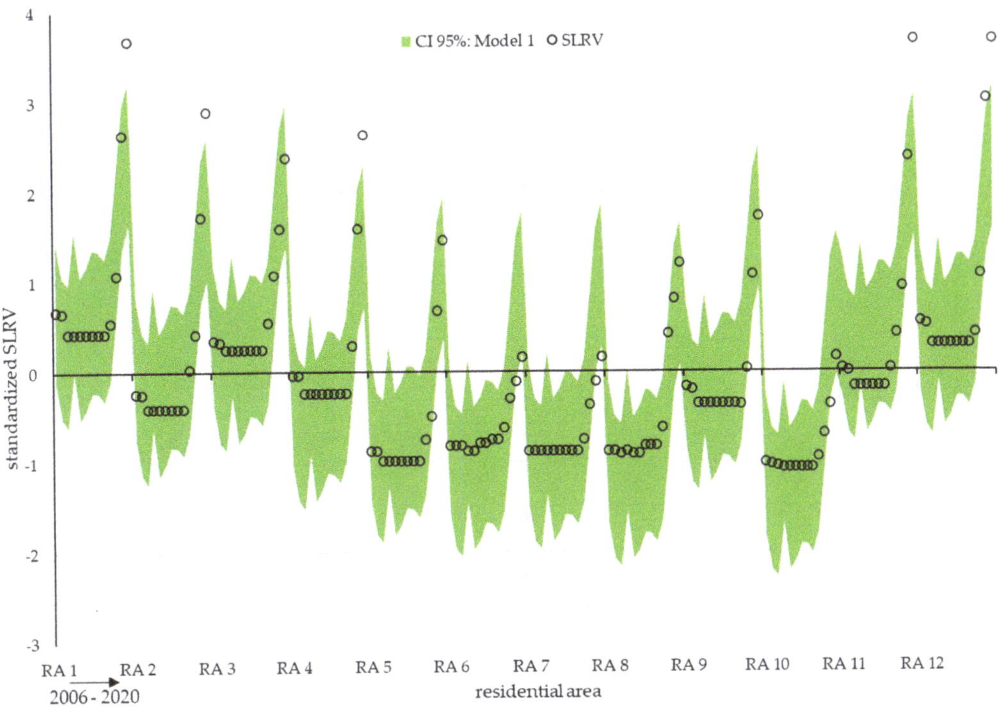

Figure 7. Presentation of Model 1 (without adjustment) based on the standardized data for the corresponding residential areas (RAi, see Figure 1).

Table 6. Results for multiple regression Model 2 Equation (25) r = 0.9192, adj. r^2 = 0.8440, se = EUR 30.31 per m^2.

Variable	Coefficienten	Standard Error	t-Statistic	*p*-Value	Lower 95%	Upper 95%
b	90.9930	3.5926	25.3282	0.0000	83.8960	98.0901
m	0.4575	0.0158	28.9734	0.0000	0.4263	0.4887

Table 7. Results for multiple regression Model 3 Equation (26) r = 0.9504, adj. r^2 = 0.9019, se = EUR 24.03 per m^2.

Variable	Coefficienten	Standard Error	t-Statistic	*p*-Value	Lower 95%	Upper 95%
b	103.7918	3.1455	32.9971	0.0000	97.5776	110.0060
m	0.1844	0.0311	5.9293	0.0000	0.1230	0.2459
n	0.0006	0.0001	9.5908	0.0000	0.0005	0.0008

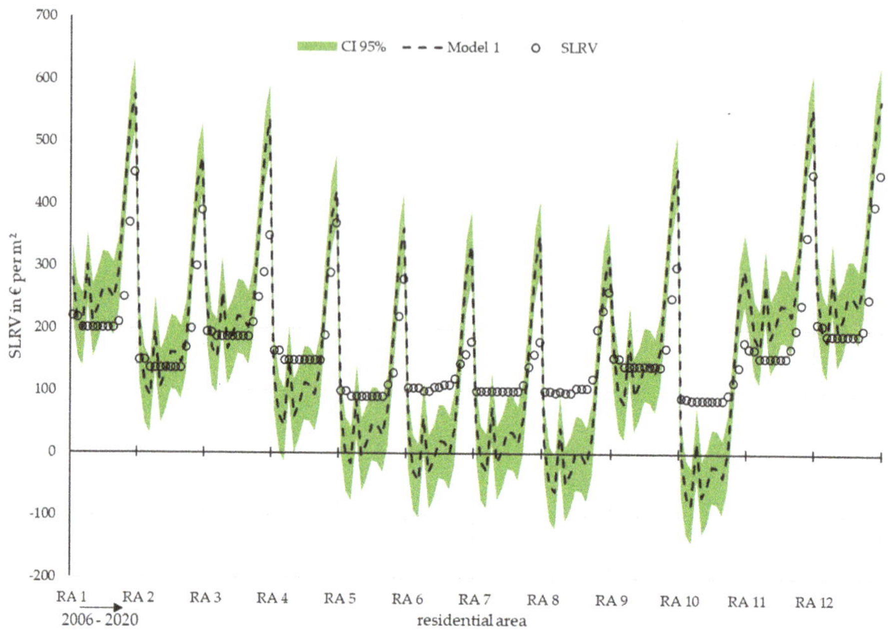

Figure 8. Presentation of Model 1 (without adjustment) based on the back-transformed data for the corresponding residential areas in EUR per m^2.

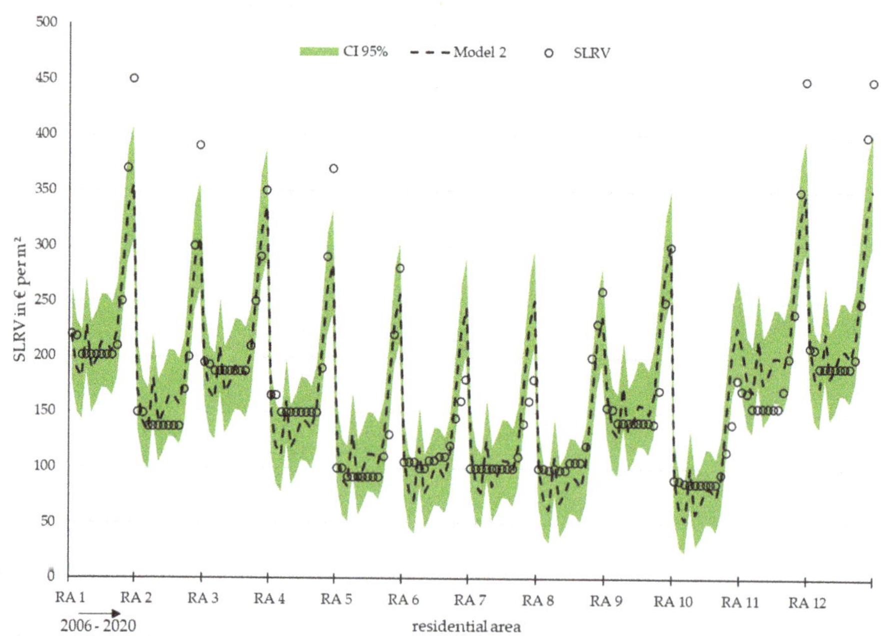

Figure 9. Presentation of Model 2 (with linear Model fitting) based on the back-transformed data for the corresponding residential areas in EUR per m^2.

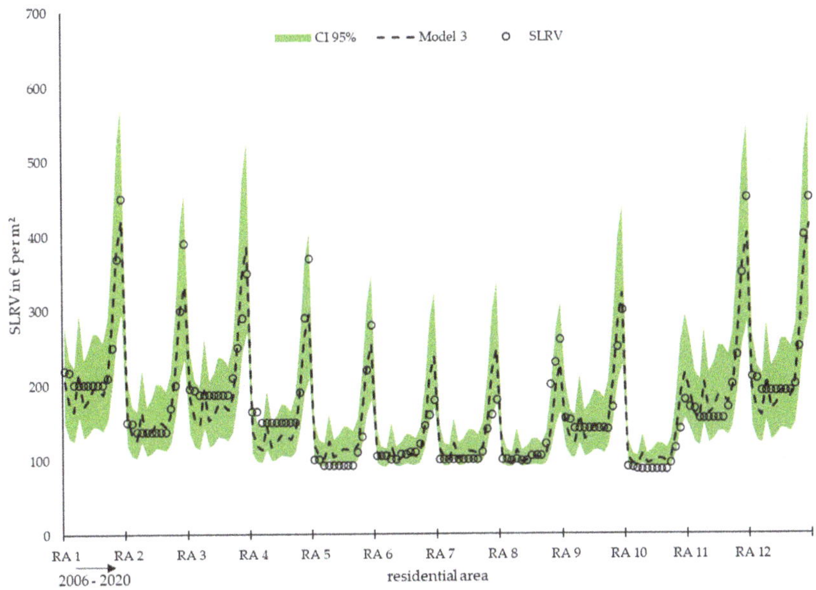

Figure 10. Presentation of Model 3 (with quadratic Model fitting) based on the back-transformed data for the corresponding residential areas in EUR per m^2.

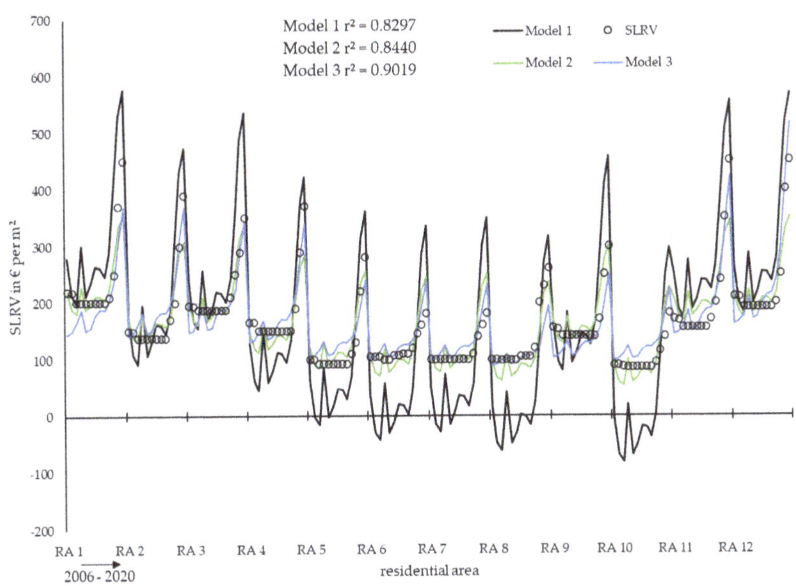

Figure 11. Presentation of the Model comparison, based on the back-transformed data for the corresponding residential areas in EUR per m^2.

3.4. Relationship between the Standard Land Values and Components of the Green Infrastructure

After the extraction of indicators and the adjustment of the model quality, the focus will now be on the effects on the SLRV that are associated with a change in variables. From the regression, it was evident that INC had the greatest influence, so to describe the interaction between SLRV, Ntree and AIG, we must out clouded effects from other variables. For this we had made the two scenarios "tree planting" and "land transformation allotment garden". It should be mentioned again that a standardized value of 0 is equivalent to the mean value of a variable. Accordingly, the interrelationships shown in the table are the initial situation for the scenarios. The baseline model is expressed in Table 8.

Table 8. Basic data for the scenario development and model interpretation based on the "Zero Model".

	SLRV (Model 3) EUR per m^2	INH Items per ha	School m	INC EUR per a	ISD % per a	IR %	Ntree Items	AIG m	min_med m	rest_pkt m
Mean	152.67	77	504	866,933,142,769	1.3	10.6	231	421	212	237
Z	0	0	0	0	0	0	0	0	0	0

The next look is at the number of trees in public green spaces. Our database ranges from a minimum of 5 to a maximum of 828, with an average of 231 trees per residential area. A residential area with only five trees therefore had an estimated increase in value compared to the mean of EUR 11.22 (CI 95%: −7.94–27.82) EUR per m^2. Meanwhile, an area with 830 trees was worth less by an estimated −25.19 (CI 95%: −45.16–1.75) EUR per m^2. Simplified, it can be stated that the SLRV per percentage change from the average of trees individuals decreases by an average of 0.11 (CI 95%: −0.54–0.36) EUR per m^2. Furthermore, a change of −52.5% in the number of trees caused a rising of 3.9 (CI 95%: −9.1–15.9) % in the SLRV. Likewise, an increase in the number of trees about 51.2% causes a decrease in SLRV of −3.6 (CI 95%: −17.1–10.7) %. Thus, the past 14 years indicate that in the evaluation of standard land values, urban trees as an important service provider for human health are considered in contrary contradiction or underrepresented. The result of the scenario "tree planting" was presented with the main parameters in the Figure 12 and with the 95% confidence interval in the Figure 13.

The results for allotments were like that for trees. In general, it could be deduced from the results that the standard land values increase with a change in percentage from the mean distance to allotment garden sites about 0.49 (CI 95%: −0.36–1.42) EUR per (m^2 × %). The distance to the allotments ranged from 196 to 916 m in our data set, with an average of 421 m distance (crow flies). Thus, 195 m to an allotment garden site causes a reduction in the standard land value by 21.62 (CI 95%: −31.72–−14.30) EUR per m^2. While an increase in the path distance to 990 m means that about 74.74 (CI 95%: 16.75–148.16) EUR per m^2 more can be achieved. Furthermore, a change of −53.7% in the distance to the allotment sites caused a reduction of 14.2 (CI 95%: −20.8–−9.4) % in the SLRV. Likewise, an increase in the distance to allotment gardens of 53.2% causes an increase in SLRV of 17.0 (CI 95%: −4.6–41.6) %. The result of the scenario "land transformation allotment garden" was presented with the main parameters in the Figure 14 and with the 95% confidence interval in the Figure 15.

In summary, when comparing the effects, a change in the number of trees for 2006 is less pronounced than a change in 2020 (see Figure 16). While the estimated SLRV increased by about 187% between 2006 and 2020, the effect increased by 1424.5% for 800 trees and for 300 m distance to allotments, the effects increased by 1334.8%.

It can be positively stated that there was no evidence of "green gentrification" in the formation of SLRV. On the negative side, it appears that the cultural values and regulatory services of green infrastructure are not being considered. Furthermore, there is a risk that important urban ecosystems or their components will fall victim to profiteering. For example, allotment gardens offer great potential to produce regional food, while at the same time providing an educational contribution for young and old, as well as being important refuges for wildlife. With increasing distance to the allotment garden sites, on the one hand, the obstacle to visit them regularly increases. Also, the trees not considered analogous to the ecosystem services can be stated. Thus, a reduction in street trees can lead to a considerable thermal load in the summer months. The reduced leaf area yields a reduced filtering function in the context of particulate air pollutants. These important aspects are not interpreted or misinterpreted when determining the SLRV. Finally, the present study indicates that (a) a uniform evaluation matrix for the SLRV should be created, and (b) the present subjective land assessments by the relevant experts should be complemented through targeting further training in the ecologically oriented planning context.

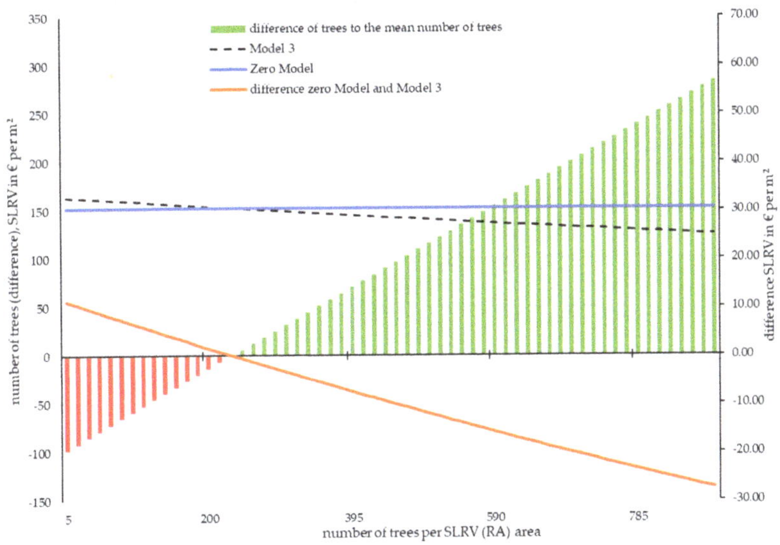

Figure 12. Presentation of the relationships between the number of street trees and standard land value. Explanation of the "difference of trees to the mean number of trees" (bar chart): increase in tree individuals compared to the mean (green), decrease in tree individuals compared to the mean (red).

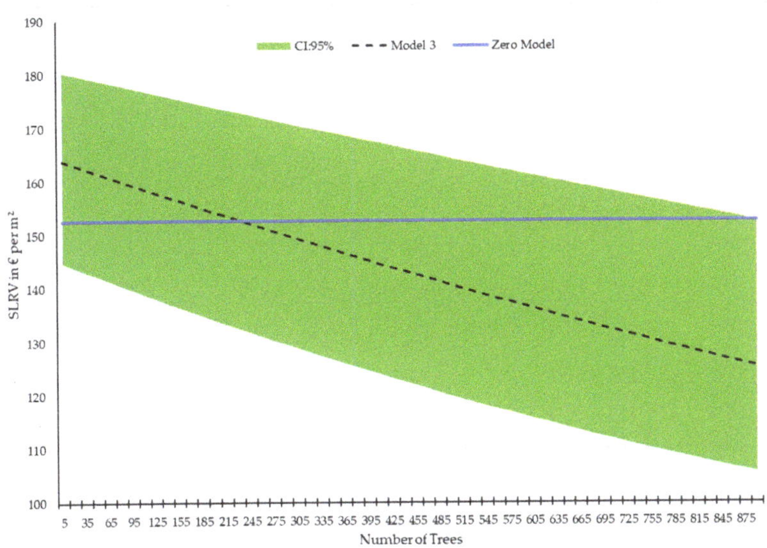

Figure 13. Presentation of the relationships between the number of street trees and standard land value with confidence interval.

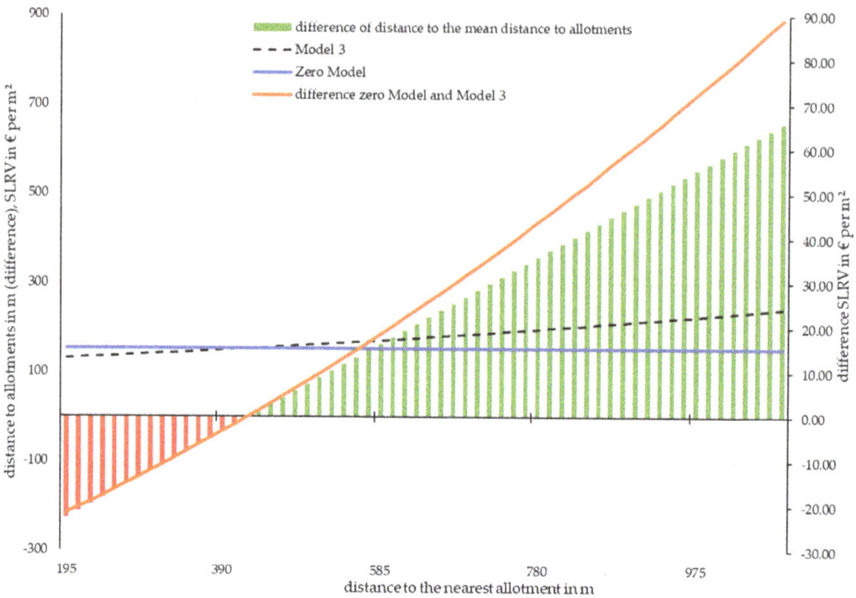

Figure 14. Presentation of the relationships between the distance to allotment gardens and standard land value. Explanation of the "difference of distance to the mean distance to allotments" (bar chart): increase in the distance to the nearest allotment garden compared to the mean (green), decrease in the distance to the nearest allotment garden compared to the mean (red).

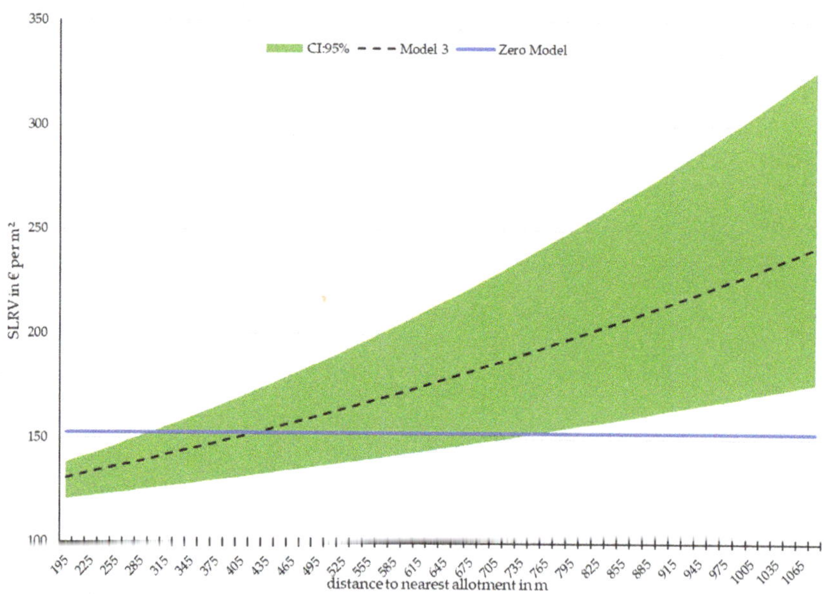

Figure 15. Presentation of the relationships between the distance to allotment gardens and standard land value with confidence interval.

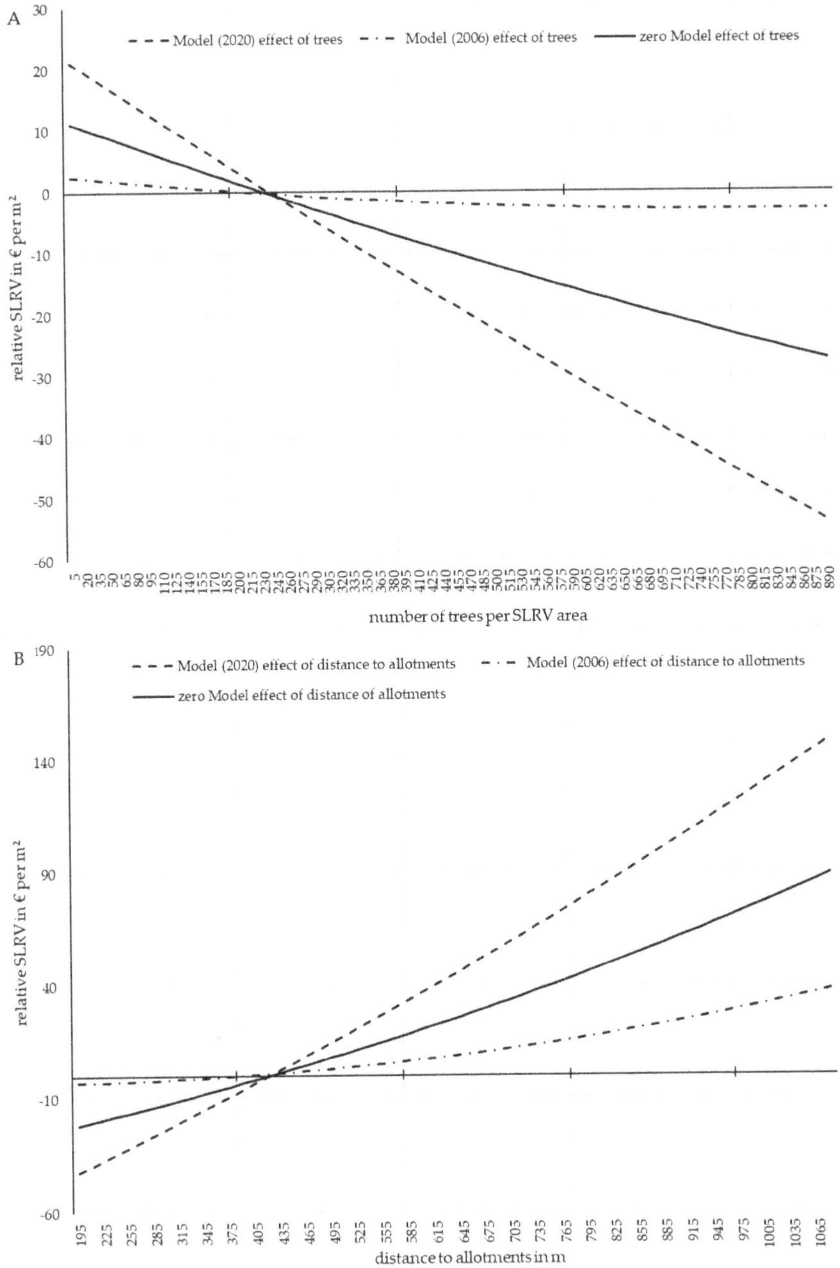

Figure 16. (**A**) Effect of trees for 2006, zero Model, 2020; (**B**) Effect of distance to allotments for 2006, zero Model, 2020 on the SLRV.

4. Discussion

The first important point to be taken up in the discussion is the underlying database. On the one hand, only a relatively short period of land use change was available in the area under consideration. An expansion of the study area or number of residential areas could contribute to improving the validity of the results and their corresponding interpretation. Furthermore, it would be important at this point to include other regionalized social factors, such as the level of education or regional income (cf. Kardan). Instead of or as a supplement to the BDI, estimates from vegetation maps could be used. Referring to Methorst et al. (2021) [51], the actual diversity and abundance of birds should also be used.

There were hardly any data available for new buildings, the closure of vacant lots or the creation and management of green spaces. Also, there was no known change in the number and distance of the medical facility, public transport stop, schools and retail outlets for the respective objects considered. The offered cold rents are based on a linear function for objects <2011; corresponding errors must be assumed. The BDI according to Gloor et al. 2018 [32] was strongly dependent on the tree species and their number, which was why there might be a relation. However, a high BDI indicates native tree species such as *Quercus robur*, *T. cordata* or the genus Malus. The test procedures were met, except for the KMO criterion for Kendall and Pearson.

The principal component analysis revealed loadings in two dimensions for all indicators, so that principal correlations were highlighted. Therefore, some significance can be assumed. Overall, however, it should be noted that the commonality for SLV is only about 0.8. Conversely, about 20% of the variance for SLRV could not be explained. This is why multiple regression was used to identify the effects of the indicators on SLRV.

The initial eleven indicators, which were supposed to describe SLRV after conversion, could be further reduced due to significance. The multiple regression showed that large effects of market economy (INC, IR and ISD) are significant for the pricing of standard land values. For green infrastructure, allotments showed the larger effect. This indicator can be interpreted as increasing densification of the settlement area. The largest distances were evident from the highly dense core areas (especially WG 11) or areas with old development structures. It can be stated based on the examined areas and the available database that for every meter distance to an allotment site, the SLRV increases by 0.49 (CI 95%: −0.36–1.42) EUR per ($m^2 \times$ %). This may well be methodical, as allotment garden sites are readily used by ecological lending in planning offices as well as local government for land development. Despite all the discussion about species extinction, loss of genetic diversity and formation of resilience for the urban ecosystem, it is not the monocultural agricultural land with almost no benefit, except for a food production, but important urban ecosystems that are required of great importance for sustainable design that are decimated. Now, if the adjacent area is "favorable", after the land use change from allotments to settlement area, the SLRV of the adjacent settlement area is used.

The tree individuals showed analogous results such as allotment gardens; here too it can be generalized that the less "green", the higher the SLRV. Highly developed urban areas usually do not offer space for many tree plantings, but seem to have value in terms of grey infrastructure. In the densely populated areas, the existing tree pits are not revitalized even after the trees have died. Furthermore, the SLRV does not provide an incentive to bring greener into the city.

If it is now assumed that public landowners want to achieve the highest possible financial benefit and are guided by the SLRV, no tree planting or other green infrastructure elements are required. However, it is imperative to note that the SLRV does not represent the asking prices; in structurally weak regions the properties are sometimes classified with, e.g., 20% of the SLRV. It is, therefore, imperative to carry out further investigations that represent a more detailed model. The focus here is on, for example, the value of ecosystem services in districts and the corresponding asking rents or the land or real estate sales prices.

5. Conclusions

Against the background of the social transformation towards a climate-neutral, ecologically oriented society, it is likely that there will be a change in mentality with regard to space marketing. Large real estate marketers are now also dealing with the ecological potential of their areas and their ecosystem services as a market argument and are having their areas ecologically upgraded. This development represents an opportunity for tasks of the century. In addition to the SLRV investigation, the determination of CO_2 binding potential, soil resources (e.g., for water retention in the area), fine dust binding potential and biomass resources represent criteria with which a quantitative upgrading of the areas can be balanced, and should be subject to future investigations.

In the future, the ecological quality of the outdoor facilities, for example, can also be an award criterion when determining the "most economical offer" for construction measures. In order to be able to allocate the costs and benefits of green spaces in the future, for example, in the award procedure, the entire life cycle costs of green spaces must be taken into account during planning, maintenance and demolition.

Sufficient areas are needed for green spaces, even in conurbations with high settlement pressure. Vertical green cannot fully replace the loss of green space. It is, therefore, important to also use the potential of unused areas for urban green spaces. Successful municipal land recycling has a high symbolic power for the establishment of a circular land use economy, in which one use becomes a prerequisite for the subsequent one and where a site, as a whole, remains in use in the long term. To the extent that fallow land is successfully reused, further concepts for saving land gain in persuasiveness (Dosch et al. 2006 [52], Jakubowski et al. 2007 [53]).

New planning processes, procedures and instruments are needed as part of green space governance. These should inform citizens and owners early and comprehensively about the future use and design of properties and at the same time address the ideas and needs, worries and concerns of the resident population. In addition, citizens are asked to contribute their energy, time and creativity to the design of communally usable properties for the benefit of the general public. This is known internationally as the Land Governance Assessment Framework (LGAF), which aims to improve the transparency of usage decisions and contributes to private owners being involved in offsetting land value increases through green spaces (called direct and indirect value capture in Anglo-Saxon countries).

Author Contributions: Conceptualization, P.S. and T.F.; methodology, T.F.; software, T.F., validation, P.S.; formal analysis, T.F., investigation, T.F.; resources, P.S.; data curation, T.F., writing—original draft preparation, P.S.; writing—review and editing, T.F.; visualization, T.F.; supervision, P.S.; project administration, P.S.; funding acquisition, P.S. All authors have read and agreed to the published version of the manuscript.

Funding: This research received no external funding. It was performed within the activities of the working group Ecological Engineering at Magdeburg-Stendal University of Applied Sciences.

Data Availability Statement: Data are available from the authors upon reasonable request.

Conflicts of Interest: The authors declare no conflict of interest.

References

1. LABO Bund/Länder-Arbeitsgemeinschaft Bodenschutz (Hrsg.) (2012): Reduzierung der Flächeninanspruchnahme. Statusbericht zu den LABO-Berichten vom 21.09.2011 sowie 30.03.2010. Bericht zur Vorlage an die Umweltministerkonferenz. Available online: https://www.labo-deutschland.de/documents/1_Anlage_LABO_Reduzierung_der_Flaecheninanspruchnahme_f11.PDF (accessed on 14 October 2015).
2. Hoymann, J.; Goetzke, R. Flächenmanagement, pp. 154–196, ARL—Akademie für Raumforschung und Landesplanung (Hrsg.): Handwörterbuch der Stadt-und Raumentwicklung Hannover 2018, ISBN 978-3-88838-559-9 (PDF-Version). 2018. Available online: http://nbn-resolving.de/urn:nbn:de:0156-55993 (accessed on 10 July 2023).
3. Ingenieurtechnischer Verband Altlasten, e.V. (ITVA), Berlin 1998, 25 S., Erarbeitet im ITVA-Fachausschuss C 5 Flächenrecycling". Available online: https://www.itv-altlasten.de/publikationen/arbeitshilfen-und-richtlinien/flaechenrecycling.html (accessed on 28 June 2023).
4. Bundesministerium für Umwelt, Naturschutz, Bau und Reaktorsicherheit (BMUB). *Grün in der Stadt—Für eine Lebenswerte Zukunft*; Grünbuch Stadtgrün: Berlin, Germany, 2015.
5. Government of Germany; Thiel, F. *Wertsteigerung Durch Grünflächen—Wer Profitiert?* Informationen zur Raumentwicklung: Bonn, Germany, 2016; pp. 691–701.

6. Schneider, P.; Fauk, T.; Mihai, F.-C. Reclaiming Urban Brownfields and Industrial Areas-Potentials for Agroecology. In *Agroecological Approaches for Sustainable Soil Management*; Prasad, M.N.V., Kumar, C., Eds.; Wiley: Hoboken, NJ, USA, 2023; pp. 409–436, ISBN 978-1-119-91198-2.
7. Debus, M. Bodenrichtwertableitung in der Praxis. In *Kreditwirtschaftliche Wertermittlungen. 8*; Pohnert, F., Ed.; Auflage: Wiesbaden, Germany, 2015.
8. Kleiber, W. *Verkehrswertermittlung von Grundstücken, Kommentar und Handbuch. 7*; Auflage: Köln, Germany, 2014.
9. The American Society of Landscape Architects. Professional Practice—Green Infrastructure. 2019. Available online: https://www.asla.org/greeninfrastructure.aspx (accessed on 22 January 2019).
10. Kardan, O.; Gozdyra, P.; Misic, B.; Moola, F.; Palmer, L.J.; Paus, T.; Berman, M.G. Neighborhood greenspace and health in a large urban center. *Sci. Rep.* **2015**, *5*, 11610. [CrossRef]
11. Gann, G.; Lamb, D. *Ecological Restoration: A Means of Conserving Biodiversity and Sustaining Livelihoods*; Society for Ecological Restoration International: Tucson, AZ, USA, 2006.
12. Fauk, T. Entwicklung eines biologisch-physikalisch basierten Ansatz zur Quantifizierung und Inwertsetzung ausgewählter Ökosystemleistungen von Dachbegrünungen als auch Dachbewässerungssystemen anhand des Stadtteils Stadtfeld Ost der Landeshauptstadt Magdeburg. Master's Thesis, Magdeburg-Stendal University of Applied Sciences, Magdeburg, Germany, 2021.
13. Schwaiger, E.; Berthold, A.; Gaugitsch, H.; Götzl, M.; Milota, E. Wirtschaftliche Bedeutung von Ökosystemleistungen Monetäre Bewertung—Risiken und Potenziale. Umweltbundesamt. 2015. Available online: http://hdl.handle.net/11159/78 (accessed on 10 July 2023).
14. Wittowsky, D.; Hoekveld, J.; Welsch, J.; Steier, M. Residential housing prices: Impact of housing characteristics, accessibility and neighbouring apartments—A case study of Dortmund, Germany. *Urban Plan. Transp. Res.* **2020**, *8*, 44–70. [CrossRef]
15. Anderson, S.T.; West, S.E. Open space, residential property values, and spatial context. *Reg. Sci. Urban Econ.* **2006**, *36*, 773–789. [CrossRef]
16. Boyle, A.; Barrilleaux, C.; Scheller, D. Does walkability influence housing prices? *Soc. Sci. Q.* **2014**, *95*, 852–867. [CrossRef]
17. Cervero, R.; Duncan, M. Neighbourhood composition and residential land prices: Does exclusion raise or lower values? *Urban Stud.* **2004**, *41*, 299–315. [CrossRef]
18. Li, W.; Joh, K.; Lee, C.; Kim, J.-H.; Park, H.; Woo, A. Assessing Benefits of Neighborhood Walkabiliy to Single-Family Property Values—A Spatial Hedonic Study in Austin, Texas. *J. Plan. Educ. Res.* **2015**, *35*, 471–488. [CrossRef]
19. Lynch, A.K.; Rasmussen, D.W. Measuring the impact of crime on house prices. *Appl. Econ.* **2001**, *33*, 1981–1989. [CrossRef]
20. Song, Y.; Knaap, G.-J. Measuring the effects of mixed land uses on housing values. *Reg. Sci. Urban Econ.* **2004**, *34*, 663–680. [CrossRef]
21. Geurs, K.T.; van Wee, B. Accessibility evaluation of land-use and transport strategies: Review and research directions. *J. Transp. Geogr.* **2004**, *12*, 127–140. [CrossRef]
22. Bourne, L. *The Geography of Housing*; Halsted Press: New York, NY, USA, 1981. [CrossRef]
23. Hanhörster, H.; Lobato, R. Migrants Access to the Rental Housing in Germany: Housing Providers and Allocation Policies. *Urban Plan.* **2021**, *6*, 7–18. [CrossRef]
24. Galvin, R. Do housing rental and sales markets incentivise energy-efficient retrofitting of western Germany's post-war apartments? Challenges for property owners, tenants, and policymakers. *Energy Effic.* **2023**, *16*, 25. [CrossRef] [PubMed]
25. Köhler, T. Bewertung ausgewählter Ökosystemleistungen im Grünzug der Goetheanlage Magdeburg, Master thesis an der Hochschule Magdeburg-Stendal, Studiengang Ingenieurökologie. Master's Thesis, Magdeburg-Stendal University of Applied Sciences, Magdeburg, Germany, 2022.
26. Ali, L.; Haase, A.; Heiland, S. Gentrification through Green Regeneration? Analyzing the Interaction between Inner-City Green Space Development and Neighborhood Change in the Context of Regrowth: The Case of Lene-Voigt-Park in Leipzig, Eastern Germany. *Land* **2020**, *9*, 24. [CrossRef]
27. Bockarjova, M.; Botzen, W.J.W.; van Schie, M.H.; Koetse, M.J. Property price effects of green interventions in cities: A meta-analysis and implications for gentrification. *Environ. Sci. Policy* **2020**, *112*, 293–304. [CrossRef]
28. Rigolon, A.; Németh, J. "We're not in the business of housing": Environmental gentrification and the nonprofitization of green infrastructure projects. *Cities* **2018**, *81*, 71–80. [CrossRef]
29. Tsai, I.-C.; Lin, C.-C. Influence of migration policy risk on international market segmentation: Analysis of housing and rental markets in the euro area. *Econ. Res. Ekon. Istraživanja* **2023**, *36*, 2121740. [CrossRef]
30. Mell, I.C.; Henneberry, J.; Hehl-Lange, S.; Keskin, B. To green or not to green: Establishing the economic value of green infrastructure investments in The Wicker, Sheffield. *Urban For. Urban Green.* **2016**, *18*, 257–267. [CrossRef]
31. Dell'Anna, F.; Bravi, M.; Bottero, M. Urban Green infrastructures: How much did they affect property prices in Singapore? *Urban For. Urban Green.* **2022**, *68*, 127475. [CrossRef]
32. Gloor, S.; Göldi Hofbauer, M. Der ökologische Wert von Stadtbäumen bezüglich der Biodiversität (The ecological value of urban trees with respect to biodiversity). In *Jahrbuch der Baumpflege*; 22. Jg.; SWILD: Zürich, Switzerland, 2018; pp. 33–48, ISBN 978-3-87815-257-6.
33. Landesvermessungsamt Sachsen-Anhalt, Germany: © GeoBasis-DE/LVermGeo ST (2023). Available online: https://www.geodatenportal.sachsen-anhalt.de/ASmobile/?appid=BORIS (accessed on 25 April 2023).

34. Kaltmieten für Magdeburg: HOMEDAY—Preisatlas. Available online: https://www.meinestadt.de/magdeburg/immobilien/mietspiegel (accessed on 25 April 2023).
35. Wohnungsboerse—Magdeburg: Tabelle Wohnungsmieten Vergleich im Jahr 2011–2022. Available online: https://www.wohnungsboerse.net/mietspiegel-Magdeburg/7772 (accessed on 25 April 2023).
36. Deutsche Bundesbank. Entwicklung des Durchschnittlichen Zinssatzes für Spareinlagen* in Deutschland in den Jahren von 1975 bis 2022 [Graph]. In *Statista*. von. Available online: https://de.statista.com/statistik/daten/studie/202295/umfrage/entwicklung-des-zinssatzes-fuer-spareinlagen-in-deutschland/ (accessed on 2 February 2023).
37. Anzahl der Erwerbstätigen in Deutschland nach dem Inlandskonzept von 1991 bis 2022 (in 1.000) [Graph], Statistisches Bundesamt. Available online: https://de.statista.com/statistik/daten/studie/214465/umfrage/erwerbstaetige-in-deutschland-nach-inlandskonzept/ (accessed on 16 May 2023).
38. Höhe des Durchschnittlichen Nettolohns/Nettogehalts im Jahr je Arbeitnehmer in Deutschland von 1991 bis 2022 [Graph], Statistisches Bundesamt. [Online]. Available online: https://de.statista.com/statistik/daten/studie/74510/umfrage/nettogehalt-im-jahr-je-arbeitnehmer-in-deutschland/ (accessed on 2 June 2023).
39. Inflationsrate in Deutschland von 1992 bis 2022 (Veränderung des Verbraucherpreisindex Gegenüber Vorjahr) [Graph], Statistisches Bundesamt. [Online]. Available online: https://de.statista.com/statistik/daten/studie/1046/umfrage/inflationsrate-veraenderung-des-verbraucherpreisindexes-zum-vorjahr/ (accessed on 22 February 2023).
40. Landesvermessungsamt Sachsen-Anhalt, Germany: © GeoBasis-DE/LVermGeo ST. 2023. Available online: https://www.lvermgeo.sachsen-anhalt.de/de/gdp-open-data.html (accessed on 10 July 2023).
41. Ziegler, C. Bevölkerung und Demografie 2021—Magdeburger Statistik (Population and Demographics 2021—Magdeburger Statistics). City of Magdeburg (eds.). 2021. Available online: https://www.magdeburg.de/PDF/Bev%C3%B6lkerung_Demografie_2021_Heft_110.PDF?ObjSvrID=37&ObjID=49447&ObjLa=1&Ext=PDF&WTR=1&_ts=1660024825 (accessed on 5 November 2022).
42. Landeshauptstadt-Magdeburg—Amt für Stadtgärten und Friedhöfe. 2022. Available online: https://opendata.unser-magdeburg.de/dataset/baumkataster-magdeburg-2021 (accessed on 10 November 2022).
43. Kalusche, D. *Ökologie in Zahlen (Ecology in Numbers)*; Springer: Berlin/Heidelberg, Germany, 2016; ISBN 978-3-662-47986-5.
44. Fauk, T.; Schneider, P. Modeling urban tree growth as a part of the green infrastructure to estimate ecosystem services in urban planning. *Front. Environ. Sci.* **2023**, *11*, 1090652. [CrossRef]
45. Tucker, C.J.; Sellers, P.J. Satellite remote sensing of primary production. *Int. J. Remote Sens.* **1986**, *7*, 1395–1416. [CrossRef]
46. Schneider, P.; Fauk, T. The Role of Allotment Gardens for Connecting Nature and People. In *Human-Nature Interactions—Exploring Nature's Values Across Landscapes*; Misiune, I., Depellegrin, D., Egarter Vigl, L., Eds.; Springer Nature: Berlin/Heidelberg, Germany, 2022; pp. 261–271. [CrossRef]
47. Bartlett, M.S. *The Effect of Standardization on a X^2 Approximation in Factor Analysis*; Biometrika—Oxford Press: Oxford, UK, 1951; pp. 337–344. [CrossRef]
48. Backhaus, K.; Erichson, B.; Gensler, S.; Weiber, R.; Weiber, T. *Multivariate Analysemethoden (Multivariate Analysis Methods). Münster, Magdeburg, München and Trier (Germany)*; Springer: Berlin/Heidelberg, Germany, 2021. [CrossRef]
49. Steiner, M.; Grieder, S.G. EFAtools: An R package with fast and flexible implementations of exploratory factor analysis tools. *J. Open Source Softw.* **2020**, *5*, 2521. [CrossRef]
50. Kassambara, A.; Mundt, F. Package: Factoextract, Version 1.0.7. 2020. Available online: https://cran.r-project.org/web/packages/factoextra/factoextra.pdf (accessed on 10 July 2023).
51. Methorst, J.; Rehdanz, K.; Mueller, T.; Hansjürgens, B.; Bonn, A.; Böhning-Gaese, K. The importance of species diversity for human well-being in Europe. *Ecol. Econ.* **2021**, *181*, 106917. [CrossRef]
52. Dosch, F.; Bergmann, E.; Jakubowski, P. Perspektive Flächenkreislaufwirtschaft, Kreislaufwirtschaft in der Städtischen/Stadtregionalen, Flächennutzung—Fläche im Kreis. Ein ExWoSt-Forschungsfeld, Band 1 Theoretische Grundlagen und Planspielkonzeption. 2006. Available online: https://www.bbsr.bund.de/BBSR/DE/veroeffentlichungen/ministerien/bmvbs/sonderveroeffentlichungen/2006/flaechenkreislaufwirtschaft.html (accessed on 28 August 2023).
53. Jakubowski, P.; Dosch, F.; Bergmann, E. Zur Theoretischen Konzeption einer Flächenkreislaufwirtschaft. *Z. Für Umweltpolit. Und Umweltr.* **2007**, *30*, 325–350.

land

MDPI

Article

Inclusion of Nature-Based Solution in the Evaluation of Slope Stability in Large Areas

Lukáš Zedek [1,*,†,‡], Jan Šembera [1,‡] and Jan Kurka [2,‡]

1 Institute of Novel Technologies and Applied Informatics, Faculty of Mechatronics, Informatics and Interedisciplinary Studies, Technical University of Liberec, Studentská 1402/2, 46117 Liberec, Czech Republic
2 AZ Consult, Spol. s r.o., Klíšská 12, 40001 Ústí nad Labem, Czech Republic
* Correspondence: lukas.zedek@tul.cz
† Current address: Institute of Novel Technologies and Applied Informatics, Faculty of Mechatronics, Informatics and Interedisciplinary Studies, Technical University of Liberec, Heydukova 1181/2, House C2, Office C02012, 46117 Liberec, Czech Republic.
‡ These authors contributed equally to this work.

Abstract: In areas affected by mining, which are undergoing reclamation, their geotechnical characteristics need to be monitored and the level of landslide risk should be assessed. This risk should preferably be reduced by nature-based solutions. This paper presents a KurZeS slope stability assessment technique based on areal data. This method is suitable for large areas. In addition, a procedure is presented for how to incorporate a prediction of the impact of nature-based solutions into this method, using the example of vegetation root reinforcement. The paper verifies the KurZeS method by comparing its results with the results of stability calculations by GEO5 software (version 5.2023.52.0) and validates the method by comparing its results with a map of closed areas in the area of the former open-cast mine Lohsa II in Lusatia, Germany. The original feature of the KurZeS method is the use of a pre-computed database. It allows the use of an original geometrical and geotechnical concept, where slope stability at each Test Point is evaluated not just along the fall line but also along different directions. This concept takes into account more slopes and assigns the Test Point the lowest safety factor in its vicinity. This could be important, especially in soil dumps with rugged terrain.

Keywords: slope failure risk prediction; slope stabilization; nature-based solutions; revegetation; root reinforcement

1. Introduction

Large-scale landslides can cause injuries, casualties, property damage and social as well as ecological problems [1]. Therefore, there is a strong need to prevent these events. This can be achieved and supported by an appropriate landslide susceptibility assessment, realized through a correctly selected evaluation method or software. The evaluation methods were intensively developed in the past decades. Nowadays, as mentioned in [2] and similarly in [3], they are usually divided into several classes: (1) qualitative methods that are based on expert experience, (2) semi-quantitative methods that are supplemented with scores/weights assigned to particularly assessed landslide susceptibility triggering aspects and (3) quantitative methods that can be further divided into geotechnical methods (physically based slope stability evaluation) [4–8], statistical analysis [2,9], neural network methods [10,11] or neuro-fuzzy logic methods [12,13].

In this paper, a physically based geotechnical model (a quantitative method) is considered. It is specifically a two- or three-layered geotechnical model with the use of the Bishop [14] (circular slip surface) and Sarma [15] methods (polygonal slip surface). In this paper, the method is referred to as KurZeS. Whereas many methods and software tools based on similar ideas use an assumption of infinite slope, e.g., [16–22], KurZeS does not, similarly to methods implemented in GRASS GIS [23–25], but it differs significantly in its geometrical concept described in Section 2 and also in its technical implementation using interpolation based on pre-computed data. The method provides an alternative procedure for slope stability assessment in large areas and an original implementation allowing calculations to be performed on desktop computers with limited computing power.

Shallow landslides, in particular, that can be assessed as described in [16,18–20,22,26–29] can be partially avoided through nature-based solutions (NBS). Root reinforcement [30,31] is the particular effect that stabilizes the slope in a natural way. Root reinforcement has not been investigated since at least 70 years ago. It is usually simulated either through additional cohesion of the rooted top soil layer [32] or through the Finite Element Method [33].

In KurZeS, the root reinforcement effect is simulated using the additional root cohesion concept. Evaluation of the additional root cohesion was described in [32,34–37]. KurZeS uses approximate values derived from values mentioned in [34,35,37].

For the visualization of the inputs and outputs of the model, we used the freely available geographic information system QGIS [38].

This article aims to analyze the influence of selected, nature-based solutions on the determination of the approximate slope stability on a large area. As typical examples, one can take areas intended for reclamation after open-pit coal mining. In the case of large areas, the prevailing reclamation method is afforestation or covering with other vegetation. Tree species recommended for reclamation of brown coal dumps are birch (*Betula*), larch (*Larix*), red oak (*Quercus rubra*) and pine (*Pinus*) [39]. In such a case, the main stabilization effect comes from the rooting of the surface layer.
The analysis is carried out in stages as follows:

1. Evaluation of the top layer shear parameter increase due to the effect of rooting. This stage was based on the available literature data.
2. The application of increased parameters in the large area surface stability model and realization of calculations using the original method created by the authors of the article.

The article is structured in five numbered sections. The first, the Introduction section, includes the background and context of the presented method. The second, the Materials and Methods section, is devoted to the description of the KurZeS method itself, its features and problems, the inclusion of nature-based solutions, the description of the testing problem and the description of the verification and validation approach. The third, the Results section, includes the results of the testing problem and results of the verification computations of specific details, and their comparison. The fourth, the Discussion section, includes specific observations concerning the KurZeS method and the inclusion of the NBS in it. The last, the Conclusions section, summarizes the main results and observations of the article.

2. Materials and Methods

2.1. Study Area

For this study, the area of the soil dump of the former open-cast mine Lohsa II in Lusatia, Germany, see Figure 1, was chosen. The map presented below was obtained from [40]. The area is interesting because several landslides have appeared there in recent times. Specifically, the slopes of the mining waste dump Scheibe at the northwest border of the closed coal mine Lohsa II moved in the spring of 2019. Even later (in the spring of 2021), landslides appeared on the slopes near the lake Knappensee.

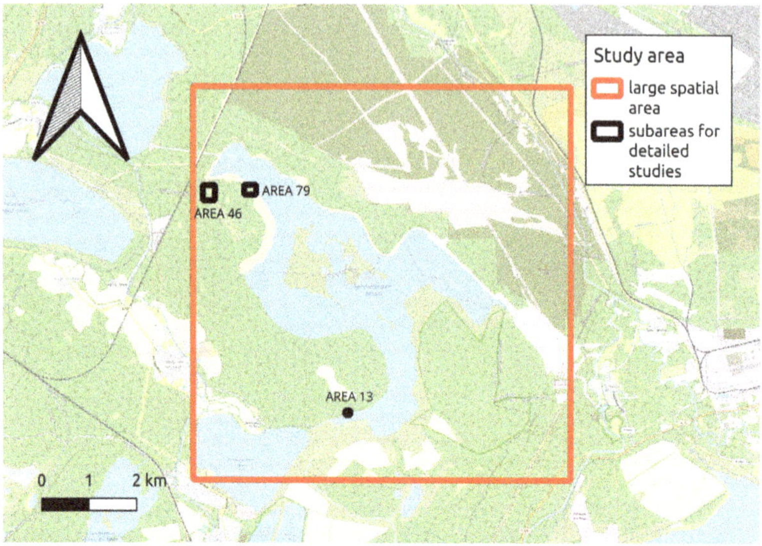

Figure 1. Study area selected for slope stability assessment (red) and three subareas for verification of results.

2.2. Geometrical Concept

The used data describing the digital elevation model (DEM) of the terrain within the study area are organized in triplets of coordinates [X; Y; Z] with a regular step of 10 m on both the X and Y axes. The safety factor is evaluated in each point of the 10 × 10 m raster (this point is referred to as the TP—the Test Point). The safety factor corresponding to the TP is derived as the minimal safety factor of all relevant slopes in the TP's neighborhood. The slopes relevant to the TP are defined by pairs of boundary points (BP_1 and BP_2) satisfying the following conditions:

- The boundary points are different points located at the points of the 10 × 10 m raster not further than 100 m from the TP.
- The TP may coincide with one of the boundary points.
- The total length of the (possibly angled) line $BP_1 - TP - BP_2$ must be at least 20 m and at most 100 m.
- The convex obtuse angle of line segments $BP_1 - TP$ and the$TP - BP_2$ shall be greater than or equal to 120°.

The safety factor of the TP is then evaluated as the minimum of the safety factors of all slopes between BP_1 and BP_2 satisfying the above-specified conditions. The algorithm KurZeS includes an optimization trick based on the obvious equivalence of the slope $BP_1 - BP_2$ and $BP_2 - BP_1$ reducing the number of safety factor estimations to a half (see Figure 2).

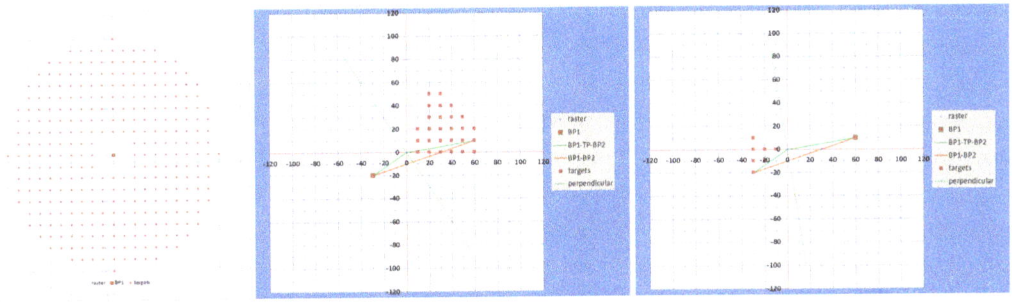

(a) (b) (c)

Figure 2. (**a**) The set of endpoints for $BP_1 \equiv TP$; (**b**,**c**) example of repetition of found pair BP_1 and BP_2 in case $BP_1 \neq TB \neq PB_2$.

2.3. *Geotechnical Concept*

The safety factor of each slope $BP_1 - BP_2$ is evaluated with interpolation from a pre-calculated database based on the following geotechnical model constructed in GEO5 [41] (version 5.2023.52.0).

2.3.1. Model

The basic model is constructed as two-layered, with a constant gradient of the terrain in variants in the range from 1 in 2 to 1 in 10. The interface between the soil layers runs parallel to the terrain at a depth of 5 m. The parameters of the top layer are exactly those of the soil it contains. The underlying layer is supposed to consist of the same type of soil but firmer, which is modeled by higher stability parameters compared to the top layer. This increase reflects the usual improvement with depth (see Figure 3a and Table 1).

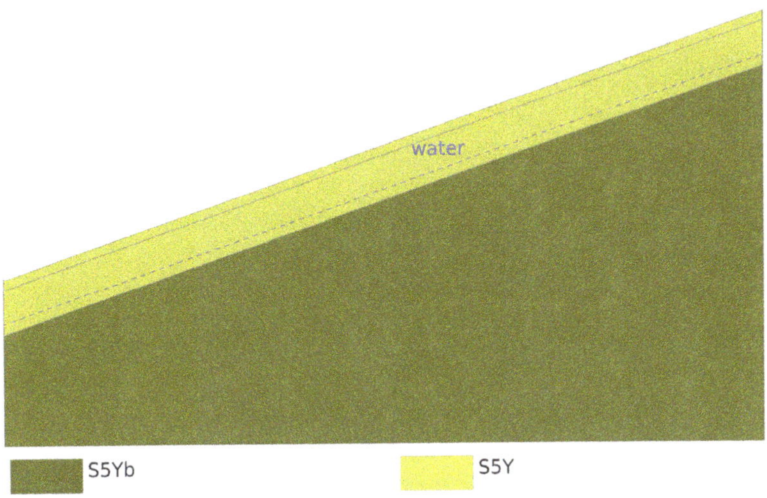

(a)

Figure 3. *Cont.*

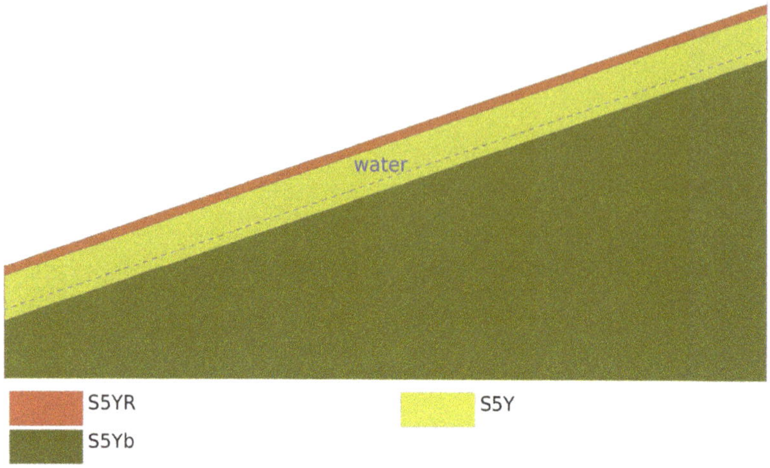

(b)

Figure 3. (a) Two-layer model; (b) three-layer model.

A more complex model is used for the case of root reinforcement. Here, the third layer of a constant thickness of 0.9 m is added to the surface (see Figure 3b and Table 1). The used soil parameters in it are also higher than in the original soil, this time due to root reinforcement. The specific values of increase are derived from the literature [34,35,37].

The groundwater level is considered parallel to the surface in variants from a depth of −20 to 0 m.

According to the method used, circular (Bishop method [14]) and polygonal (Sarma method [15]) slip surfaces with a fixed length from 20 to 100 m are considered.

2.3.2. Resulting Database

The result of an individual stability calculation in the given model is the search for the least favorable slip surface (either circular or polygonal, using the optimization procedure) and the determination of the safety factor F_s by the limit equilibrium method (LEM). F_s values were calculated for all options including 8 slope inclinations, 6 slope lengths, 7 groundwater levels and 2 cases with and without a rooted topsoil layer. For each selected soil type, the results database consists of 672 F_s values.

The safety factor of a specific slope $BP_1 - BP_2$ is derived from this database so that the value of the corresponding slope inclination, slope length and groundwater level of the slope $BP_1 - BP_2$ is specified and the F_s value is interpolated from neighboring values tabulated in the database.

It is typical for the particular two- or three-layer model with parallel construction described above that the type and shape of the slip surface changes as the least favorable slip surface lengthens. Short slip surfaces with the lowest F_s are typically circular in shape (B—Bishop) confined in the uppermost layer, or in the central and rooted layer, not extending into the lowest firmer layer with an interface 5 m below the surface (Figure 4a). The intermediate slip surfaces are most commonly polygonal (S—Sarma) following the top of the interface at 5 m depth. In terms of type, they most closely resemble the infinite slope model (Figure 4b). In some cases, usually with the groundwater level between depths of 3 and 6 m below ground level, long slip surfaces of circular shape cross the aforementioned interface at a depth of 5 m below the surface and extend deep into the firmer layer (BD—Bishop, deep)—Figure 4c.

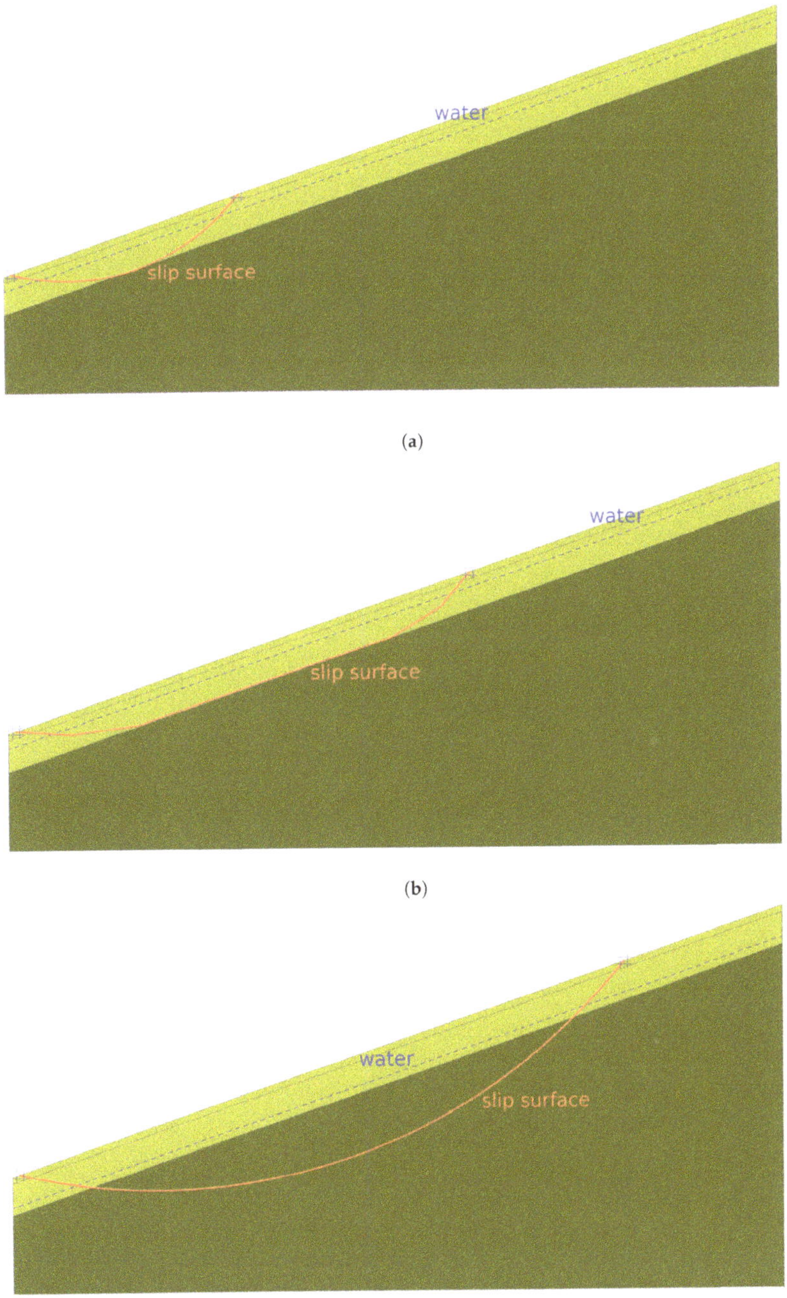

(c)

Figure 4. (**a**) Short circular slip surface; (**b**) medium-length polygonal slip surface; (**c**) long deep circular slip surface.

It is not clear in advance which of the slip surfaces will be the least favorable. The result is mainly influenced by the geometry of the model, soil parameters and the position of the water table or surface water level. It should be emphasized repeatedly that the results are based on the chosen model; for a differently designed model, the composition of the least favorable slip surfaces could be different.

2.4. *Quantification of the Effect of Rooting*

Figure 5 shows an example from the resulting database of F_s for one selected soil type denoted as S5Y (sandy dump). The left half of the table applies to the model without a rooted topsoil layer, and the other to the model with a rooted layer. In the header of the table, the parameters used for effective strength (effective angle of internal friction (°) and cohesion (kPa)) and bulk weight (kN/m^3) are typed. The first column shows the depth of the water table, the second column shows the slope inclination and the first row under the heading shows the fixed length of the slip surface. The types of slip surfaces in the sense of the text above are distinguished by color for cells with the resulting safety factors F_s (short circular slip surface in blue, medium-length polygonal slip surface in green and deep circular slip surface in red).

S5Y		Φ_{eff}	c_{eff}	γ	γ_{sat}		
	1	25	2	19	19.5		
	1plus	29	4	20	20.5		
		25	18	19	19.5		
gndwater	1 in ?	20	30	40	60	80	100
-20	3	1.75	1.66	1.62	1.56	1.54	1.52
-6	3	1.75	1.67	1.62	1.56	1.53	1.49
-4	3	1.75	1.66	1.57	1.49	1.43	1.35
-3	3	1.75	1.53	1.44	1.35	1.29	1.27
-2	3	1.58	1.36	1.28	1.18	1.13	1.12
-1	3	1.31	1.16	1.08	1.01	0.99	0.96
0	3	**0.97**	**0.90**	**0.86**	**0.82**	**0.80**	**0.79**
-20	4	2.33	2.22	2.15	2.08	2.04	2.02
-6	4	2.33	2.22	2.15	2.08	2.04	2.00
-4	4	2.33	2.22	2.09	1.96	1.90	1.80
-3	4	2.33	2.05	1.92	1.78	1.73	1.69
-2	4	2.11	1.82	1.70	1.55	1.51	1.50
-1	4	1.76	1.54	1.44	1.34	1.30	1.27
0	4	**1.29**	**1.19**	**1.14**	**1.09**	**1.06**	**1.05**
-20	5	2.90	2.76	2.69	2.60	2.55	2.52
-6	5	2.90	2.76	2.69	2.60	2.55	2.50
-4	5	2.90	2.76	2.62	2.46	2.39	2.25
-3	5	2.90	2.57	2.40	2.23	2.14	2.11
-2	5	2.63	2.28	2.12	1.98	1.89	1.86
-1	5	2.18	1.93	1.80	1.68	1.62	1.60
0	5	**1.60**	**1.49**	**1.42**	**1.36**	**1.32**	**1.30**
-20	6	3.48	3.32	3.22	3.12	3.06	3.02

S5Y root		Φ_{eff}	c_{eff}	γ	γ_{sat}		
	1	25	2	19	19.5		
	1plus	29	4	20	20.5		
		25	18	19	19.5		
gndwater	1 in ?	20	30	40	60	80	100
-20	3	2.01	1.82	1.72	1.62	1.58	1.55
-6	3	2.01	1.82	1.72	1.62	1.58	1.50
-4	3	2.02	1.77	1.66	1.55	1.44	1.36
-3	3	1.93	1.63	1.53	1.40	1.33	1.28
-2	3	1.71	1.46	1.36	1.23	1.17	1.14
-1	3	1.46	1.25	1.17	1.06	1.02	0.98
0	3	**1.18**	**1.02**	**0.95**	**0.89**	**0.84**	**0.81**
-20	4	2.70	2.43	2.30	2.17	2.10	2.08
-6	4	2.70	2.42	2.30	2.17	2.10	2.01
-4	4	2.70	2.36	2.21	2.02	1.93	1.81
-3	4	2.57	2.17	2.03	1.84	1.78	1.71
-2	4	2.30	1.94	1.81	1.61	1.56	1.54
-1	4	1.96	1.66	1.55	1.41	1.35	1.31
0	4	**1.56**	**1.34**	**1.26**	**1.16**	**1.16**	**1.09**
-20	5	3.35	3.03	2.87	2.69	2.62	2.57
-6	5	3.35	3.04	2.87	2.69	2.62	2.51
-4	5	3.35	2.95	2.76	2.54	2.41	2.27
-3	5	3.20	2.73	2.54	2.32	2.20	2.13
-2	5	2.86	2.42	2.24	2.06	1.94	1.90
-1	5	2.42	2.08	1.92	1.77	1.67	1.65
0	5	**1.92**	**1.68**	**1.57**	**1.45**	**1.39**	**1.35**
-20	6	4.05	3.63	3.42	3.23	3.14	3.09

Figure 5. Example from the resulting database of F_s for S5Y soil (sandy dump).

2.5. *Data Sets*

All data sets were obtained from freely available sources and they were transformed into the form of georeferenced rasters, where the centers of gravity of particular raster cells are the nodes of the network described in Section 2.2 dedicated to the geometrical concept of the model.

2.5.1. Digital Elevation Model

The main data source for performed simulations was the digital elevation model (DEM) DGM1 (German DEM [42]). This DEM (multiple shape files) was rasterized with the use of QGIS [38]. The selected cell edge length was 10 m. This selection of the cell edge length brought a compromise between acceptable computation times (huge number of elements/nodes) and sufficient accuracy of the slope description, see Figure 6.

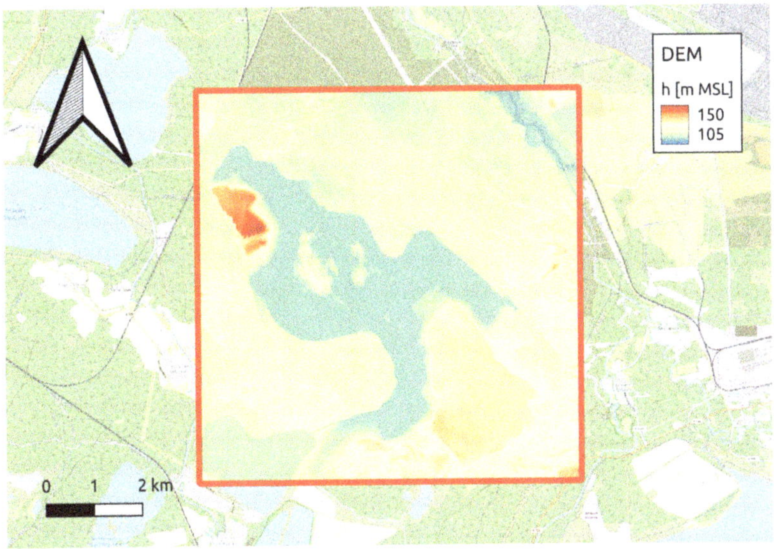

Figure 6. Digital elevation model rasterized with cell edge length 10.0 m.

2.5.2. Ground Water Table

Because of missing data, the groundwater table was derived from the DEM. For this purpose, raster calculator "gdal_calc", which is a part of GDAL-extension [43] for QGIS, was used together with Equation (1), where W means water elevation, A represents a surface elevation in DEM-raster, 115 stands for the selected lake water level, 0.75 defines the "speed" of water table increase and 20 and 0 are the maximal and minimal groundwater table depths, respectively.

$$W = \begin{cases} 115 & \text{if } A \leq 115 \\ A - 20 & \text{if } (A - 115) \cdot (1 - 0.75) > 20 \\ 115 + (A - 115) \cdot 0.75 & \text{otherwise} \end{cases} \quad (1)$$

The resulting water table raster is shown in Figure 7.

2.5.3. Type of Soils

A pedological map was obtained from LfULG (Bodenkarte BK50 [44]). Shapefiles from the original map were merged and divided into two or four (non-rooted soil and rooted soil, respectively) different groups, and they were rasterized in the same spatial extent and with the same cell size (10 × 10 m) as the DEM, see Figure 8a. For the case of rooted soils, two further soil types appeared, see Figure 8b. The distribution of rooted and non-rooted soil was based on the difference between the DEM and the selected lake water level (115 m MSL). Rooted soils were considered for positive values (above water) and non-rooted soils for negative (below water).

The strength parameters of soils can be found in Table 1.

Table 1. The strength parameters of used soil types. φ_{eff} denotes effective angle of internal friction and c_{eff} stands for effective cohesion of soil. The thickness of the topsoil layer is 5 m and there is no rooted soil layer in case of no measures. In the case of 0.9 m rooting depth, the thickness of the rooted soil layer is 0.9 m and the thickness of topsoil layer is 4.1 m. Parameter values are based on expert judgment combined with information from the literature [34,35,37].

	Basal Soil Layer		Topsoil Layer		Rooted Soil Layer	
	φ_{eff}	c_{eff}	φ_{eff}	c_{eff}	φ_{eff}	c_{eff}
S45 *	31°	16 kPa	27°	8 kPa	27°	24 kPa
S5Y *	29°	4 kPa	25°	2 kPa	25°	18 kPa

* The soil type notation comes from the Czech Technical Standard ČSN P 73 1005 (731005) Engineering geological survey [45]; the Y symbol denotes the relocated material in non-original placement.

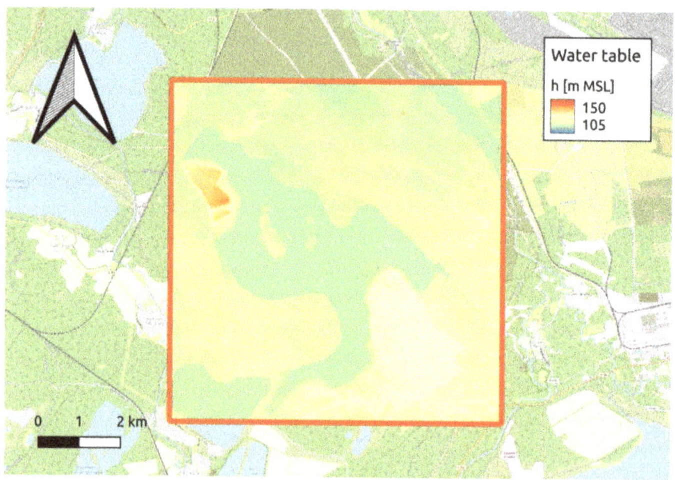

Figure 7. Linearly increasing water table.

(**a**)

Figure 8. *Cont.*

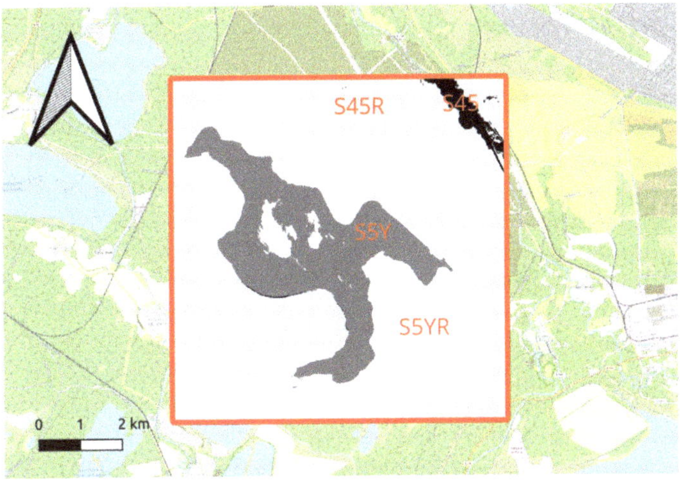

(b)

Figure 8. Simplified categorization of soil types. (**a**) Non-rooted soil types. (**b**) Soil types without (under the water) and with roots.

2.6. Implementation

Slope stability was derived from interpolation tables (resulting database, see Section 2.3.2) pre-computed in the GEO5 software [41] (version 5.2023.52.0). derived the safety factor from the slope length (in range 20 to 100 m), slope inclination (in range 1:10 to 1:2), soil type (2 types of rooted soil and 2 types without reinforcement) and depth of the water table (in range 0 to 20 m).

Two variants of interpolation were tested: linear interpolation and polynomial interpolation. As they had comparable results and linear interpolation was noticeably faster, the presented computations were performed using multidimensional linear interpolation with the use of the linearNDInterpolator implemented in SciPy [46].

The calculations work with a circular or polygonal slip surface of general depth. Such a slip surface crosses the root-reinforced layer only along the possible landslide scarp. Thus, its area is the only place where the improvement in soil properties is accounted for by this method. In this way, the application of root reinforcement differs significantly from commonly used infinite slope stability models, e.g., [47].

For the presented computations, a computer with the following parameters was used:

- Intel® Core™ i7-3770K CPU; 3.50 GHz × 4;
- RAM 15.5 GiB;
- Ubuntu 22.04.4 LTS;
- Python 3.8.16.

Computation for such a large-scale area was not possible at once because of extensive memory demands. The results were obtained by dividing the area into 36 overlapping subareas (see Figure 9), which effectively decreased the memory requirements. Eventually, the computation could be realized in four parallel threads, i.e., in four subareas at the same time. A 100 m overlapping of subareas was necessary because of the geometrical concept of the model, see Section 2.2. One turn (4 subareas) took 14 min.

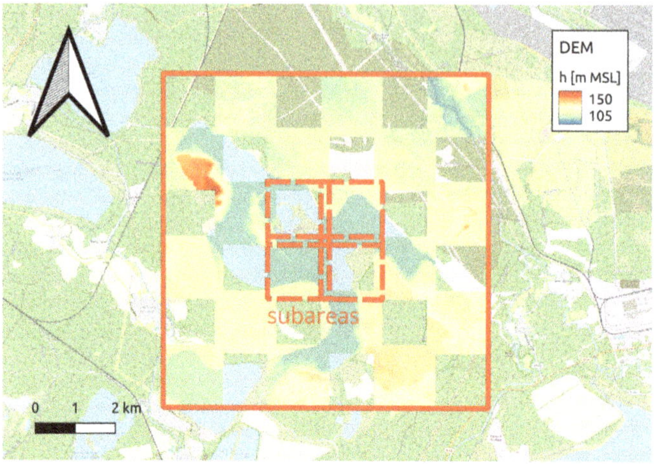

Figure 9. Division of the study area enabled parallel computation.

2.7. Verification

The mechanism for verification of the results can be found in Figure 10. Two safety factor computations were performed—one without root reinforcement and another with root reinforcement under consideration. Two hypotheses were tested as follows:

- Because all F_s in the pre-calculated database are higher for root-reinforced soils than for the soils without vegetation, the safety factor should be higher or equal in each Test Point of prediction considering root reinforcement than in prediction without root reinforcement.
- The results of the prediction should be comparable with F_s predicted by GEO5 at each slope in the area.

Several slopes in the area were taken, the surface and groundwater table cuts were taken from QGIS and included in GEO5 software and the following observations were made:

- Consideration of root reinforcement brought either the same or higher value of the predicted safety factor in every Test Point.
- For selected cuts, the predicted safety factor corresponded closely to the values obtained from GEO5.

For more details, see Section 3.

The two above-mentioned points verified the expected behaviour of the simulation code.

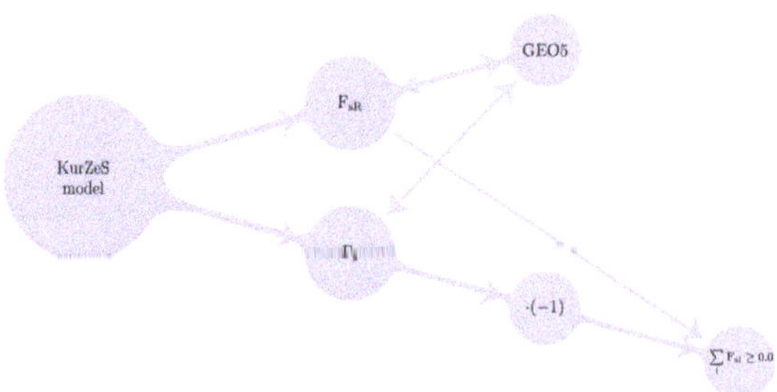

Figure 10. Verification diagram.

2.8. Validation

The approach for validation of the results is shown in Figure 11. The computed safety factor for all the cells in the study area with and without additional root cohesion under consideration was imported to QGIS [38] and compared with the shape file map of the area closed to the public because of safety reasons. The closed area correlates well with the slopes with a predicted F_s very close to one, which validates the results with only independent accessible public data.

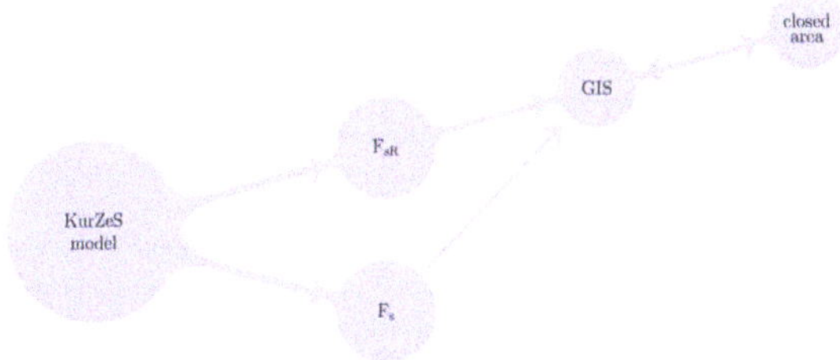

Figure 11. Validation diagram.

Another method of validation was the comparison of landslide risk maps with orthophoto maps of the study area and also with maps showing the slope gradient there. A noticeable correspondence was observed between these data, but due to the unclear, precise dating and spatial location of the landslides, these comparisons were not included in the study.

3. Results

3.1. Slope Stability Improvement Predicted by GEO5

To get a better idea about the expectable predicted safety factor improvement caused by additional root cohesion, selected data sets from the resulting database were visualized and analyzed. Safety factors of two selected soil types present in the study area, sandy clay (S45) and sandy dump (S5Y), each for one slope inclination, are depicted with and without vegetation cover (root reinforcement) in Figure 12a,b.

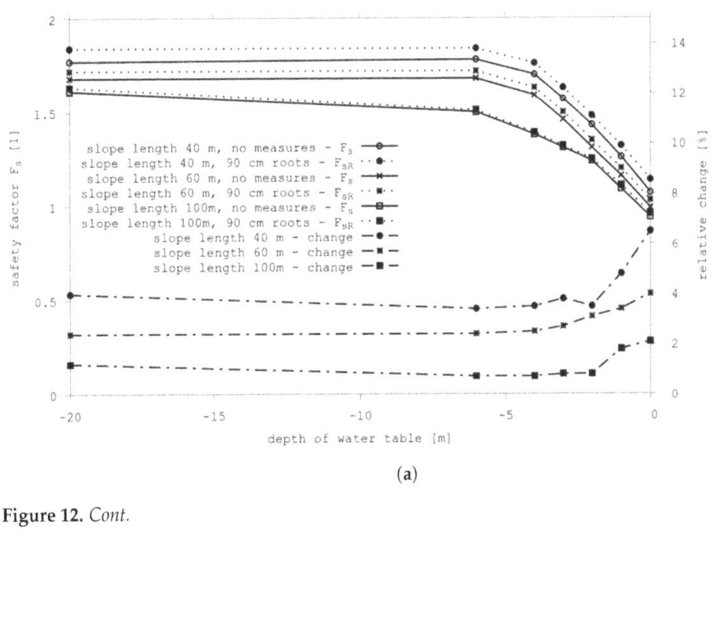

(a)

Figure 12. *Cont.*

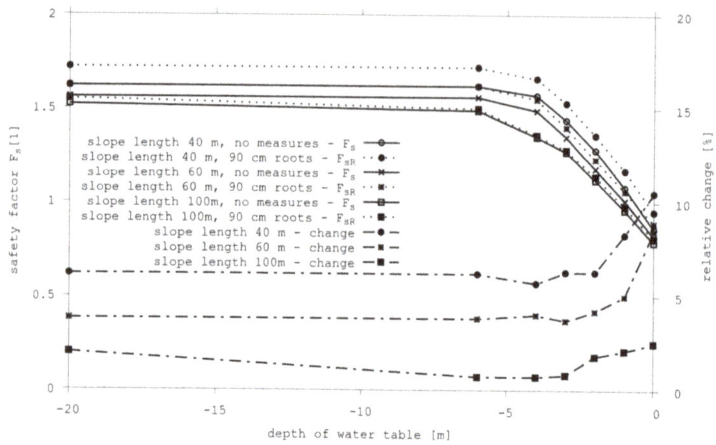

(**b**)

Figure 12. Comparison of predicted slope stability. (**a**) Sandy clay (S45), slope inclination 1 in 2.5 and three different slope lengths. (**b**) Sandy dump (S5Y), slope inclination 1 in 3 and three different slope lengths.

Both Figure 12a,b show the dependency of the absolute and relative improvement of the predicted safety factor on the depth of the water table as well as on the root reinforcement under consideration. Safety factor values (left-hand axis), as well as relative improvement (right-hand axis), decrease with growing slope length. The effect of root reinforcement is stronger in the case of sandy dump. However small the absolute improvement may seem, it can be crucial for the cases where driving and resisting forces are near to equilibrium (i.e., for F_s around 1). The most significant relative improvement is predicted for slopes where groundwater reaches near the earth's surface, i.e., the least stable slopes.

3.2. Slope Stability Evaluation with KurZeS

As a first step for the testing of KurZeS, the predicted safety factor F_s without the root reinforcement (additional cohesion) under consideration was evaluated, see Figure 13. The most landslide-threatened slopes appeared on the lake shore (e.g., "AREA 79" on the Figures 1) and on the steep slopes in the northwestern part of the waste dump Scheibe ("AREA 46" on the Figure 1).

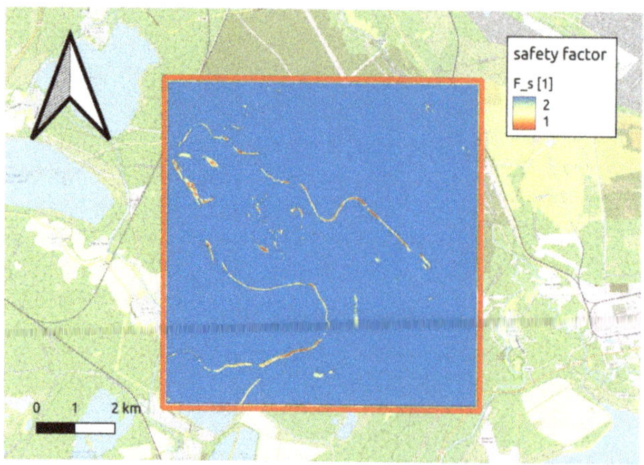

Figure 13. Slope stability evaluation without root reinforcement under consideration.

Figure 14 shows the comparison of the predicted safety factor of slopes for non-rooted soils with a shape file map of the area closed to the public. It shows good correspondence between both data sets and it shows the beneficial effect of the KurZeS approach for the identification of unstable areas suggested for further in situ investigation.

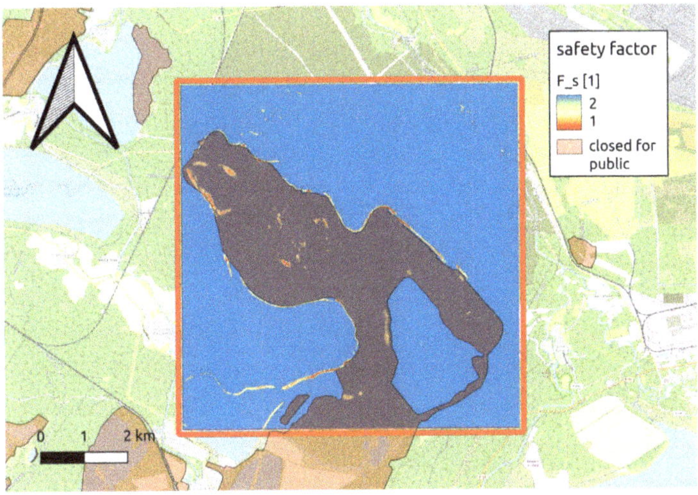

Figure 14. Comparison of results with shape file mapping area closed for public.

Another prediction was made with the addition of the root reinforcement under consideration. The resulting prediction of the safety factor is denoted as F_{sR}. The result can be seen in Figure 15. The values of F_{sR} are generally higher (better) than the values of F_s (without root reinforcement). It will be further analyzed in the following Section.

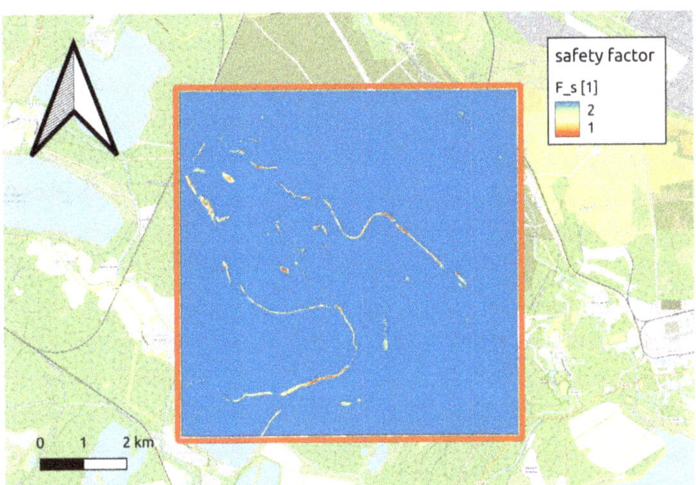

Figure 15. Slope stability evaluation with root reinforcement under consideration.

3.3. *Improvement in Slope Stability Predicted by KurZeS Due to Root Reinforcement*

From raster outputs of slope stability analyses, absolute and relative improvement in the predicted safety factor obtained with or without the roots under consideration was evaluated. The absolute improvement $F_{sR} - F_s$ reaches values in the range $\langle 0.0, 0.25 \rangle$. Relative improvement of the predicted safety factor $F_{sR}/F_s - 1$ depicted in Figure 16 reaches values in the range $\langle 0.0, 17.7\% \rangle$. Both the maximal values overreach the expectations constructed on the base from Section 3.1. The explanation coming from further analysis resides in the fact that the best improvement appears for extremely short slopes, which were not considered in Section 3.1.

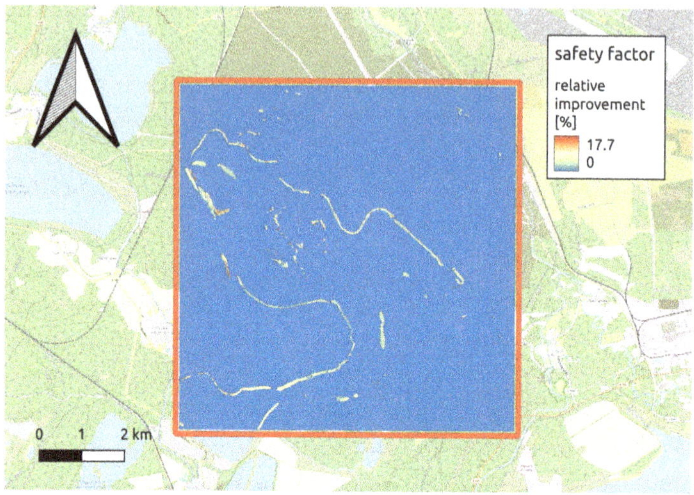

Figure 16. Relative improvement of slope stability due to root reinforcement.

3.4. Comparison of Results from KurZeS and GEO5

To verify the results of slope stability evaluation by the KurZeS approach, several further studies of slope stability, in particular mesh cells, were performed. The KurZeS results were compared with those obtained from GEO5 for the particular slopes. In this section, we present two representative examples as well as data sets prepared for the following research.

For nine Test Points, the boundary points BP_1 and BP_2 realizing the minimal associated F_s were identified. The connection $BP_1 - BP_2$ was prolonged in QGIS [38] and on such a cross-section, the model input data (DEM, depth of water table, soil type) were extracted from intersected raster cells. These cross-section data served as inputs to GEO5 software for cross-sectional safety factor evaluations. The Test Points were selected from three areas that were somehow interesting. The areas were named "AREA 13" (it included Test Points 1 to 3 and corresponding cross-sections 1 to 3), "AREA 46" (it included Test Points 4 to 6 and corresponding cross-sections 4 to 6) and "AREA 79" (which included Test Points 7 to 9 and corresponding cross-sections 7 to 9). Here, we present the results of cross-sections 5 and 8 that represent all nine cross-sections well.

3.4.1. Cross-Section 5

In Figure 17, you can see the map from [40] of "AREA 46" that includes very steep slopes that are not at the lake shore with well-defined slope direction (almost planar slopes of a long straight valley). Test Point TP_5 that is not a part of the cross-section was selected. In Figure 17, in red, 13 sampling points (SP_i, j) formed the cross-section model in GEO5.

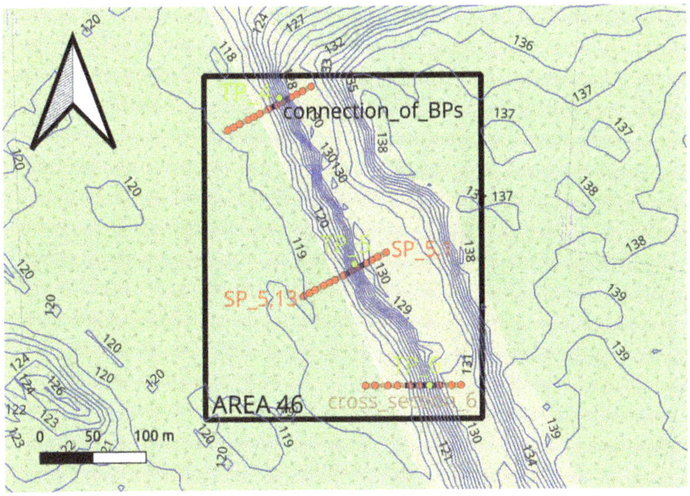

Figure 17. Map of "AREA 46".

For the non-rooted soils, KurZeS evaluated the safety factor $F_s(TP_5) = 1.10$. Results for the whole "AREA 46" are shown in Figure 18.

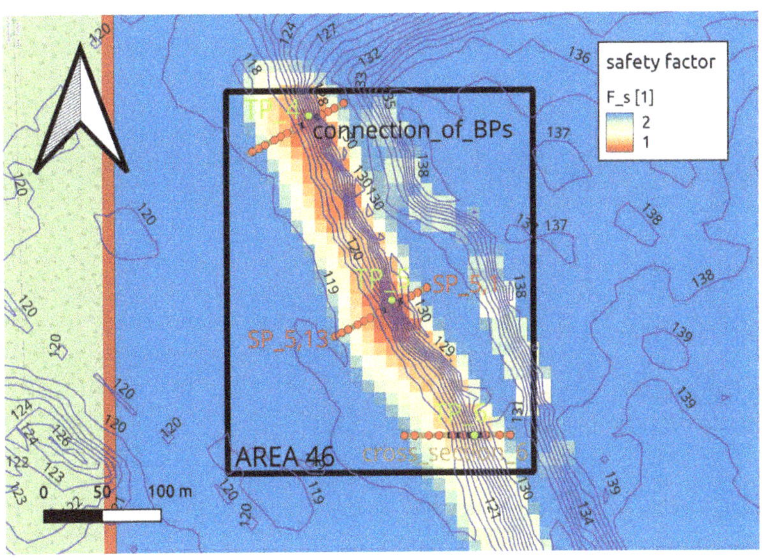

Figure 18. Safety factor predicted by KurZeS for non-rooted soil.

For the scenario with non-rooted soil, GEO5 evaluated the safety factor $F_s(TP_5) = 1.08$. Figure 19 shows a 2D cross-section with visualization of the data extracted for complementary safety factor assessment.

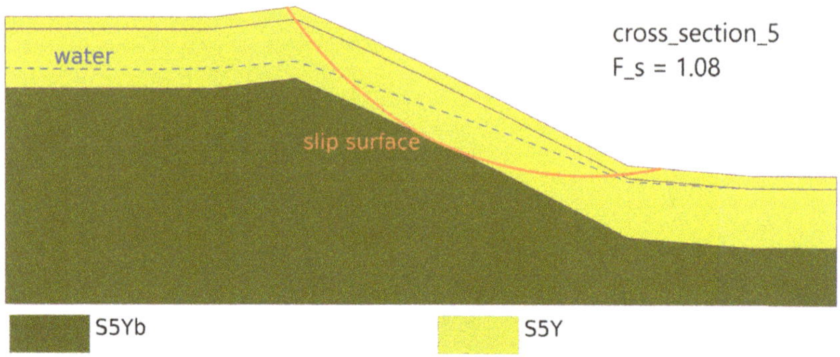

Figure 19. Safety factor predicted by GEO5 for non-rooted soil.

For the model with rooted soil, the situation is very similar. KurZeS predicted a safety factor $F_s(TP_5) = 1.18$. Figure 20 depicts the results evaluated for the whole "AREA 46". Safety factor $F_{sR}(TP_5)$ in Figure 20 evinces evident improvement in comparison with Figure 18.

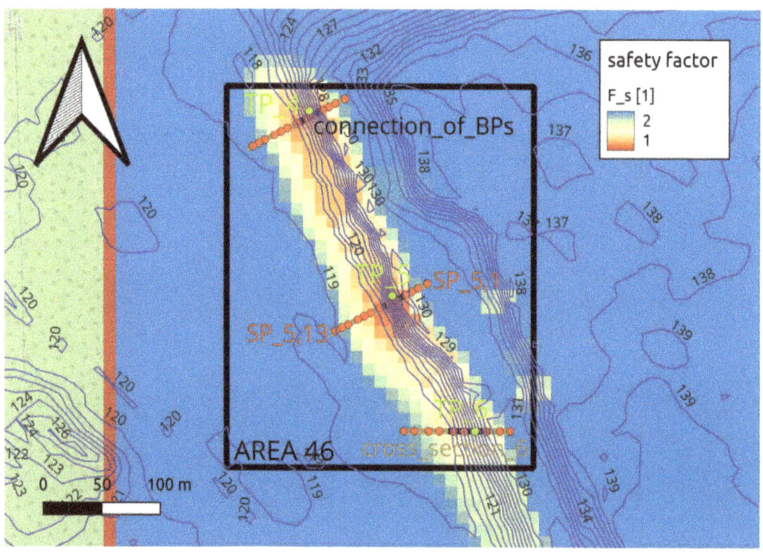

Figure 20. Safety factor predicted by KurZeS for rooted soil. TP_i denotes i-th Testing Point and SP_i, j denotes j-th Sampling Point lying on the i-th cross-section.

GEO5 evaluated the safety factor $F_{sR}(TP_5) = 1.16$ for the rooted soils. Visualization of the input data extracted along cross_section_5 is shown in Figure 21.

The difference between the resulting safety factors obtained from KurZeS and GEO5 can be denoted as negligible.

This comparison of two very close results from two different software tools verified the function of KurZeS for the case of a model without surface water.

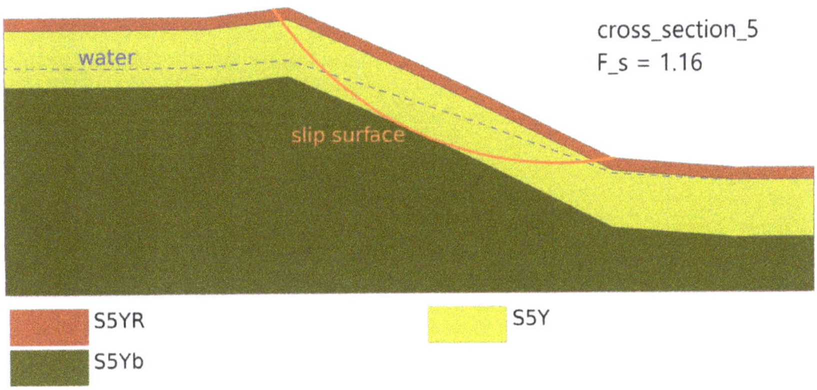

Figure 21. Safety factor predicted by GEO5 for rooted soil.

3.4.2. Cross-Section 8

Another subarea selected for verification of the results takes place at the northeastern slope of the mining waste dump Scheibe and represents very steep slopes finishing in surface water, see Figure 22. Here, we present the comparison of safety factor predictions at Test Point TP_8, which lies directly on cross-section 8.

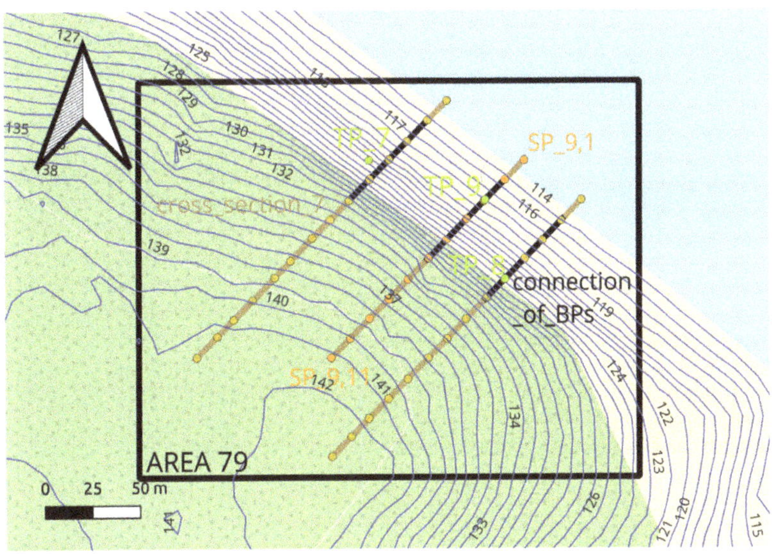

Figure 22. Map of AREA 79. TP_i denotes i-th Testing Point and SP_i, j denotes j-th Sampling Point lying on the i-th cross-section.

The KurZeS approach predicted the safety factor $F_s(TP_8) = 1.24$ for slopes covered by non-rooted soil. You can see the results for the whole "AREA 79" in Figure 23. The direction of cross-section 8 is very close to fall lines (almost perpendicular to contour lines).

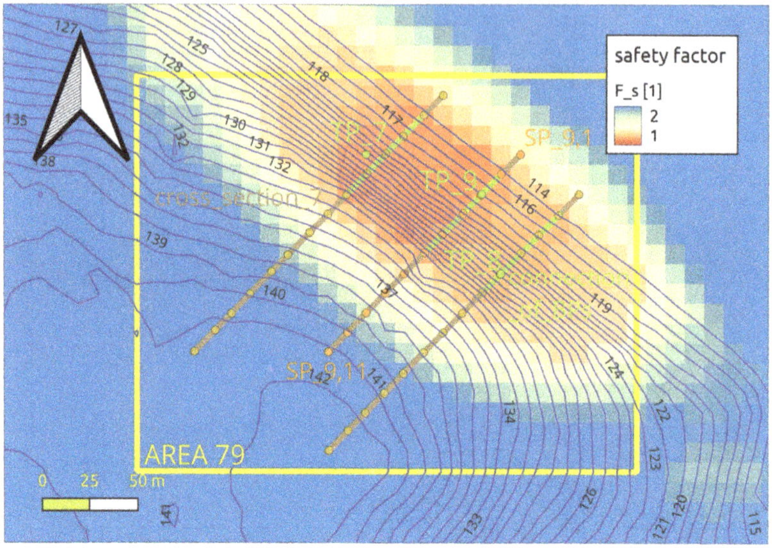

Figure 23. Safety factor predicted by KurZeS for non-rooted soil. TP_i denotes i-th Testing Point and SP_i, j denotes j-th Sampling Point lying on the i-th cross-section.

The safety factor of the slope covered by non-rooted soil and evaluated by GEO5 has a value $F_s(TP_8) = 1.25$. In Figure 24, you can see a 2D cross_section_8, which was extracted from input files for KurZeS and evaluated in GEO5.

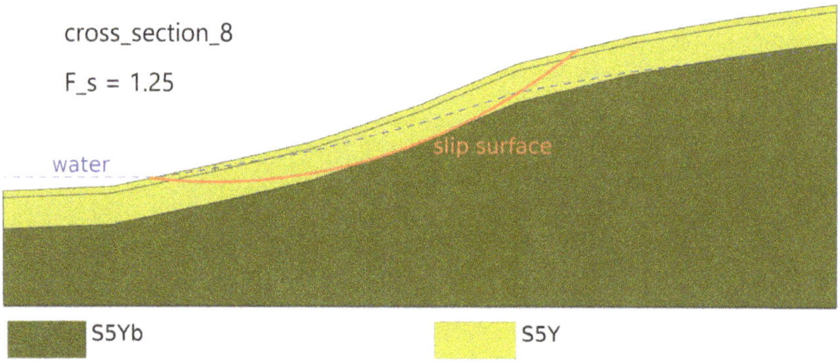

Figure 24. Safety factor predicted by GEO5 for non-rooted soil.

The difference between safety factors predicted by KurZeS and GEO5 for slopes formed by non-rooted soils is even smaller than in the case evaluated in "AREA 46" and it can be denoted as negligible.

For the case of a slope formed by rooted soil, KurZeS predicted safety factor $F_{sR}(TP_8) = 1.30$. The resulting raster for the whole "AREA 79" can be seen in Figure 25. Lighter colors than in Figure 23 indicate an improvement in slope stability due to root reinforcement.

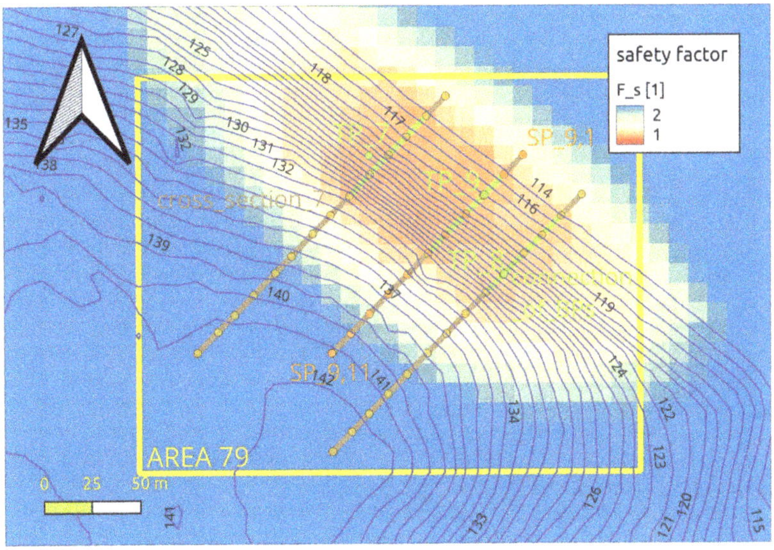

Figure 25. Safety factor predicted by KurZeS for rooted soil. TP_i denotes i-th Testing Point and SP_i, j denotes j-th Sampling Point lying on the i-th cross-section.

Verification result obtained for the case of rooted soil from GEO5 has the value $F_s(TP_8) = 1.27$ and the appropriate cross_section_8 with rooted soil is depicted in Figure 26.

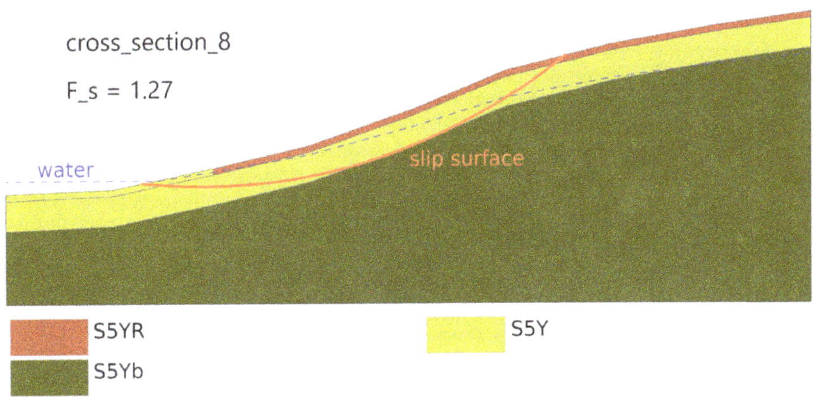

Figure 26. Safety factor predicted by GEO5 for rooted soil.

For the case of a slope formed by rooted soil, the difference between safety factors obtained from KurZeS and GEO5 is slightly higher than in previously evaluated cases and it makes for approximately 2.3%, which we feel to be still very low.

This comparison of two very close results from two different software tools verified the function of KurZeS for the case of a model including surface water.

The above-mentioned results show that the results of KurZeS using the pre-calculated database are very comparable with the results of GEO5 for the same slopes. The analysis was conducted both for slopes finishing in surface water and slopes finishing in land, and both for cross-sections including their Test Point and cross-sections that passed it. The comparability of the results verifies the approach using the pre-calculated database as a good simplification of the problem. This approach efficiently allows many stability factor evaluations for each Test Point and finds the weakest slope in its vicinity.

4. Discussion

Commercial software tools for slope stability evaluation have offered the "Method of Slices" to compute safety factors in 2D slope cross-sections. The KurZeS approach uses the "Method of Slices" implemented in GEO5 for pre-computed interpolation tables for selected soil types. These tables are used to approximate the slope stability of large areas.

The use of a pre-computed database allows the KurZeS approach presented in this article to bring an original geometrical and geotechnical concept, where slope stability in each Test Point is not evaluated just along the fall line but also along rather different directions. This concept takes into account more slopes and assigns the TP the lowest safety factor in its vicinity. This "looking around" the TP approach could be important, especially in places where surface water or material interface appear. In such a place, the weakest slope may not be the one that runs exactly along the fall line passing the Test Point.

As in all applications, the accuracy must be balanced with simplicity. The price that should be paid for achievable large-scale models is a simplified two- (case without NBS) or three- (case with NBS) layer stratigraphy, which has been evaluated as sufficient detail for the soil dump slopes under consideration.

In most cases, the simplifications and inaccuracies of the method are related to the mesh discretization density and schematization of the model. In the geometric part, slip surfaces shorter than 20 m are not allowed.

The method allows the $BP_1 - BP_2$ line to pass the TP. The effect of this feature on the accuracy of the result is largely limited by the fact that the least favorable slip surfaces described by both pairs of line segments $BP_1 - TP$ and $TP - BP_2$, or the line segment $BP_1 - BP_2$, usually have the minimum angular deviation from the slope gradient and are close to each other in shape according to the experience analysis. This means that the angle between $BP_1 - TP$ and $TP - BP_2$ is close to 180°, and therefore, the distance between TP and $BP_1 - BP_2$ is small. Examples of cases with different probabilities of finding the least favorable slip surface with a comparison of the degree of inaccuracy are shown in Figure 27.

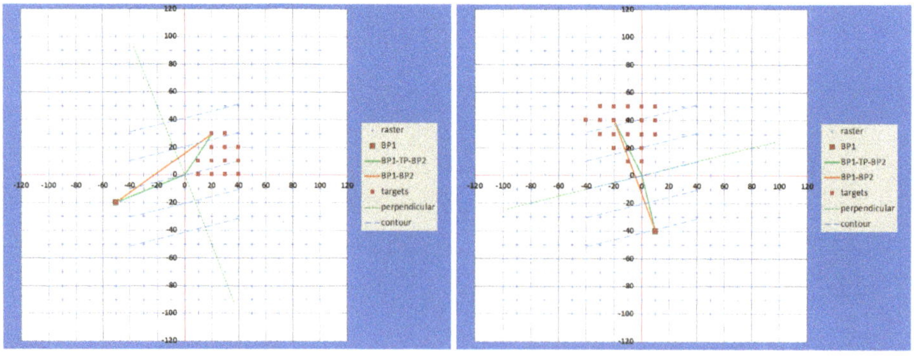

Figure 27. (**a**) Low probability that the link would represent the area with the lowest F_s; (**b**) a high probability of the same.

The inaccuracies of the method resulting from the influence of the mutual rotation of the 10 × 10 m mesh and the gradient direction increase with the decreasing length of the slip surfaces. Therefore they are analyzed in greater detail for the cases of the shortest slip surfaces with a changing angle between the contour lines and the orientation of the 10 × 10 m mesh. For this analysis, a completely homogeneous model is considered, without the influence of groundwater. Therefore, only the slope of the terrain on the line and its length have influence.

For TP $\neq$ BP$_1$ and the 120° angle, there are only two patterns with three lines (Figure 28, the first and second columns of three) with a matching BP$_2$, that are repeated four times. In the case of TP $\equiv$ BP$_1$, there is one pattern repeated four times (third column).

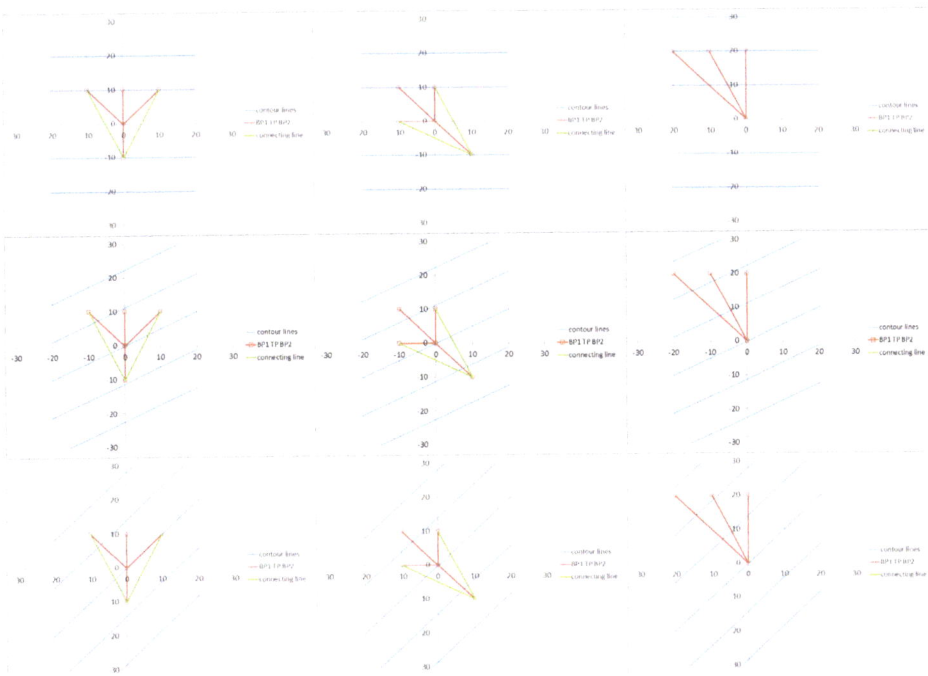

Figure 28. Analysis of the influence of the mesh rotation relative to the slope for short links, variants for 120° angle.

In the first triad, the contour lines have the same direction as the horizontal X axis and the slope as the vertical Y axis. In the second triplet, these lines form an angle of 26, 6° and in the third, 45°. One of the possible links for TP $\neq$ BP$_1$ and TP $\equiv$ BP$_1$ is always perpendicular to the contour lines, and its steepness is maximal under the given conditions, but its length changes by leaps from 20 m to 22.36 m ($10 \times \sqrt{5}$) and 28.28 m ($10 \times 2 \times \sqrt{2}$)—Figure 29.

In favor of the adherence of the method to reality, the regularities in terrain modelling, especially in the case of reclamation, the "inverse proportion" of the slope gradient and length are at work. The longest slope cannot be the steepest one.

Further, the potential slope stability improvement achieved through root reinforcement represented by additional root cohesion was studied. Tree species recommended for reclamation of brown coal dumps are birch (*Betula*), larch (*Larix*), red oak (*Quercus rubra*) and pine (*Pinus*) [39]. In the presented simulations, a special type of slope with specific soil composition was considered. This kind of slope is common in soil dumps that are present in brown coal mining areas, e.g., in the Northern Bohemian Brown Coal Basin or Saxony and Lusatia, Germany.

The values of the stability parameter included in the presented model are rather conservative values coming from the literature and an expert's experience. Such a choice of parameters can avoid damage caused by an overestimation of the predicted slope stability.

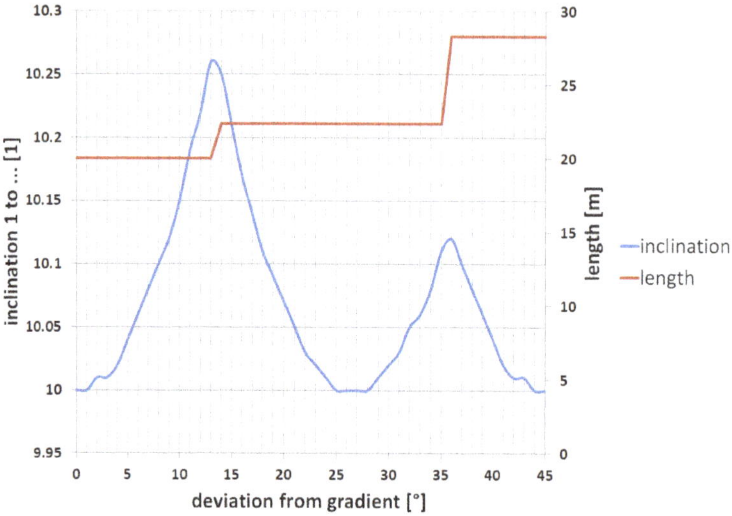

Figure 29. Effect of mesh direction rotation relative to fall line for short links.

Predicted slope stability improvements may seem to reach small relative values; however, in very unstable areas where the safety factor approaches the value of 1, the improvement may mean an important difference between the risk and safety.

In this study, neither rainfall [16,18,48,49] nor groundwater flow [50] were considered. These effects could be taken into account by coupling the KurZeS method with different appropriate model tools.

A probable plant-induced water table decrease was not considered. The expected stabilization effect is in the order of a few percent. However, this small improvement may play a crucial role in the boundary cases where the slope is near to collapse.

Temporal dependency of the stabilizing effect, which may reach its optimum after a couple of decades, was also not considered. It could be implemented using time-dependent stability parameters of the soils.

For the sake of simplicity, we have also not considered in the model, the effect of earthquakes, as in [51,52].

In this study, the volume of soil removed by landslides was not predicted. The description of this phenomenon can be found in [8,47,53] and possible methods of its evaluation are described in [54,55].

Before the stabilization effects of the industrial forest fully appear, another NBS (fascines, grass, shrub) may be applied to avoid the risk of shallow landslides and erosion. On a small scale, at specific places, NBS may be complemented with standard technical solutions used for slope stabilization. The same holds for the cases where NBS are not able to stabilize the slope in sufficient measures and where the residual risk of landslides remains too high.

5. Conclusions

The presented method allows the geotechnical stability of large areas to be analyzed with generally accessible data on a usual computer in a reasonably short time, which was demonstrated on a specific area of soil dump of the former open-cast mine Lohsa II. It also allows evaluation of the influence of possible natural-based or technological measures, which was demonstrated in the case of root reinforcement.

The two most original aspects of the presented approach are the use of a pre-computed database of safety factors of slopes of various parameters and the testing of the stability of the whole neighborhood of each Test Point (the associated slopes do not need to cross their

TPs). Both these aspects are somehow connected since the pre-computed database shortens computational time so taking more possible slopes into account is allowed.

A compromise between the detail and the large area of the study area is the choice of a square network with a node spacing (TP) of 10 m and a limitation of the shortest slip surface length to 20 m. However, the method includes, to a controlled extent, a rather detailed examination of the immediate vicinity of the TP by also considering links that do not directly pass through the TP. By taking into account slopes, directions, slip surface lengths and other parameters, the method is suitable for rugged terrain with variable soil composition or in contact with water bodies, despite the limitations.

The possibility of the inclusion of NBS was presented in the study of the influence of possible safety factor change due to vegetation (root reinforcement) that can be performed on a computer to help identify the slopes that can be stabilized by root reinforcement and do not need other (specifically technological) measures.

The presented approach was verified and validated in the area of soil dump of the former open-cast mine Lohsa II in Lusatia, Germany. Verification was performed by comparing several slope stability computations obtained from the KurZeS method and GEO5 software. The validation was performed by comparing the KurZeS results with the published map of areas closed to the public.

The maximum predicted relative increase in the stability factor due to the consideration of nature-based solutions was 17% and was observed for extremely short slopes with a length of 20 m or less. Shorter slopes are not accounted for by the model due to its geometrical concept.

However small this improvement may appear, it can be crucial in cases where the stability factor is close to 1.0, i.e., when the driving and motion-resisting forces are close to equilibrium.

The predicted maximum improvement exceeded expectations based on reference calculations from GEO5. This was due to the choice of the minimum tested slope length in the reference calculations being too high.

In further research, the authors will focus on the generalization of quantification of slope stabilization effects achieved through various NBS for various soil compositions. In the application, the focus will be directed towards the simulation of the stability of selected, large-scale mining waste dump slope areas. This will enable us not just to assess risks and threats but also to evaluate the potential to avoid shallow landslides using NBS.

Author Contributions: Conceptualization, L.Z.; Methodology, J.K.; Software, L.Z. and J.K.; Validation, J.Š.; Resources, L.Z.; Writing—original draft, L.Z., J.Š. and J.K.; Writing—review & editing, J.Š. and J.K.; Visualization, J.Š.; Supervision, L.Z.; Project administration, L.Z. All authors have read and agreed to the published version of the manuscript.

Funding: This research received no external funding.

Data Availability Statement: The original contributions presented in the study are included in the article, further inquiries can be directed to the corresponding author.

Conflicts of Interest: The authors declare no conflict of interest.

Abbreviations

BP	Boundary point
DEM	Digital elevation model
F_s	Safety Factor
GIS	Geographic Information System
LfULG	Landesamt für Umwelt, Landwirtschaft und Geologie
MSL	Mean Sea Level
NBS	Nature-based solutions
SWB	Soil and Water Bioengineering
TP	Test Point

References

1. Li, F.; Torgoev, I.; Zaredinov, D.; Li, M.; Talipov, B.; Belousova, A.; Kunze, C.; Schneider, P. Landslide-Induced Mass Transport of Radionuclides along Transboundary Mailuu-Suu River Networks in Central Asia. *Remote Sens.* **2021**, *13*, 698. [CrossRef]
2. Chalkias, C.; Ferentinou, M.; Polykretis, C. GIS-Based Landslide Susceptibility Mapping on the Peloponnese Peninsula, Greece. *Geosciences* **2014**, *4*, 176–190. [CrossRef]
3. Torgoev, A. Assessment of Landslide Hazard in the Environmental Hotspot Areas of the Kyrgyz Tien-Shan: Spatial Analysis and Numeric Modelling. Ph.D. Thesis, Université de Liège, Liège, Belgium, 2016.
4. Xie, M.; Esaki, T.; Cai, M. GIS-Based Implementation of Three-Dimensional Limit Equilibrium Approach of Slope Stability. *J. Geotech. Geoenviron. Eng.* **2006**, *132*, 656–660. [CrossRef]
5. Jia, N.; Yang, Z.; Xie, M.; Mitani, Y.; Tong, J. GIS-based three-dimensional slope stability analysis considering rainfall infiltration. *Bull. Eng. Geol. Environ.* **2015**, *74*, 919–931. [CrossRef]
6. Yu, G.; Xie, M.; Liang, J.; Farooq, A.; Williams, E.J. A GIS-based 3D slope stability analysis method based on the assumed normal stress on the slip surface. *Sci. Rep.* **2020**, *10*, 4442. [CrossRef]
7. Ullah, S.; Khan, M.U.; Rehman, G. A brief review of the slope stability analysis methods. *Geol. Behav.* **2020**, *4*, 73–77. [CrossRef]
8. Li, F.; Torgoev, I.; Zaredinov, D.; Li, M.; Talipov, B.; Belousova, A.; Kunze, C.; Schneider, P. Influence of Earthquakes on Landslide Susceptibility in a Seismic Prone Catchment in Central Asia. *Appl. Sci.* **2021**, *11*, 3768. [CrossRef]
9. Juliev, M.; Mergili, M.; Mondal, I.; Nurtaev, B.; Pulatov, A.; Hübl, J. Comparative analysis of statistical methods for landslide susceptibility mapping in the Bostanlik District, Uzbekistan. *Sci. Total Environ.* **2019**, *653*, 801–814. [CrossRef]
10. Sakellariou, M.G.; Ferentinou, M.D. A study of slope stability prediction using neural networks. *Geotech. Geol. Eng.* **2005**, *23*, 419–445. [CrossRef]
11. Soranzo, E.; Guardiani, C.; Chen, Y.; Wang, Y.; Wu, W. Convolutional neural networks prediction of the factor of safety of random layered slopes by the strength reduction method. *Acta Geotech.* **2023**, *18*, 3391–3402. [CrossRef]
12. Fattahi, H. Prediction of slope stability using adaptive neuro-fuzzy inference system based on clustering methods. *J. Min. Environ.* **2016**, *8*, 163–177. [CrossRef]
13. Safa, M.; Sari, P.A.; Shariati, M.; Suhatril, M.; Trung, N.T.; Wakil, K.; Khorami, M. Development of neuro-fuzzy and neuro-bee predictive models for prediction of the safety factor of eco-protection slopes. *Phys. A Stat. Mech. Appl.* **2020**, *550*, 124046. [CrossRef]
14. Bishop, A.W. The use of the Slip Circle in the Stability Analysis of Slopes. *Géotechnique* **1955**, *5*, 7–17. [CrossRef]
15. Sarma, S.K. Stability analysis of embankments and slopes. *Géotechnique* **1973**, *23*, 423–433. [CrossRef]
16. Rosso, R.; Rulli, M.C.; Vannucchi, G. A physically based model for the hydrologic control on shallow landsliding. *Water Resour. Res.* **2006**, *42*, 2005WR004369. [CrossRef]
17. Günther, A. SAGA-GIS Module Library Documentation (Module SHALSTAB). 2012. Available online: https://saga-gis.sourceforge.io/saga_tool_doc/2.1.3/ta_slope_stability_2.html (accessed on 12 March 2024).
18. Rossi, G.; Catani, F.; Leoni, L.; Segoni, S.; Tofani, V. HIRESSS: A physically based slope stability simulator for HPC applications. *Nat. Hazards Earth Syst. Sci.* **2013**, *13*, 151–166. [CrossRef]
19. Escobar-Wolf, R.; Sanders, J.; Vishnu, C.; Oommen, T.; Sajinkumar, K. A GIS tool for infinite slope stability analysis (GIS-TISSA). *Geosci. Front.* **2021**, *12*, 756–768. [CrossRef]
20. Wang, S.; Zhang, K.; Van Beek, L.; Tian, X.; Bogaard, T. Physically-based landslide prediction over a large region: Scaling low-resolution hydrological model results for high-resolution slope stability assessment. *Environ. Model. Softw.* **2020**, *124*, 104607. [CrossRef]
21. LISEM. SlopeFailure. 2022. Available online: https://lisemmodel.com//docs/slopefailure/ (accessed on 6 December 2023).
22. Thomas, J.; Gupta, M.; Srivastava, P.K.; Petropoulos, G.P. Assessment of a Dynamic Physically Based Slope Stability Model to Evaluate Timing and Distribution of Rainfall-Induced Shallow Landslides. *ISPRS Int. J. Geo-Inf.* **2023**, *12*, 105. [CrossRef]
23. Mergili, M.; Marchesini, I.; Rossi, M.; Guzzetti, F.; Fellin, W. Spatially distributed three-dimensional slope stability modelling in a raster GIS. *Geomorphology* **2014**, *206*, 178–195. [CrossRef]
24. Mergili, M.; Marchesini, I.; Alvioli, M.; Metz, M.; Schneider-Muntau, B.; Rossi, M.; Guzzetti, F. A strategy for GIS-based 3-D slope stability modelling over large areas. *Geosci. Model Dev.* **2014**, *7*, 2969–2982. [CrossRef]
25. Palacio Cordoba, J.; Mergili, M.; Aristizábal, E. Probabilistic landslide susceptibility analysis in tropical mountainous terrain using the physically based r.slope.stability model. *Nat. Hazards Earth Syst. Sci.* **2020**, *20*, 815–829. [CrossRef]
26. Sciarra, M.; Coco, L.; Urbano, T. Assessment and validation of GIS-based landslide susceptibility maps: A case study from Feltrino stream basin (Central Italy). *Bull. Eng. Geol. Environ.* **2017**, *76*, 437–456. [CrossRef]
27. Shirzadi, A.; Soliamani, K.; Habibnejhad, M.; Kavian, A.; Chapi, K.; Shahabi, H.; Chen, W.; Khosravi, K.; Thai Pham, B.; Pradhan, B.; et al. Novel GIS Based Machine Learning Algorithms for Shallow Landslide Susceptibility Mapping. *Sensors* **2018**, *18*, 3777. [CrossRef] [PubMed]
28. Bulzinetti, M.A.; Segoni, S.; Pappafico, G.; Masi, E.B.; Rossi, G.; Tofani, V. A Tool for the Automatic Aggregation and Validation of the Results of Physically Based Distributed Slope Stability Models. *Water* **2021**, *13*, 2313. [CrossRef]
29. Murgia, I.; Giadrossich, F.; Mao, Z.; Cohen, D.; Capra, G.F.; Schwarz, M. Modeling shallow landslides and root reinforcement: A review. *Ecol. Eng.* **2022**, *181*, 106671. [CrossRef]
30. Masi, E.B.; Segoni, S.; Tofani, V. Root Reinforcement in Slope Stability Models: A Review. *Geosciences* **2021**, *11*, 212. [CrossRef]

31. Mao, Z. Root reinforcement models: Classification, criticism and perspectives. *Plant Soil* **2022**, *472*, 17–28. [CrossRef]
32. Bischetti, G.B.; Chiaradia, E.A.; Epis, T.; Morlotti, E. Root cohesion of forest species in the Italian Alps. *Plant Soil* **2009**, *324*, 71–89. [CrossRef]
33. Chok, Y.H.; Jaksa, M.B.; Kaggwa, W.S.; Griffiths, D.V. Assessing the influence of root reinforcement on slope stability by finite elements. *Int. J. Geo-Eng.* **2015**, *6*, 12. [CrossRef]
34. Schmidt, K.M.; Roering, J.J.; Stock, J.D.; Dietrich, W.E.; Montgomery, D.R.; Schaub, T. The variability of root cohesion as an influence on shallow landslide susceptibility in the Oregon Coast Range. *Can. Geotech. J.* **2001**, *38*, 995–1024. [CrossRef]
35. Schwarz, M.; Preti, F.; Giadrossich, F.; Lehmann, P.; Or, D. Quantifying the role of vegetation in slope stability: A case study in Tuscany (Italy). *Ecol. Eng.* **2010**, *36*, 285–291. [CrossRef]
36. Schwarz, M.; Cohen, D.; Or, D. Spatial characterization of root reinforcement at stand scale: Theory and case study. *Geomorphology* **2012**, *171–172*, 190–200. [CrossRef]
37. Zydron, T.; Gruchot, A. Influence of root systems of deciduous trees on soil reinforcement—A case study from the Carpathians, Poland. *Environ. Eng. Manag. J.* **2021**, *20*, 419–433. [CrossRef]
38. QGIS Development Team. QGIS Geographic Information System. 2023. Available online: https://www.qgis.org (accessed on 6 December 2023).
39. Krümmelbein, J.; Bens, O.; Raab, T.; Anne Naeth, M. A history of lignite coal mining and reclamation practices in Lusatia, eastern Germany. *Can. J. Soil Sci.* **2012**, *92*, 53–66. [CrossRef]
40. OpenStreetMap Contributors. 2017. Planet Dump. Available online: https://planet.osm.org (accessed on 12 March 2024).
41. Geotechnical Software GEO5. Available online: https://www.finesoftware.eu/geotechnical-software/ (accessed on 6 December 2023).
42. Referat Kommunikation und Öffentlichkeitsarbeit. Downloadbereich Digitale Höhenmodelle—Offene Geodaten. Available online: http://www.geodaten.sachsen.de/downloadbereich-digitale-hoehenmodelle-4851.html (accessed on 6 December 2023).
43. Rouault, E.; Warmerdam, F.; Schwehr, K.; Kiselev, A.; Butler, H.; Łoskot, M.; Szekeres, T.; Tourigny, E.; Landa, M.; Miara, I.; et al. GDAL. 2024. Available online: https://zenodo.org/records/10472126 (accessed on 12 March 2024).
44. Referat Kommunikation und Öffentlichkeitsarbeit. Fachbereich Boden—LUIS—Landwirtschaft- und Umweltinformationssystem für Geodaten—sachsen.de. Available online: https://luis.sachsen.de/fachbereich-boden.html (accessed on 6 December 2023).
45. *ČSN P 73 1005 (731005)*; Inženýrskogeologický Průzkum. 2016. Available online: https://www.technicke-normy-csn.cz/csn-p-73-1005-731005-222271.html# (accessed on 6 December 2023).
46. Virtanen, P.; Gommers, R.; Oliphant, T.E.; Haberland, M.; Reddy, T.; Cournapeau, D.; Burovski, E.; Peterson, P.; Weckesser, W.; Bright, J.; et al. SciPy 1.0: Fundamental algorithms for scientific computing in Python. *Nat. Methods* **2020**, *17*, 261–272. [CrossRef] [PubMed]
47. Rossi, L.; Rapidel, B.; Roupsard, O.; Villatoro-sánchez, M.; Mao, Z.; Nespoulous, J.; Perez, J.; Prieto, I.; Roumet, C.; Metselaar, K.; et al. Sensitivity of the landslide model LAPSUS_LS to vegetation and soil parameters. *Ecol. Eng.* **2017**, *109*, 249–255. [CrossRef]
48. Zieher, T.; Rutzinger, M.; Schneider-Muntau, B.; Perzl, F.; Leidinger, D.; Formayer, H.; Geitner, C. Sensitivity analysis and calibration of a dynamic physically based slope stability model. *Nat. Hazards Earth Syst. Sci.* **2017**, *17*, 971–992. [CrossRef]
49. Van Zadelhoff, F.B.; Albaba, A.; Cohen, D.; Phillips, C.; Schaefli, B.; Dorren, L.; Schwarz, M. Introducing SlideforMAP: A probabilistic finite slope approach for modelling shallow-landslide probability in forested situations. *Nat. Hazards Earth Syst. Sci.* **2022**, *22*, 2611–2635. [CrossRef]
50. Marino, P.; Santonastaso, G.F.; Fan, X.; Greco, R. Prediction of shallow landslides in pyroclastic-covered slopes by coupled modeling of unsaturated and saturated groundwater flow. *Landslides* **2021**, *18*, 31–41. [CrossRef]
51. Rao, B.; Burman, A.; Roy, L.B. An Efficient Box Search Method for Limit Equilibrium Method-Based 3D Slope Stability Analysis. *Transp. Infrastruct. Geotechnol.* **2023**, *11*, 327–358. [CrossRef]
52. Wan, Y.; Gao, Y.; Zhang, F. Stability Analysis of Three-Dimensional Slopes Considering the Earthquake Force Direction. *Math. Probl. Eng.* **2018**, *2018*. [CrossRef]
53. Marchesini, I.; Cencetti, C.; De Rosa, P. A preliminary method for the evaluation of the landslides volume at a regional scale. *GeoInformatica* **2009**, *13*, 277–289. [CrossRef]
54. Reid, M.E.; Christian, S.B.; Brien, D.L.; Henderson, S.T. Scoops3D: Software to Analyze 3D Slope Stability throughout a Digital Landscape. 2015. Available online: https://doi.org/10.3133/tm14a1 (accessed on 6 December 2023).
55. Welcome to the LISEM Documentation. 2022. Available online: https://lisemmodel.com//docs/home/ (accessed on 6 December 2023).

land

MDPI

Article

Nature Through Young Eyes: Exploring Children's Understanding of Nature in Urban Landscapes in Beijing, China

Jianjiao Liu and Raymond James Green *

Faculty of Architecture, Building and Planning, University of Melbourne, Parkville, VIC 3010, Australia; jianjiao.liu@student.unimelb.edu.au
* Correspondence: rjgreen@unimelb.edu.au

Abstract: Rapid urbanization over the last few decades has resulted in children growing up in cities increasingly disconnected from the natural world. A better understanding of children's conceptions of nature in cities is needed to try to address this unfortunate trend. This study was conducted within four core districts of Beijing, China, and involved interviewing 81 children aged 8 to 12 years to explore their general, conceptual understanding of nature and accessible nature and how it manifests in their daily lives. Some of these children ($N = 54$) were involved in a follow-up photo Q-sort task aimed at investigating their perceptions of nature in the city, which required them to sort 33 stimuli photographs in piles according to their perceived "naturalness". The interview data were content analyzed, while the photo-sorting data were subjected to various statistical analyses, including Multidimensional Scaling (MDS). These analyses identified three distinct ways children perceive nature in the city, reflecting ecological, emotional, and visual values for nature in the city. The findings highlight the influence of personal experiences, education, the media, and evolutional predispositions on children's constructs of nature. The findings can guide environmental design and education actions aimed at integrating nature in urban settings that align better with the children's conceptions of nature.

Keywords: children; nature; personal construct theory; built environment; environmental education; evolutionary predisposition

1. Introduction

Children growing up in cities today are increasingly losing their sensitivity to and connection with the natural world, which is being driven primarily by rapid urbanization and land use changes, resulting in reduced access to natural environments [1,2]. Changes in lifestyles, and the transformative effects of technology on children's conceptions of nature, also play an important role. Children's leisure time has increasingly shifted from direct experiences with nature gained through outdoor play and exploration, an effective way for children to develop meaningful connections with nature, to more sedentary, indoor activities. This shift has been correlated with a decline in children's physical and psychological well-being, as well as their cognitive and social development—what has been referred to as "Nature-Deficit Disorder" [3]. The potential consequences of this trend highlight the necessity for a better understanding of the relationship between children and nature—how they understand, perceive, and interact with nature, especially within built urban environments where direct access to nature may be limited, in order to reverse this trend. The study reported here is an effort toward achieving this understanding.

Many studies have explored the relationship between children and nature. A key focus has been the impact of nature exposure on children's health and well-being, aligning

https://doi.org/10.3390/land14030624

with the notion of biophilia, which represents an innate emotional connection between humans and other living organisms [4]. Two major theories have then been proposed in the literature of environmental psychology—Stress Recovery Theory (SRT) and Attention Restoration Theory (ART). SRT proposes that (visual) exposure to nature could quickly elicit positive emotions and block negative ones, thereby reducing blood pressure and stress hormones [5]. ART, pioneered by Kaplan, distinguishes two types of attention—directed and involuntary attention. Directed attention is needed to maintain a high level of performance in people's day-to-day functioning and decreases when overused, while involuntary attention requires no effort. In this theory, nature is often regarded as rich in (soft) fascination that quickly elicits involuntary attention and simultaneously allows directed attention to be recharged [6]. Subsequent research has employed quantitative indexes of nature exposure derived from satellite, parent-reported, or self-reported data or by evaluating the quality of natural features through on-site measurements, providing empirical support for the positive impacts of passively viewing, immersing oneself in, or actively interacting with environments rich in natural elements on children's cognitive development and psychological well-being [7–12]. However, the increasing disconnection of children from nature in cities has deprived them of this source of soft fascination and the benefits it provides.

Children's sense of environmental stewardship is another important area of research. Some studies have examined the impact of natural/environmental education on children's pro-environmental behaviors, with many educational programs incorporating hands-on, nature-based activities [13]. Some research has also explored the influence of connection to nature and the mediating effects of environmental knowledge and attitudes in these processes, with varying effects ranging from medium to weak being identified [2,14–16]. These variables have primarily been assessed using standardized scales that include abstract statements, such as *"I often feel a sense of oneness with the natural world around me"*, in assessing children's connectedness with nature [2]. However, few studies have investigated how children conceptualize and interpret the notion of nature itself—how they define the concepts of "nature" or "natural", how they reason about nature, and how they make sense of human–nature interactions.

There is also a lack of research on the impacts of urbanization and human activities on children's perception and understanding of nature. As Orr [17] (p. 282) put it, children *"on average can recognize over 1000 corporate logos but only a handful of plants and animals native to their places"*, reflecting a worrisome trend toward what has been termed "biotic impoverishment" [18]. For many children living in cities today, direct exposure to nature is primarily limited to nearby parks, green/blue spaces, and other types of places where "urbanized nature" can be found, including artificial and semi-natural environments that nonetheless may possibly be perceived as natural [19]. However, opportunities for virtual contact with imagined and exotic manifestations of nature have increased dramatically due to technology, leading to a neglect of perceptual, affective, cognitive, and behavioral interactions with actual and local sources of contact with nature, or what has been called the "extinction of experience" [20]. This can impact children's cognitive and affective responses toward nature, which, in turn, can influence their motivation for engaging in nature-based activities and adopting pro-environmental attitudes and actions [21].

Reintegrating nature into children's daily lives in cities is essential for reversing this trend, but first, it is necessary to better understand how children conceptualize and perceive nature in urban contexts. The environmental psychologist Wohlwill [22] (p. 7) succinctly defined the notion of "nature" as broadly dealing

"... with the landscape rather than with the built environment. It includes the world of rock and sand, of shoreline, desert, woods, mountains, etc., and the diverse manifestations of plant and

animal life that are encountered there. It excludes the man-made world: our cities and towns, our houses and factories, along with the diverse implements devised by mankind, for transport, recreation, commerce, and other human needs."

However, children's understanding of nature might differ due to their cognitive development, direct experiences of nature, and the type of environment in which they grew up. Personal construct theory, initially proposed by the psychologist George Kelly, provides a valuable lens in this regard, emphasizing that individuals actively interpret and make sense of the world through a system of personal constructs developed through experiences, typically through templates of bipolar dimensions of connotative meaning, such as, for example, nasty–nice, clean–dirty, natural–unnatural, and other dimensions of meanings [23]. Rather than relying on fixed definitions of nature, children develop their own constructs associated with their understanding of what nature is and is not. From this perspective, two research questions are addressed in this study:

(1) How do children growing up in urban areas conceptualize nature, including *conceptual nature* (nature in the abstract) and *accessible nature* (that they encounter in their daily lives that they consider to be natural)?

(2) How do children perceive the nature that is accessible to them in their daily lives?

In exploring these questions, this study sought to address the abovementioned gaps and deepen the understanding of how children, particularly in urban environments, conceptualize and perceive nature by employing both verbal (interview) and perception-based (photo-sorting) methods. In addition to the academic contribution, this study also provides insights that can inform environmental design actions and educational interventions for children to foster more meaningful connections with the natural world.

2. Methods

2.1. Research Site and Sampling

This study focused on children aged 8 to 12 years living in four core administrative districts of Beijing—Haidian, Chaoyang, Dongcheng, and Xicheng. These areas had the highest level of Gross Domestic Product (GDP, Figure 1a) and density of primary school children (Figure 1b) while also having the lowest amounts of public open green spaces per capita (Figure 1c), according to the Beijing Statistical Yearbook 2021. This disparity underscores tensions between urbanization, economic development, population density, and children's access to nature. Children involved in this study are in an age group where they increasingly become curious about their home range environments and acquire better comprehension and communication skills and, for many children, an increasing appreciation for the natural world [24]. These factors also enabled them to participate more effectively in interviews and photo-sorting tasks. The children were recruited via social media advertisements, posters, and a snowball sampling approach through personal networks. Parental consent was obtained before involving the children in data collection, and the study protocol was approved by the University of Melbourne's Human Research Ethics Committee (Reference No. 2021-22465-24141-3).

A total of 81 children participated in the first part of the study, which was focused on exploring their understanding of conceptual and accessible nature in their daily lives. The child respondents included a nearly even balance of genders (42 girls, 39 boys) and were evenly distributed by age. For anonymity, each child was assigned a unique identification number prefixed with "**C**". In total, 54 children (32 girls and 22 boys) who participated in the first part of the study also participated in the second part. Figure 1d shows the geographic locations of the research site and the child participants.

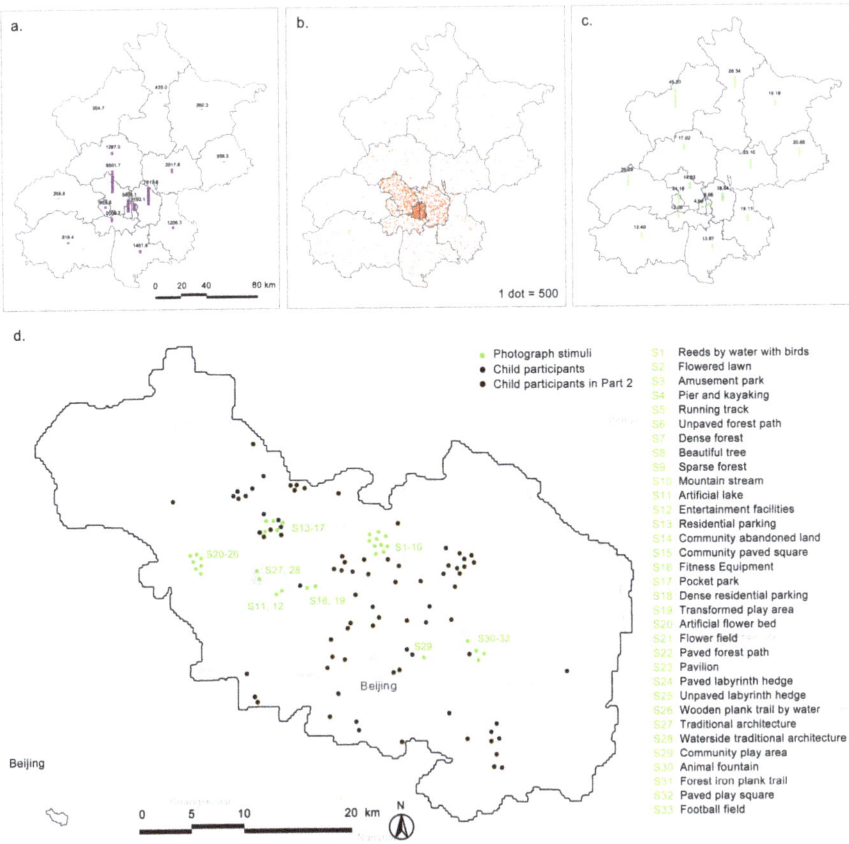

Figure 1. (**a**) The Gross Domestic Product (GDP, billion) in Beijing in 2021. (**b**) The density of primary school children in Beijing in 2021. (**c**) The public open green spaces per capita (m^2) in Beijing in 2021. (**d**) The geographic locations of the research site and the participants.

2.2. Data Collection and Analysis

2.2.1. Part 1—Understanding of Nature

This part of the research investigated children's understanding of nature by allowing them to freely express themselves about what nature is to them without the researchers imposing any predefined definitions of the notion of nature. Data collection involved two tasks incorporated in one-on-one semi-structured interviews, where children were firstly asked to close their eyes and visualize what nature is or what elements of the natural world they have in their minds—*conceptual nature*. The interview questions included, for example, *"Can you describe what you think of when you hear the word nature?"* and *"What kinds of places or things are part of nature?"*. Probing techniques for interviews proposed by Zeisel were also used, which have been widely used in environmental and behavioral research [25]. These probes were aimed at encouraging the children to fully express themselves and elaborate on their responses (depth probes) without the researchers biasing their answers. For example, when a child mentioned "forest", the researcher would follow up with questions, such as *"What can you find in the forest?"* or *"Can you think of anything else?"*, to encourage the child to provide more detailed and in-depth responses.

The second task required them to describe what, among the many things they encounter in their daily lives, they consider to be nature—*accessible nature*. These questions included, for example, *"Do you think there is anything in your daily life that you consider to be part of nature?"*, *"Are there any things or places around your home or school that you would call nature?"*, and *"When was the last time you felt like you were in nature and what was there?"*. A similar probing technique was also adopted as used in the first task.

The interviews took approximately 15 to 20 min per participant. Qualitative data obtained through interviews were input into the NVivo 20 program for conducting content analysis using inductive and deductive coding processes to identify similarities and differences between these two aspects.

2.2.2. Part 2—Perceptions of Nature in the City

A photo-based Q methodology was used to explore the second research question concerning children's perception of nature in their daily life. The Q methodology has primarily been employed in small-sample studies to explore people's attitudes, perceptions, and values [26–28]. It was traditionally used by psychologists, performed using a set of verbal statements (the Q set), in which participants were asked to sort them along a continuum of the construct of interest being tested, with their responses analyzed to identify patterns. Pitt and Zube [29] were the first to adapt this method for use in assessing people's perceptions of landscapes using photographs as stimuli, with the reliability and validity of the results being well-supported by subsequent studies [30–34]. This photo-sorting also has the advantage of eliciting personal constructs and related connotative meanings associated with people's perceptions of complex landscapes, which aligns well with the purpose of this study [35].

The photo Q set. To develop the set of photographs used as stimuli for the Q methodology, the researchers initially took numerous photographs across diverse sites in Beijing, including residential areas, city and national-level parks, and a variety of other types of environments children might encounter in the course of their daily lives. All photographs were taken with the same Canon EOS 200D II camera (Canon Inc., Tokyo, Japan) under similar weather and lighting conditions (clear sky between 10 a.m. and 11 a.m. or 3 p.m. and 4 p.m.) in May 2023. The photo selection criteria included context relevance and diversity of depicted content elements to capture a range of environmental settings the children might be likely to encounter. The researchers also consulted various environmental psychology and landscape design experts in selecting the final set of 33 photographs that were used as stimuli (labeled **S**) in the study. All of the photographs were printed in color and mounted on 13 cm × 9 cm cards—large enough for children to observe details yet easy enough to handle and be manageable to administer in the field.

Procedure. The 54 child participants were asked to independently sort the 33 photographs into nine piles arranged from natural to unnatural. The numbers of photographs in each pile were prescribed to reflect a normal distribution [29]. A diamond-shaped normal distribution template was adopted, where the children were asked to place the most natural environments at the top and the least natural ones on the bottom [36]. The template, which was printed on durable fabric, defines a specific number of photographs (1, 2, 4, 6, 7, 6, 4, 2, 1) to the nine piles, which are ranked in values from −4 to +4. Before each interview, the photographs were randomly distributed in front of the child respondent, and they were instructed to initially categorize the 33 photographs into three groups: most natural, most unnatural, and neutral. Next, they had to place the photographs in the different piles, working from the top and bottom ends to the center. They were then asked to verbally explain the environments depicted in the seven photographs they had placed in the top and bottom places that corresponded to what they thought were the most and least natural environments depicted in the photographs.

Analysis. Multidimensional Scaling (MDS) was used to analyze the naturalness rating data from photo Q-sorting method. MDS is an exploratory analytical method well-suited for analyzing sorting task data, as it can reveal latent perceptual structures without imposing predefined categories. It also provides a visual representation through plotting spatial configurations of the stimuli elements that helps with visualizing cognitive maps of children's environmental concepts in a way that reflects both proximity and relational meaning compared to clustering methods, which often create discrete, non-overlapping categories [37,38]. Therefore, MDS was considered well-suited for exploring the underlying dimensions of meaning, or personal constructs, in which the children discriminated between various environmental stimuli settings as depicted in the photographs and associated attributes [30,39]. The data matrix created from the photo-sorting data was subjected to Classical MDS (CMDS) using the alternating least squares scaling algorithm (ALSCAL) in SPSS 22 software [40]. Because the data in the matrix are ordinal, the Euclidean distance was used. Several MDS models, from 2 to 4 dimensions, were initially generated. An examination of the fitness values of each model suggested that the three-dimensional solution was the best based on its low stress value of 0.051, high RSQ of 0.989, and better interpretability [41].

The data matrix was also subjected to a hierarchical cluster analysis using the between-group method and the stimuli photographs as variables. Examination of the cluster analysis dendrogram allowed the photographs of environmental settings to be categorized into groups of similarly perceived elements. Six meaningful clusters were identified based on interpretability and labeled as "closed vegetative landscapes", "open vegetative landscapes", "nature in orderly frames", "mixed built landscapes", "urban leftover spaces", and "human-dominated landscapes". Combined with content analysis of the after-sorting interviews, it further facilitated the interpretation of the underlying perpetual dimensions apparent in the MDS stimulus configuration plot [30,42].

3. Results

3.1. Understanding of Nature—Conceptual and Accessible

This section reports the results of the first part of the study, which examined how the children conceptualize and understand nature in general (conceptual) and nature in the city (accessible). Their answers were divided into living and non-living manifestations of nature and scenarios.

3.1.1. Living Organisms

Almost all of the children mentioned living organisms when describing both conceptual nature ($N = 78$) and accessible nature ($N = 79$). Their descriptions of living organisms were categorized into three groups, flora, fauna, and fungi, broadly corresponding to the roles of producers, consumers, and decomposers within ecosystems (Table 1).

Flora. The children most frequently mentioned plants when defining nature, including both conceptual ($N = 74$) and accessible nature ($N = 75$). Interestingly, "tree" was the most frequently mentioned feature, both for conceptual ($N = 54$) and accessible nature ($N = 64$). They typically used terms for plants/trees more generally in their descriptions, with few of them being able to name particular plant types or species. This, to some extent, reflected a form of plant blindness [43], especially when compared to their understanding of animals, as discussed below. When describing conceptual nature, only four children mentioned specific plants, such as apple and pine trees. Interestingly, one child referenced the Corpse Flower (*Rafflesia arnoldii*), an exotic species with unique biological, ecological, and sensory properties. In comparison, 21 children named specific plant types in describing accessible nature, including willows, poplars, pine trees, gingko, and other trees commonly seen in Beijing. They also identified environmental settings where trees grew, such as at their schools, residential areas, or parks, with which they were familiar. Some children described experiences of climbing trees, collecting leaves, and other activities, or simply seeing them, bringing them into contact with nature. As **C67** (a 10-year-old girl) said: *"I especially like the ginkgo woods in the park. Every autumn, the ginkgo trees turn yellow and fall. The earth seems to be covered with a blanket, a thick yellow blanket. I always want to pick up ginkgo leaves in the park."*

Table 1. The proportions of child participants who mentioned each category of nature.

Categories	Sub-Categories		Conceptual Nature	Accessible Nature
Living organisms	Total		(87.5%, 100%]	(87.5%, 100%]
	Flora	General flora	(87.5%, 100%]	(87.5%, 100%]
		Specific flora	(0%, 12.5%]	(25%, 37.5%]
	Fauna	General fauna	(67.5%, 75%]	(25%, 37.5%]
		Specific fauna	(50%, 67.5%]	(67.5%, 75%]
	Fungi		(0%, 12.5%]	
Non-living entities	Total		(75%, 87.5%]	(50%, 67.5%]
	Natural elements		(75%, 87.5%]	(50%, 67.5%]
	Natural phenomena		(12.5%, 25%]	(0%, 12.5%]
Scenarios	Total		(37.5%, 50%]	(37.5%, 50%]
	Human-involved		(0%, 12.5%]	(25%, 37.5%]
	Nonhuman		(25%, 37.5%]	(0%, 12.5%]

Note: = 0%, = (0%, 12.5%], = (12.5%, 25%], = (25%, 37.5%], = (37.5%, 50%]; = (50%, 67.5%], = (67.5%, 75%], = (75%, 87.5%], = (87.5%, 100%].

Fauna. Animals were another frequently mentioned category in describing conceptual (N = 68) and accessible nature (N = 67). Children often used more general terms to describe animals in their notions of conceptual nature, while significantly more children could identify different types of animals when describing daily accessible nature (N = 60). Birds, mammals, and insects were the most frequently mentioned animals, although there were subtle differences in mentions of animals between conceptual and accessible nature.

Birds are visually appealing to children and contribute to sensory experiences of nature (e.g., the sounds of birdsongs), creating a multisensory connection to nature. The birds mentioned most in describing accessible nature were sparrows, magpies, woodpeckers, swallows, ducks, and pigeons, which were accessible to them in their everyday environments. This is in contrast to the more abstract depiction of conceptual nature, in which penguins, hummingbirds, and other exotic animals were mentioned, in addition to the more commonly seen birds.

Insects were mentioned in the descriptions of conceptual (N = 20) and accessible nature (N = 27). Half of the children identified specific types of insects, and some reported experiences of catching butterflies, dragonflies, and other insects and expressed a fondness for certain insects. As **C74** (an 8-year-old boy), for example, said: *"I like insects because they've been on the earth for over a billion years, and there are so many of them. I often observe insects... I especially like praying mantises because they can capture insects much larger than themselves, even birds."* In contrast, some children lacked knowledge about insects or feared them, sometimes feeling disgust toward insects, a biophobic response that may be rooted in an evolutionarily innate predisposition to avoid harmful creatures or negative experiences with them (e.g., painful bites or stings) [4,44]. Their first-hand observations and sensory experiences likely contribute to the perception of insects as integral parts of nature.

Mammals stood out as the category of animals where children most diverged when describing conceptual versus accessible nature. While slightly more children mentioned mammals when describing accessible nature, the diversity of mammals referenced in conceptual nature was three times greater. When describing accessible nature, they frequently mentioned animals common in the city—cats and dogs (primarily kept as pets), as well as other urban wildlife, including hedgehogs, weasels, squirrels, and rabbits that can be seen in parks and open spaces in residential neighborhoods. In contrast, larger mammals, such as elephants, rhinos, dolphins, and whales, were frequently mentioned in their descriptions of conceptual nature, which contained iconic types of "wild animals" that they are likely exposed to through the internet, cultural narratives, educational materials, and various forms of media that portray a version of nature that is more diverse, larger, and wilder than what is accessible to the children in Beijing. The children can experience some of the animals they mentioned in zoos and museums, as well as by visiting national parks and wild areas.

Some of the children described reptiles. Only one child mentioned turtles in accessible nature, while several referenced crocodiles and snakes, and one child even mentioned dinosaurs when describing conceptual nature, highlighting the potential influence of media and education on their perceptions of what nature means to them.

Fungi. Fungi are often relatively small, primarily found underground or low to the ground, and visible at certain times of the year; hence, they are not easily observed or as captivating as plants and animals. Only one child mentioned fungi when describing conceptual nature.

3.1.2. Non-Living Entities

The children also mentioned non-living entities when describing conceptual ($N = 67$) and accessible nature ($N = 45$), categorized into natural elements and phenomena. Natural elements are relatively stable and long-lasting, such as the sun and mountains. Natural phenomena, such as weather, involve dynamic processes or transient events. More of the children mentioned natural elements than natural phenomena.

Natural elements. Non-organic natural elements were mentioned by 66 and 44 children in describing conceptual and accessible nature, respectively. These included land-based geomorphological features (e.g., mountains and deserts) and water-based elements (e.g., rivers and seas). These are visually distinctive features, and while they were frequently mentioned, they would not be easily accessible to the children in their everyday experiences in the areas in Beijing that were the focus of this study. For example, as **C71** (a 10-year-old girl) mentioned, *"I absolutely love the ocean. It gives me a very unique and dreamy feeling, though I haven't had the opportunity to visit the seaside."* Their identification of these landscape features was likely due to the influence of education and media on their conceptions of nature. Nonetheless, many children also mentioned direct experiences with some of these natural elements through visiting a forest, hiking in the mountains, or playing by a river, given that Beijing is surrounded to the west by mountain ranges featuring lakes and rivers. There are some large parks in Beijing, such as the Olympic Forest Park and the Summer

by closed vegetative landscapes that create a lush and immersive natural setting with minimal to no visible traces of human intervention. Many of these photographs also display other natural elements, including water bodies, rocks, and birds, conveying a sense of thriving and diverse ecosystems. For example, **C35** (an 11-year-old girl) said in describing **S1** (depicting waterside trees and reeds with birds) that *"There are animals, and animals usually live in natural environments. The surrounding shrubs and other vegetation also appear very abundant and natural, along with a waterbody"*. Analysis of the open-ended responses suggested that these environmental settings were most often described as being "green", "wild", "ecological", and "like primeval forests", noting the absence of "artificial traces".

Cluster 2: Open vegetative landscapes. Two photos depicting scenes of open vegetative landscapes, including a lawn and a field with flowers, are included in this group. The aggregate mean value of 1.88 means they were perceived as highly natural. These environments were often described as spacious, aesthetically pleasing with open views, and full of life. **C35**, for instance, remarked, *"I feel the lawn is spacious, natural, and very beautiful. I like this place very much at first sight"*. The children frequently mentioned trees, grass, flowers, and animals, including birds, bees, and butterflies, that they felt contributed to a lively and vibrant atmosphere. While noting the absence of artificial structures, some features, such as neatly arranged flowers, which were artificially planted, were suggested to display *"such orderliness [they] would not typically occur in a wild natural setting"*, as **C26** (an 11-year-old boy) said.

Cluster 3: Nature in "orderly frames". This group is represented by photographs displaying natural environments and artificially framed elements (e.g., by roads and hedges) defining orderly spaces. The children generally emphasized the natural elements present in these settings, such as trees, water, and stones, but the presence of roads, bridges, and buildings that frame these elements were also apparent in the photographs. These artificial structures do not dominate the natural scenery but suggest thoughtful integration within the environment and what has been referred to as "cues to care", the presence of human intervention that shows the environment is being maintained and cared for, as described in Nassauer's seminal article *Messy ecosystems orderly frames* [45]. As **C26** described, *"It's like placing some human-made things within nature"*. The children often used terms including "harmonious" and "congruous" to describe these scenes. For example, **C53** (an 11-year-old boy) described the pavilion depicted in **S23** as follows: *"It feels like it belongs there; the scene is very harmonious, so it doesn't feel out of place"*. Similarly, **C15** (a 12-year-old girl) described the trees, grass, sunshine, and a kitten in **S14** as *"a very natural and harmonious scene"*.

Cluster 4: Mixed built landscapes. Similar to Cluster 3, this set of photographs depicts scenes where natural and artificial elements are equally represented. However, the key difference between these two clusters lies in the prominence of each element depicted. In Cluster 3, natural elements are the focus, while artificial traces subtly blend into orderly frameworks. In contrast, this cluster has more artificial structures that dominate the scenes, often positioned prominently at the center of the scene, which likely affects their perceptions of their degree of naturalness. As **C13** (a 10-year-old boy) commented, *"The flower bed artificially confines the flowers, and I feel it would be better if it were more like this photo [P21]"*, suggesting a preference for more organic, less controlled arrangements of features where natural elements dominate.

Cluster 5: Urban leftover spaces. This cluster is represented by only one photograph, **S19**, which depicts an urban leftover space in a residential area that had once been transformed into a play area. Most children noted that artificial structures, including buildings, occupied a large area in the photograph, while the presence of trees was sparse. Although some children acknowledged that wooden play structures were made from natural materials, it did little to influence their perceptions of the naturalness of the scene. Beyond these descriptions of physical features, some of the children also expressed their attitudes toward the depicted environment, describing it as old, desolate, less vibrant, and too cluttered.

Cluster 6: Human-dominated landscapes. This cluster received the lowest aggregate mean rating for perceived naturalness (M = −1.73). It primarily contains photographs displaying environments dominated by human-made elements, including residential areas, playgrounds, sports fields, and other environmental settings the children may encounter in their daily lives in the city. They frequently used the terms "artificial", "built for human use", and "unsuitable for animals" to describe the buildings, parking lots, and fitness facilities made of plastic and metal visible in these photographs, while emphasizing the scarcity of greenery, particularly trees. For example, **C74** (an 8-year-old boy) said in describing **S32** (paved play square): *"This place is for playing, and there are many people here... There are only a few trees. Places with many people are not suitable for animals"*. Interestingly, despite the limited presence of natural elements, some children remarked that these environments felt familiar to them as they are like those they encounter in their daily lives, evoking a sense of familiarity and comfort similar to feelings associated with home environments. As **C11** (a 9-year-old girl) noted when describing **S13**, **S15**, and **S18**, *"Places that feel like home are more natural"*. This suggests that for some children, the familiarity of their everyday surroundings can influence their perception of naturalness, highlighting the role of personal experiences and emotional connections to known spaces in shaping how children interpret and value the environment.

Cluster 1 Closed vegetative landscapes
(aggregate group M = 1.98, SD = 1.31)

S10 - Mountain stream
M = 2.57 SD = 1.16

S1 - Reeds by the water with birds
M = 2.09 SD = 1.10

S7 - Dense forest
M = 2.07 SD = 1.76

S6 - Unpaved forest path
M = 1.65 SD = 1.07

S8 - Beautiful tree
M = 1.52 SD = 1.09

Cluster 2 Open vegetative landscapes
(aggregate group M = 1.88, SD = 1.36)

S2 - Flowered lawn
M = 1.98 SD = 1.37

S21 - Flower field
M = 1.78 SD = 1.36

Figure 2. *Cont.*

Cluster 3 Nature in "orderly frames"
(aggregate group M = 0.55, SD = 1.07)

S22 - Paved forest path
M = 1.33 SD = 1.06

S26 - Wooden plank trail by the water
M = 1.07 SD = 0.99

S24 - Paved labyrinth hedge
M = 0.98 SD = 0.94

S25 - Unpaved labyrinth hedge
M = 0.91 SD = 0.83

S11 - Artificial lake
M = 0.76 SD = 0.73

S31 - Forest iron plank trail
M = 0.70 SD = 1.00

S9 - Sparse forest
M = 0.33 SD = 1.06

S17 - Pocket park
M = 0.30 SD = 0.74

S23 - Pavilion
M = 0.28 SD = 0.92

S28 - Waterside traditional architecture
M = 0.20 SD = 1.07

S5 - Running track
M = −0.20 SD = 0.94

S14 - Community abandoned land
M = −0.02 SD = 1.32

Cluster 4 Mixed built landscapes
(aggregate group M = −0.48 , SD = 1.29)

S20 - Artificial flower bed
M = −0.30 SD = 1.34

S4 - Pier and kayaking
M = −0.41 SD = 1.25

S27 - Traditional architecture
M = −0.74 SD = 1.26

Figure 2. *Cont.*

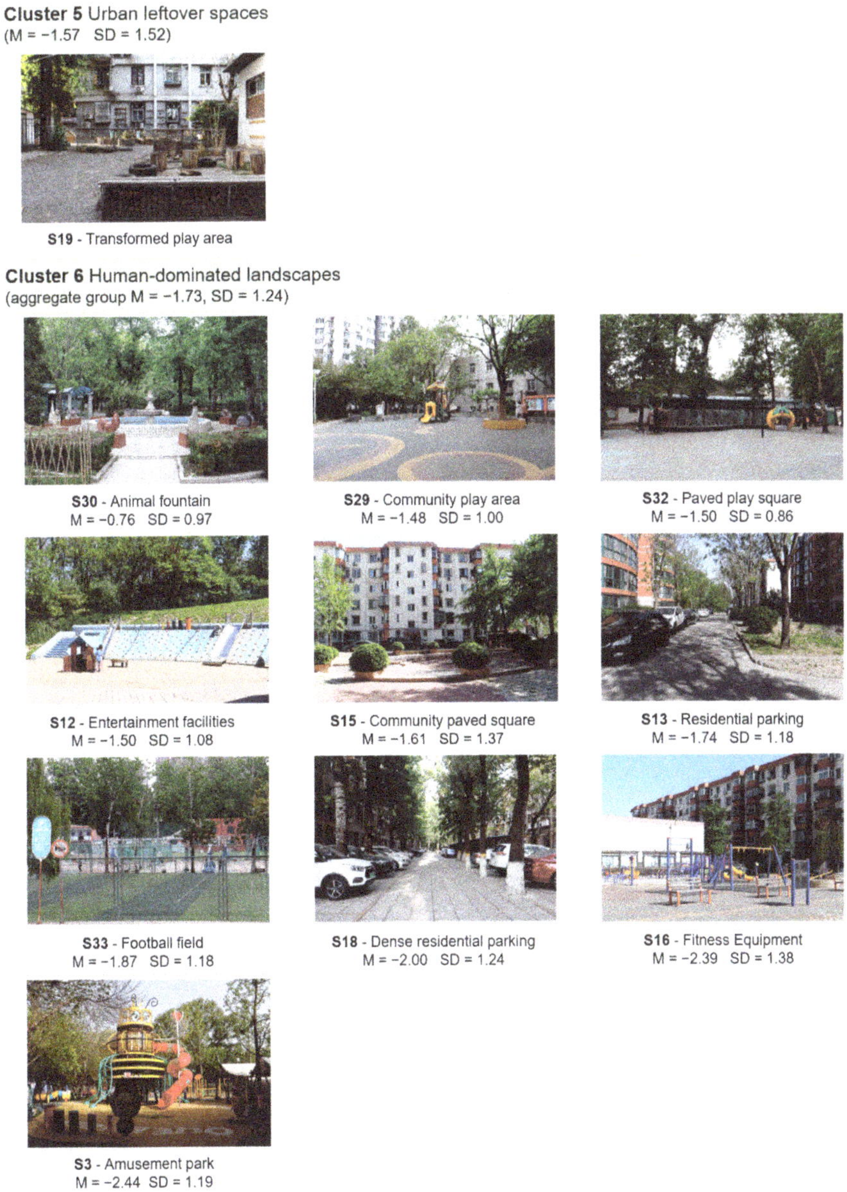

Figure 2. Clusters of photograph stimuli arranged through hierarchical cluster analysis with associated naturalness mean ratings (M = mean value, SD = standard deviation).

3.2.3. Underlying Perceptual Dimensions

The results of the MDS analysis revealed three underlying bipolar dimensions of meaning, as revealed in the stimulus configuration plots (see Figures 3 and 4), which visually map the perceptual relationships and distances between the various stimuli elements (photographs of environments). These dimensions represent the key discriminating factors that influence how children perceive the naturalness of the different environments tested in this study and categorize them accordingly in clusters, as previously described.

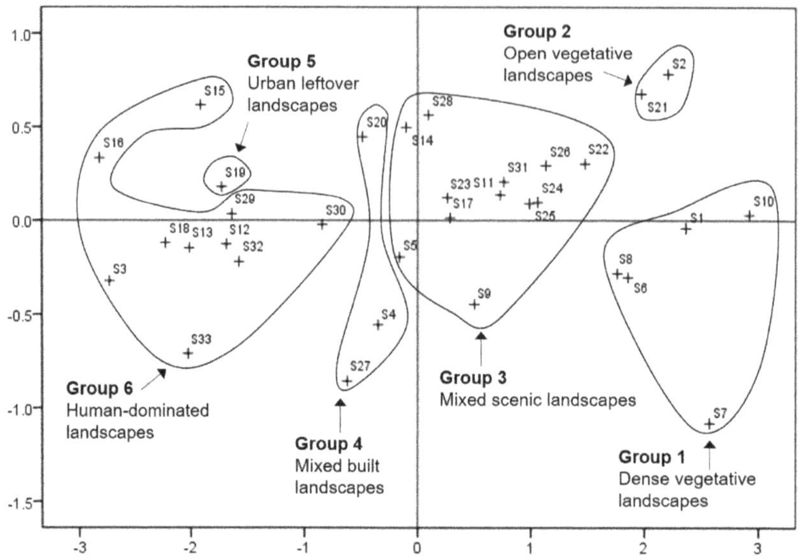

Figure 3. MDS stimulus configuration of Dimensions 1 and 2. Dimension 1: organic to inorganic (horizontal right to left); Dimension 2: congruous to incongruous (vertical top to bottom).

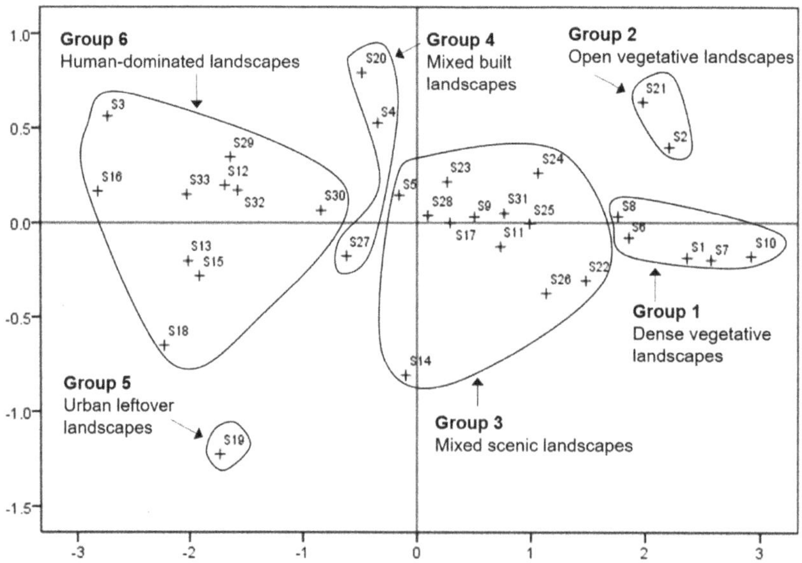

Figure 4. MDS stimulus configuration of Dimensions 1 and 3. Dimension 1: organic to inorganic (horizontal right to left); Dimension 3: colorful to drab (vertical top to bottom).

Dimension 1: organic to inorganic (natural to artificial). In this dimension, Clusters 1 and 2 are positioned on the far right of the plot, indicating that children perceive these environments as the most natural. These clusters are characterized by landscapes dominated by plants with almost no visible signs of artificial elements and were described as "green" and "ecological", along with references to animals, such as birds and insects. In contrast, Clusters 5 and 6, which represent environments dominated by artificial elements, are located on the far left of the plot, suggesting that Dimension 1 represents a spectrum from organic/natural to inorganic/artificial based on the interpretation of the environments included in these four clusters, highlighting this dimension as mainly related to objective evidence and signs of naturalness. Additionally, the positioning of Clusters 3 and 4 further reinforces this interpretation. Cluster 3 integrates orderly frames within predominantly natural settings and is located closer to the right than Cluster 4, which features artificial structures as the main focus of the environment.

Dimension 2: congruous to incongruous (fitting to non-fitting). The interpretation of Dimension 2, like that of Dimension 1, is revealed by examining the relative positions of the clusters and, in this case, the vertical distribution of photographs and the associated clusters. Cluster 2 is at the top of the plot, for which the children frequently used terms including "spacious", "comfortable", and "beautiful" to describe the depicted environments. Similarly, **S14** (community abandoned land) and **S28** (traditional architecture at the water's edge) in Cluster 3 are also positioned near the top, with the children describing them as being "harmonious". In contrast, **S9** (sparse forest) was described as a place that felt "dark" and "depressing", with the stones resembling tombstones and being "incongruous". Similar descriptions were given of **S7** (a dense forest-like setting), which was perceived as being the least natural along this dimension. Interestingly, **S15** (community paved square) in Cluster 6 is also positioned near the top, and, in fact, most photographs in this cluster are positioned relatively high or at least around the middle along this dimension compared to the first dimension, where the depicted environments were perceived as the least natural. These photographs predominantly depict residential areas, community parks, and similar environments familiar to children in their daily lives. A common feature among the photographs positioned higher in these clusters is that children described these environments as being "comfortable", "fitting", and "harmonious", suggesting a subjective evaluative dimension associated with emotional responses to the depicted environments.

Dimension 3: colorful to drab (vibrant to dull). By comparing the plot shown in Figure 3 with the one shown in Figure 4, Cluster 2 remains in a relatively high position on the plot; however, **S21** (a field of flowers) is positioned higher than **S2** (a flowered lawn). In Dimension 3, the photographs in Cluster 4 have generally shifted upwards, with **S20** (an artificial flower bed) positioned at the very top. In Cluster 6, **S3** (an amusement park) has moved to the highest position. What these photos have in common is that they depict the most colorful environments, suggesting a strong association with vibrancy and visual appeal. In contrast, Cluster 5 shows the most significant change and is rated the least natural along this dimension. Similarly, **S14** (community abandoned land) and **S18** (a dense residential parking area) were also perceived as being less natural, with a common trait of being perceived as drab and dull in color. This shift indicates that environments with more vivid colors and visually engaging elements tend to be perceived as more natural along this dimension.

4. Discussion

The findings of this study indicate that plants, particularly trees, are the dominant elements of children's constructs of nature. This is evident from the interviews on what conceptual and accessible nature means to them and the criteria they used to rate the stimuli photographs regarding the degree of perceived naturalness, especially regarding the first dimension that differentiates organic from inorganic. Plants, in general, were found to be highly instrumental in defining and characterizing natural settings and features, especially in urban open space areas, including parks and reserves. Trees are among the most conspicuous natural elements, given their large scale, towering presence, distinct shapes, and seasonal changes, and they also provide habitats and sustenance for other biological organisms (e.g., birds and small mammals), potentially appearing as an independent system to children. Children's experiences with plants (e.g., observing plant landscapes, climbing trees, and collecting leaves) provide them with the city's most tangible signs of nature, thereby shaping their constructs of plants/trees as manifestations of nature [46].

The prominence of plants/trees in the children's conceptualization of nature could also be traced back to human evolution, as early humans relied on trees and other plants as a primary source of food, shelter, and other resources necessary for their survival. For example, trees provide elevated locations to obtain views (prospect) over the landscape and find refuge from predators [47]. The frequently mentioned grass and flowers and the children's fondness for them might be explained by their resemblance to savanna-type landscapes where early humans are believed to have lived during a crucial period of evolution, and flowers would have also signaled prospective food sources available to our ancient ancestors [48,49]. Non-living entities, especially water bodies, were also frequently mentioned in the interviews and as key criteria that define what are perceived as being natural settings, as gleaned from the sorting tasks, with many children showing fondness for waterscapes, which aligned with existing research [50]. To some extent, this reflects the enduring influence of biophilic learning rules in modern societies, in which human–nature connections are correlated with multiple meanings and expressions of satisfaction—aesthetically, intellectually, cognitively, and even spiritually [4].

The children's conceptualization and perception of nature could also be influenced by media and educational materials. One example is the notion of "plant blindness", as suggested by the findings of this study, where few children could name specific plant species, especially when compared to their ability to identify animals, which is consistent with other studies [51–53]. Despite the relatively weak visual cues for identifying plants, especially when they are not in bloom or bearing fruit, a possible explanation lies in the influence of media and educational materials that portray plants as more of a collective presence that contributes to the overall ambiance of settings rather than independent plant types and species [54,55], compared to animals, which have more distinctive features and resemblance to human traits (e.g., the ability to move and make sounds) and therefore often play key roles in children's stories [56]. The differences in children's descriptions of animals between conceptual and accessible personal constructs of nature may also reflect the impact of media and education, as well as their daily experiences in the city, on their personal constructs. More children mentioned animals commonly encountered in their everyday lives in describing accessible nature, particularly cats, dogs, birds, and insects, compared to conceptual nature, where children referred to a broader range of animals, often iconic, large, distinctive, or endangered types of wildlife, such as elephants, rhinos, and cheetahs.

However, how nature is presented in children's books and media is also likely to influence how they perceive nature in their everyday environment. Children tend to spend more time indoors engaging with electronic devices, from which they more frequently encounter exotic species of plants and animals virtually at the expense of awareness of everyday biodiversity in their surroundings; thus, they are likely to be desensitized to everyday nature, which they may find to be ordinary and boring in comparison [18,57]. This phenomenon is reflected in the children's discrimination of real versus fake nature, as discussed in the results section, which highlights children's tendency to romanticize nature, and it can lead them to perceive the relationship between humans and nature in a somewhat distorted manner. This may, in turn, further exacerbate their disconnection with nature, impacting their motivation to engage in efforts at conserving nature, and may even develop in them a conservation bias—a selective focus on the conservation of certain charismatic, "cute", or endangered species at the expense of species they may encounter in their everyday city lives [58].

In addition, it is interesting to note that despite the first dimension of organic to inorganic, an objective evaluative dimension consistent with the interview results and the definition of nature provided by Wohlwill, two subjective dimensions were also identified in children's perceptions of accessible nature in urban settings tested in this study. These dimensions focused more on non-living and artificial elements and were associated with particular affective responses (congruous to incongruous, feelings of comfort, harmony, and familiarity) and aesthetic qualities (colorful to drab, visually vibrant features) of the environments, indicating the impact of their daily experiences and surroundings on their conceptions of nature. These findings may differ from those of adults or those with specialized knowledge, such as ecologists, who might place more emphasis on biodiversity and ecosystem functions and services. These insights highlight the influence of the design of urban environments and human activities on children's perceptions of nature, suggesting that both environmental design and environmental education should seek to provide children with more meaningful and direct experiences of nature that they can encounter in their everyday surroundings.

An important step in doing this is to encourage children to spend more time in outdoor natural environments over engaging in indoor activities and the use of digital and electronic devices. It is, therefore, crucial to provide them with accessible natural environments, particularly blue (water) and green (vegetated) spaces in proximity to residential neighborhoods and schools where they live and study. These environments should incorporate a diverse array of native plant species and create suitable habitats that can accommodate common types of urban wildlife, such as birds, insects, and small mammals. Because the children tended to associate vibrant, colorful, and harmonious environments with nature, these environments should also be visually appealing and engaging, and drawing on biophilic design principles (e.g., shapes, colors, and sounds of nature) helps to create immersive environments that can facilitate regular, unstructured outdoor activities and exposure to elements of the natural world. In addition, environmental educational interventions should incorporate elements of local nature and encourage study of nature in the curriculum that involves hands-on interaction with nature, such as bird observation and insect surveys, that emphasize greater understanding of local biodiversity.

There are several limitations that should be acknowledged. Firstly, the sample of children used in the study was not large or varied enough to fully represent the population of a megacity like Beijing, as most participants shared similar socioeconomic backgrounds, which could limit the generalizability of the findings to other populations and geographic contexts. Secondly, while studies on environmental perception often employ a somewhat larger number of photographs as stimuli in using the photo Q-sort method, this study used only 33 photos considering the cognitive and attentional capacities of children compared to adults, which could restrict the scope and variety of environments represented. However, using children as respondents and highly urban environmental settings as stimuli for this method is innovative and helps to fill a gap in the current literature addressing this topic. Future research could, however, address these limitations by including a larger and more diverse sample of child participants and environmental contexts. Comparative studies between urban and rural children across different age groups or cultural contexts may also provide deeper insights into how children understand and perceive nature.

5. Conclusions

The study reported here first explored children's understanding of conceptual and accessible nature and identified similarities and differences between the two. The results revealed that plants, particularly trees, and vegetated landscapes were perceived by the children to be the most natural elements, followed by animals. They also frequently mentioned non-living natural elements and phenomena, which was consistent across both conceptual and accessible modes of conceptualizing nature. The most notable difference between the two is that when describing nature conceptually, the children tended to romanticize it, considering exotic and dramatic elements as "real nature" while seemingly being desensitized to ordinary, accessible, everyday nature.

Further exploration of children's perceptions of accessible nature in urban settings shows that they distinguish between natural and non-natural environments primarily through an objective evaluative dimension, contrasting nature with artificial/inorganic elements. Interestingly, the other two dimensions revealed through the MSD analysis—congruous to incongruous and colorful to drab—highlight the influence of the design of urban environments and associated human activities on children's understanding of nature in cities. This study also highlighted the impacts of personal experiences, education, media, and possibly evolutionarily ingrained predispositions on their personal constructs of nature. This understanding can provide useful insights for urban and landscape design and environmental educational curricula and activities aimed at engaging children in more meaningful contact with nature that will allow them to benefit from cultural ecosystems services, which can both promote their physical and psychological well-being and health and influence them in adopting pro-environmental attitudes, which, in turn, can help to ensure for them a more sustainable future.

Author Contributions: J.L.: conceptualization, resources, methodology, investigation, data curation, formal analysis, writing—original draft and editing. R.J.G.: conceptualization, resources, methodology, writing—review and editing, supervision. All authors have read and agreed to the published version of the manuscript.

Funding: Research reported in this article was partially funded by the China Scholarship Council (CSC)—grant number: 201906350021.

Data Availability Statement: The original contributions presented in this study are included in the article. Further inquiries can be directed to the corresponding author.

Conflicts of Interest: The authors declare no conflicts of interest.

References

1. Chawla, L. Childhood Nature Connection and Constructive Hope: A Review of Research on Connecting with Nature and Coping with Environmental Loss. *People Nat.* **2020**, *2*, 619–642. [CrossRef]
2. Cheng, J.C.H.; Monroe, M.C. Connection to Nature: Children's Affective Attitude toward Nature. *Environ. Behav.* **2012**, *44*, 31–49. [CrossRef]
3. Louv, R. *Last Child in the Woods: Saving Our Children From Nature-Deficit Disorder (Updated and Expanded)*; Algonquin Books of Chapel Hill: New York, USA, 2008; ISBN-13: 978-1-56512-605-3.
4. Kellert, S.R.; Wilson, E.O. *The Biophilia Hypothesis*; Island Press: Washington, DC, USA, 1993; ISBN 9780415475976.
5. Ulrich, R.S.; Simon, R.F.; Losito, B.D.; Fiorito, E.; Miles, M.A.; Zelson, M. Stress Recovery during Exposure to Natural and Urban Environments. *J. Environ. Psychol.* **1991**, *11*, 201–230. [CrossRef]
6. Kaplan, S. The Restorative Benefits of Nature: Toward an Integrative Framework. *J. Environ. Psychol.* **1995**, *15*, 169–182. [CrossRef]
7. Bagot, K.L.; Allen, F.C.L.; Toukhsati, S. Perceived Restorativeness of Children's School Playground Environments: Nature, Playground Features and Play Period Experiences. *J. Environ. Psychol.* **2015**, *41*, 1–9. [CrossRef]
8. Chawla, L.; Keena, K.; Pevec, I.; Stanley, E. Green Schoolyards as Havens from Stress and Resources for Resilience in Childhood and Adolescence. *Health Place* **2014**, *28*, 1–13. [CrossRef] [PubMed]
9. Weeland, J.; Moens, M.A.; Beute, F.; Assink, M.; Staaks, J.P.C.; Overbeek, G. A Dose of Nature: Two Three-Level Meta-Analyses of the Beneficial Effects of Exposure to Nature on Children's Self-Regulation. *J. Environ. Psychol.* **2019**, *65*, 101326. [CrossRef]
10. Liu, J.; Green, R.J. The Effect of Exposure to Nature on Children's Psychological Well-Being: A Systematic Review of the Literature. *Urban For. Urban Green.* **2023**, *81*, 127846. [CrossRef]
11. Shu, S.; Ma, H. The Restorative Environmental Sounds Perceived by Children. *J. Environ. Psychol.* **2018**, *60*, 72–80. [CrossRef]

12. Hartley, K.; Perazzo, J.; Brokamp, C.; Gillespie, G.L.; Cecil, K.M.; LeMasters, G.; Yolton, K.; Ryan, P. Residential Surrounding Greenness and Self-Reported Symptoms of Anxiety and Depression in Adolescents. *Environ. Res.* **2021**, *194*, 110628. [CrossRef]
13. Liu, J.; Green, R.J. Children's pro-Environmental Behaviour: A Systematic Review of the Literature. *Resour. Conserv. Recycl.* **2024**, *205*, 107524. [CrossRef]
14. Dunlap, R.E.; Van Liere, K.D.; Mertig, A.G.; Jones, R.E. Measuring Endorsement of the New Ecological Paradigm: A Revised NEP Scale. *J. Soc. Issues* **2000**, *56*, 425–442. [CrossRef]
15. Frick, J.; Kaiser, F.G.; Wilson, M. Environmental Knowledge and Conservation Behavior: Exploring Prevalence and Structure in a Representative Sample. *Pers. Individ. Dif.* **2004**, *37*, 1597–1613. [CrossRef]
16. Otto, S.; Pensini, P. Nature-Based Environmental Education of Children: Environmental Knowledge and Connectedness to Nature, Together, Are Related to Ecological Behaviour. *Glob. Environ. Chang.* **2017**, *47*, 88–94. [CrossRef]
17. Orr, D.W. Political Economy and the Ecology of Childhood. In *Children and Nature: Psychological, Sociocultural and Evolutionary Investigations*; Peter, H., Kahn, J., Kellert, S.R., Eds.; The MIT Press: Cambridge, MA, USA, 2002; pp. 279–303.
18. Pyle, R.M. Eden in a Vacant Lot: Special Places, Species and Kids in the Neighborhood of Life. In *Children and Nature*; Peter, H., Kahn, J., Kellert, S.R., Eds.; The MIT Press: Cambridge, MA, USA, 2002; pp. 305–327.
19. Van Aart, C.J.C.; Michels, N.; Sioen, I.; De Decker, A.; Bijnens, E.M.; Janssen, B.G.; De Henauw, S.; Nawrot, T.S. Residential Landscape as a Predictor of Psychosocial Stress in the Life Course from Childhood to Adolescence. *Environ. Int.* **2018**, *120*, 456–463. [CrossRef]
20. Miller, J.R. Biodiversity Conservation and the Extinction of Experience. *Trends Ecol. Evol.* **2005**, *20*, 430–434. [CrossRef]
21. Soga, M.; Evans, M.J.; Yamanoi, T.; Fukano, Y.; Tsuchiya, K.; Koyanagi, T.F.; Kanai, T. How Can We Mitigate against Increasing Biophobia among Children during the Extinction of Experience ? *Biol. Conserv.* **2020**, *242*, 108420. [CrossRef]
22. Wohlwill, J.F. The Concept of Nature: A Psychologist's View. In *Behavior and the Natural Environment*; Altman, I., Wohlwill, J.F., Eds.; Springer: Berlin/Heidelberg, Germany, 1983; ISBN 978-1-4613-3541-2.
23. Kelly, G.A. *A Theory of Personality: The Psychology of Personal Constructs*; Norton: New York, USA, 1955.
24. Kahn, P.H.; Kellert, S.R. *Children and Nature: Psychological, Sociocultural, and Evolutionary Investigations*; Kahn, P.H., Kellert, S.R., Eds.; The MIT Press: Cambridge, MA, USA, 2002; ISBN 9780262112671.
25. Zeisel, J. *Inquiry by Design: Tools for Environment-Behavior Research*; Cambridge University Press: Cambridge, UK, 1984.
26. Wijngaarden, V. Q Method and Ethnography in Tourism Research: Enhancing Insights, Comparability and Reflexivity. *Curr. Issues Tour.* **2017**, *20*, 869–882. [CrossRef]
27. Mckeown, B.; Thoma, D.B. *Q Methodology*, 2nd ed.; SAGE Publications, Inc.: Thousand Oaks, CA, USA, 1988.
28. Stephenson, W. *The Study of Behaviour: Q-Technique and Its Methodology*; University of Chicago Press: Chicago, IL, USA, 1953.
29. Pitt, D.G.; Zube, E.H. The Q Sort Method: Use in Landscape Assessment Research and Landscape Planning. In *Proceedings on Our National Landscape: A Conference on Applied Techniques for Analysis and Management of the Visual Resource*; Elsner, G.H., Smardon, R.C., Eds.; USDA Forest Service: Berkeley, CA, USA, 1979; pp. 227–234.
30. Green, R. Community Perceptions of Environmental and Social Change and Tourism Development on the Island of Koh Samui, Thailand. *J. Environ. Psychol.* **2005**, *25*, 37–56. [CrossRef]
31. D'Souza, M.; Johnson, M.F.; Ives, C.D. Values Influence Public Perceptions of Flood Management Schemes. *J. Environ. Manage.* **2021**, *291*, 112636. [CrossRef]
32. Fairweather, J.R.; Swaffield, S.R. Visitor Experiences of Kaikoura, New Zealand: An Interpretative Study Using Photographs of Landscapes and Q Method. *Tour. Manag.* **2001**, *22*, 219–228. [CrossRef]
33. Naspetti, S.; Mandolesi, S.; Zanoli, R. Using Visual Q Sorting to Determine the Impact of Photovoltaic Applications on the Landscape. *Land Use Policy* **2016**, *57*, 564–573. [CrossRef]
34. Zube, E.H.; Pitt, D.G.; Evans, G.W. A Lifespan Developmental Study of Landscape Assessment. *J. Environ. Psychol.* **1983**, *3*, 115–128. [CrossRef]
35. Harrison, J.N.; Sarre, P. Personal Construct Theory in the Measurement of Environmental Images. *Environ. Behav.* **1975**, *7*, 3–58. [CrossRef]
36. Milcu, A.I.; Sherren, K.; Hanspach, J.; Abson, D.; Fischer, J. Navigating Conflicting Landscape Aspirations: Application of a Photo-Based Q-Method in Transylvania (Central Romania). *Land Use Policy* **2014**, *41*, 408–422. [CrossRef]
37. Berman, M.G.; Hout, M.C.; Kardan, O.; Hunter, M.C.R.; Yourganov, G.; Henderson, J.M.; Hanayik, T.; Karimi, H.; Jonides, J. The Perception of Naturalness Correlates with Low-Level Visual Features of Environmental Scenes. *PLoS ONE* **2014**, *9*, e114572. [CrossRef]
38. Detrinidad, E.; López-Ruiz, V.R. The Interplay of Happiness and Sustainability: A Multidimensional Scaling and K-Means Cluster Approach. *Sustainability* **2024**, *16*, 10068. [CrossRef]
39. Dobbie, M.; Green, R. Public Perceptions of Freshwater Wetlands in Victoria, Australia. *Landsc. Urban Plan.* **2013**, *110*, 143–154. [CrossRef]

40. Takane, Y.; Young, F.W.; de Leeuw, J. Nonmetric Individual Differences Multidimensional Scaling: An Alternating Least Squares Method with Optimal Scaling Features. *Psychometrika* **1977**, *42*, 7–67. [CrossRef]
41. Kruskal, J.B. Multidimensional Scaling by Optimising Good- Ness of Fit to a Nonmetric Hypothesis. *Psychometrika* **1964**, *29*, 1–27. [CrossRef]
42. Real, E.; Arce, C.; Sabucedo, J.M. Classification of Landscapes Using Quantitative and Categorical Data, and Prediction of Their Scenic Beauty in North-Western Spain. *J. Environ. Psychol.* **2000**, *20*, 355–373. [CrossRef]
43. Wandersee, J.H.; Schussler, E.E. Preventing Plant Blindness. *Am. Biol. Teach.* **1999**, *61*, 82–86. [CrossRef]
44. Zhang, W.; Goodale, E.; Chen, J. How Contact with Nature Affects Children's Biophilia, Biophobia and Conservation Attitude in China. *Biol. Conserv.* **2014**, *177*, 109–116. [CrossRef]
45. Nassauer, J.I. Messy Ecosystems, Orderly Frames. *Landsc. J.* **1995**, *14*, 161–170. [CrossRef]
46. Collado, S.; Íñiguez-Rueda, L.; Corraliza, J.A. Experiencing Nature and Children's Conceptualizations of the Natural World. *Child. Geogr.* **2016**, *14*, 716–730. [CrossRef]
47. Appleton, J. *The Experience of Landscape*; John Wiley & Sons: Chichester, UK, 1996.
48. Orians, G.H. An Ecological and Evolutionary Approach to Landscape Aesthetics. In *Landscape Meanings and Values*; Routledge: Berkeley, CA, USA, 1986; pp. 3–25.
49. Balling, J.D.; Falk, J.H. Development of Visual Preference for Natural Environments. *Environ. Behav.* **1982**, *14*, 5–28. [CrossRef]
50. Bozkurt, M.; Woolley, H. Let's Splash: Children's Active and Passive Water Play in Constructed and Natural Water Features in Urban Green Spaces in Sheffield. *Urban For. Urban Green.* **2020**, *52*, 126696. [CrossRef]
51. Bebbington, A. The Ability of A-Level Students to Name Plants. *J. Biol. Educ.* **2005**, *39*, 63–67. [CrossRef]
52. Patrick, P.; Tunnicliffe, S.D. What Plants and Animals Do Early Childhood and Primary Students' Name? Where Do They See Them? *J. Sci. Educ. Technol.* **2011**, *20*, 630–642. [CrossRef]
53. Balas, B.; Momsen, J.L. Attention "Blinks" Differently for Plants and Animals. *CBE Life Sci. Educ.* **2014**, *13*, 437–443. [CrossRef]
54. Schussler, E.E.; Link-Pérez, M.A.; Weber, K.M.; Dollo, V.H. Exploring Plant and Animal Content in Elementary Science Textbooks. *J. Biol. Educ.* **2010**, *44*, 123–128. [CrossRef]
55. Honey, J.N. Where Have All the Flowers Gone?—The Place of Plants in School Science. *J. Biol. Educ.* **1987**, *21*, 185–189. [CrossRef]
56. Link-Pérez, M.A.; Dollo, V.H.; Weber, K.M.; Schussler, E.E. What's in a Name: Differential Labelling of Plant and Animal Photographs in Two Nationally Syndicated Elementary Science Textbook Series. *Int. J. Sci. Educ.* **2010**, *32*, 1227–1242. [CrossRef]
57. Clucas, B.; McHugh, K.; Caro, T. Flagship Species on Covers of US Conservation and Nature Magazines. *Biodivers. Conserv.* **2008**, *17*, 1517–1528. [CrossRef]
58. Rosalino, L.M.; Gheler-Costa, C.; Santos, G.; Gonçalves, M.T.; Fonseca, C.; Leal, A.I. Conservation Priorities for Elementary School Students: Neotropical and European Perspectives. *Biodivers. Conserv.* **2017**, *26*, 2675–2697. [CrossRef]

land

MDPI

Article

Advancing Urban and Extra-Urban Afforestation: A Case Study of the Italian National Urban Forestry Plan in the Metropolitan City of Genoa

Adriano Magliocco * and **Paola Sabbion**

Department of Architecture and Design, University of Genoa, 16123 Genoa, Italy; paola.sabbion@unige.it
* Correspondence: adriano.magliocco@unige.it

Abstract: Urban afforestation is a key strategy for combating climate change and enhancing urban living by reducing carbon emissions and improving air quality. Reflecting these priorities, Italy's National Ecological Transition Plan (NETP), led by the Ministry of Ecological Transition (MiTE), introduced the National Urban Forestry Plan to accelerate sustainability and resilience in urban areas. This plan focuses on reducing carbon emissions, enhancing natural capital through reforestation, and promoting Nature-Based Solutions (NBS) for climate adaptation. Our study aligns with this framework by exploring how local urban forestry initiatives support the NETP's objectives, providing a concrete example of policy implementation. Genoa (Italy) serves as a compelling case study. In 2023, the city received PNRR funding under the "Protection and Enhancement of Urban and Extra-Urban Greenery" initiative to improve air quality, restore biodiversity, and strengthen ecological corridors. Our research identifies three key areas for improvement: ecological planning, citizen engagement, and impact monitoring. We present initial findings and quantitative data, evaluating CO_2 sequestration potential, communication strategies, and consider the strengths and weaknesses of the specific context. These findings highlight the need for further investigation to optimize and communicate urban greening efforts.

Keywords: urban afforestation; carbon sequestration; green infrastructure; nature-based solutions; climate change mitigation

1. Introduction

Nature-based solutions play a vital role in reducing carbon emissions and enhancing carbon sequestration, significantly contributing to achieving climate adaptation objectives. These solutions encompass reforestation, afforestation, and the conservation of forests, wetlands, coastal habitats, and grasslands. Forests are among the most effective carbon sinks [1]. Urban forests represent a complex system that includes forests, groups of trees, street trees, and individual trees situated in urban and peri-urban areas.

According to the FAO Guidelines on Urban and Peri-Urban Forestry [2], urban forests are considered the "backbone" of Green Infrastructure (GI), which is defined as a "strategically planned and managed network of natural land, landscapes, and other open spaces that protect the values and functions of ecosystems while providing a wide range of benefits to human populations and wild nature" [3]. Urban forests (UF) serve as a critical connection between rural and urban areas, significantly enhancing a city's environmental sustainability by delivering essential ecosystem services and improving urban quality of life. Recognizing this, FAO has urged cities worldwide to invest in forest-based solutions to foster more sustainable and resilient urban development [4]. These initiatives not only address climate challenges but also support biodiversity, mitigate urban heat island effects, and contribute to public health and well-being, underscoring the multifaceted benefits of integrating urban forestry into strategic planning.

Urban forests significantly contribute to atmospheric purification by absorbing gaseous pollutants such as ozone, nitrogen oxides, and sulfur chlorides, intercepting particulate matter—including dust, ash, pollen, and smoke—and releasing oxygen via photosynthesis. Additionally, trees play a pivotal role in mitigating urban heat through shading and transpiration processes, which, in turn, can suppress ozone formation. Beyond these ecological functions, urban forestry is critical for carbon sequestration, energy demand reduction, and the promotion of sustainable urban development frameworks [5]. Urban forestry also facilitates essential ecosystem services such as carbon storage, erosion control, landslide risk mitigation, and biodiversity conservation. These functions align closely with international climate objectives, enabling local governments to adopt localized, bottom-up strategies that contribute to global sustainability targets [6,7].

Urban areas account for over two-thirds of global energy consumption and generate more than 70% of greenhouse gas (GHG) emissions, making afforestation a critical strategy for combating climate change in these settings [8]. Evidence suggests that increasing urban tree canopy cover can reduce summer temperatures by up to 3 °C, thereby lowering peak ozone levels and improving air quality [9]. Urban afforestation is therefore a pivotal strategy for mitigating climate change and improving urban living conditions. It contributes directly to the decarbonization of metropolitan areas and aligns with the objectives of the Paris Agreement. As urban populations continue to expand, the integration of urban forestry into climate strategies becomes increasingly essential for fostering resilient and sustainable urban environments capable of both adapting to and mitigating the impacts of climate change.

Municipalities play a central role in the effective implementation of urban forestry as part of climate mitigation efforts. Sustainable Energy and Climate Action Plans (SECAPs) frequently incorporate urban forestry as a key component of broader initiatives aimed at reducing carbon footprints and meeting climate objectives. The strategic inclusion of urban forestry into municipal planning and execution is critical for achieving long-term decarbonization goals. By facilitating these efforts, local governments contribute significantly to the broader global commitment to addressing climate change through scientifically informed, localized action [10].

UF, which include all trees within urban realms such as parks, streets, private properties, and greenways, play a critical role in carbon sequestration. Forests act as a temporary carbon sink, delaying the release of stored carbon until the wood is harvested and reintroduced into the carbon cycle. This process helps offset emissions from urban activities, particularly in cities, where transportation, industry, and energy production are major sources of greenhouse gases [11]. Despite growing public interest in urban forestry, its role in climate mitigation has received comparatively less scientific attention than rural reforestation studies [12].

According to literature, urban forestry demonstrates a lower climate mitigation potential—estimated at 82.4 $\pm$ 25.7 $MtCO_2$ per year—compared to rural reforestation, which sequesters approximately 1600 $MtCO_2$ per year. Nevertheless, UF can offset more than 25% of carbon emissions within city boundaries, making it a vital strategy for local-level climate change mitigation [12]. Urban forests currently cover approximately 6.0 $\pm$ 1.5 million hectares globally, representing 9.7% of total urban areas, with estimates suggesting that 100,000 urban trees can sequester around one million tons of CO_2 [13].

UF, parks, and street trees are particularly effective for carbon sequestration due to their high woody biomass. However, their sequestration capacity varies widely, influenced by factors such as biome type, plant species, biophysical parameters, location, climate, and maintenance requirements [14,15]. To optimize carbon sequestration in urban areas, strategies include the protection, rehabilitation, or creation of green spaces and the introduction of vegetation with high carbon storage potential. Overall, the global implementation of urban vegetation strategies could improve annual carbon sequestration by 0.1 to 0.3 $GtCO_2$, highlighting their significant potential for supporting climate mitigation targets [16]. Heavily urbanized areas exhibit the highest emissions per unit area (57.95 kg CO_2e/m^2), followed by suburban areas (7.99 kg CO_2e/m^2), whereas afforested areas emit the least (0.73 kg CO_2e/m^2) [7]. Expanding green spaces, particularly on brownfields, presents an opportunity to enhance carbon sequestration while simultaneously delivering additional ecosystem services.

In addition to direct carbon sequestration, afforestation contributes indirectly to decarbonization by reducing energy consumption. Urban trees provide shade, mitigate the urban heat island (UHI) effect, and lower ambient temperatures, which in turn decreases the demand for air conditioning. This natural cooling effect reduces energy use and the associated greenhouse gas emissions from power plants. Furthermore, energy cost savings are realized through the combined effects of shading, evapotranspiration, and wind speed regulation, which help conserve energy in buildings. Urban green spaces also encourage active transportation, such as walking and cycling, reducing dependency on motorized vehicles and thereby contributing to lower emissions. This shift not only supports decarbonization but also aligns with broader sustainability goals by improving public health and reducing air pollution [17].

In alignment with international commitments, the European Union has introduced the European Green Deal, aiming to achieve net-zero emissions by 2050 and a 55% reduction in emissions by 2030. Complementing the Green Deal is the Next Generation EU initiative, developed to stimulate economic recovery following the COVID-19 pandemic. This program operates through two primary mechanisms: the Recovery and Resilience Facility, which mandates member states to submit detailed plans for investments and reforms, and the REACT-EU package, which provides recovery assistance to support cohesion and regional development across Europe.

Given the growing political interest and significant public investment in Nature-Based Solutions (NBS) in Italy, the National Urban Forestry Plan has been introduced as a strategic framework for sustainable urban development. Italy's national policies, including the National Ecological Transition Plan (NETP), emphasize the enhancement of natural capital and the expansion of urban green networks. Spearheaded by the Ministry of Ecological Transition (MiTE), the NETP provides a comprehensive approach to fostering urban resilience, with key objectives such as reducing carbon emissions, increasing green infrastructure, and promoting NBS for climate adaptation.

Our study aligns with this framework by examining how local urban forestry initiatives support the NETP's strategic goals, demonstrating how national policies translate into tangible, city-level actions.

2. Materials and Methods

Urban forestry plays a crucial role in climate change mitigation and adaptation, particularly in Mediterranean cities, where rising temperatures and extreme heat events are increasingly frequent. Vegetation cover significantly influences urban microclimates, reducing urban heat island (UHI) effects through evapotranspiration, shading, and surface cooling. This relationship is particularly relevant in coastal cities like Genoa, where urban density, topography, and meteorological conditions contribute to temperature amplification, exacerbating thermal discomfort and energy demand for cooling. The success of urban reforestation initiatives relies on the careful selection of tree species adapted to urban microclimates and a comprehensive understanding of their ecological physiology. Addressing the specific environmental challenges of urban areas enhances the effectiveness of reforestation efforts, making significant contributions to climate change mitigation and urban resilience.

Genoa was selected as a case study due to its specific conditions, dense urban fabric, and ongoing urban greening initiatives, which together create an ideal setting to assess both the benefits and challenges of implementing nature-based solutions. Local authorities have been actively incorporating urban greenery into city planning to enhance ecosystem services, such as air quality improvement, temperature regulation, and biodiversity conservation. Despite spatial limitations and challenging topography, the city has pioneered urban forestry projects, including reforestation efforts, green corridors, and the revitalization of underutilized areas.

Due to the ongoing nature of the urban afforestation projects in Genoa and the limited availability of detailed implementation data from local authorities, this study adopted a mixed-methods approach based on policy analysis, literature review, and preliminary quantitative estimates.

First, we conducted a policy coherence analysis by comparing the urban reforestation projects currently underway in Genoa with the directives and strategic goals outlined by the National Urban and Peri-Urban Forestry Plan, as defined by Italy's Ministry of Ecological Transition (MiTE). The analysis assessed compliance according to key criteria derived from national guidelines, including: (1) targeted increase in tree canopy cover; (2) contribution to ecological connectivity and biodiversity; (3) prioritization of native species adapted to urban and peri-urban environments; and (4) alignment with Nature-Based Solutions (NBS) principles aimed at climate adaptation and resilience.

Second, to quantify the potential carbon sequestration of urban forestry initiatives, we estimated the CO_2 capture capacity based on species-specific carbon sequestration rates derived from existing scientific literature and market values recognized internationally (e.g., IPCC guidelines, scientific publications, and forestry databases). We considered predominant species selected for afforestation in Genoa, incorporating tree growth rates, lifespan, and climatic suitability for Mediterranean urban environments.

Third, we evaluated the strengths and weaknesses of urban afforestation specific context, explicitly addressing the challenge posed by Genoa's dense urban fabric, steep topography, and limited availability of space within urban boundaries. Given these constraints, reforestation initiatives were primarily implemented in peri-urban zones.

Given the preliminary stage of the projects analyzed, our evaluation primarily focused on verifying compliance with certain ecological criteria explicitly set by the National Urban and Peri-Urban Forestry Plan, such as the use of native species, species diversity, and structural composition of urban forests (proportion of shrubs to trees). Due to the lack of finalized quantitative data from ongoing projects, we performed a qualitative assessment based on available documentation, administrative reports, and literature-based estimations.

The evaluation criteria include: ecological benefits (e.g., biodiversity conservation, habitat restoration, climate regulation); socio-economic factors (e.g., recreational opportunities, public acceptance, aesthetic improvement); implementation feasibility (availability of land, maintenance requirements, adaptability of selected tree species), and potential drawbacks such as limited accessibility by citizens, higher costs of maintenance due to remote locations, and delayed impacts on urban microclimates.

The data sources utilized in this study included official documents and reports from national and local authorities (e.g., MiTE, PNRR documentation), peer-reviewed scientific literature on urban forestry, and semi-structured interviews conducted with key stakeholders involved in urban greening initiatives in Genoa. Literature sources included policy analyses regarding urban forestry efforts across major Italian metropolitan cities (e.g., Milan, Rome, Bologna, Turin), providing a comparative framework for interpreting Genoa's initiatives.

The methodological framework of this study was developed within the PRIN2022 project "Better Policy" (Building Environmental Tools To Empower Responsive Policies Outreaching LIfeCYcle: Guidelines and protocols to enable Public Administrations-driven processes in the Italian construction sector), coordinated by the author and focused on decarbonization strategies within public administration processes. Although primarily concerned with reducing emissions in the construction sector, exploring complementary strategies such as urban reforestation was deemed relevant given the broader climate resilience objectives of the project.

3. Results

3.1. National Urban and Peri-Urban Forestry Plan

Recently, there has been a growing nationwide interest in urban reforestation to counteract the effects of climate change and enhance biodiversity [18–21]. Several major Italian metropolitan cities have initiated significant urban forestry projects, including Milan ("ForestaMi" project), Turin, Rome ("Roma Resiliente" plan), Bologna ("Città 30 e Verde" project), Florence, Naples, Bari, and Catania.

In Italy, climate and ecological transition policies are shaped by a comprehensive framework that includes the National Climate Change Adaptation Strategy (SNAC) and its implementation plan, the National Climate Change Adaptation Plan (PNACC). Additional guiding policies include the National Sustainable Development Strategy (SNSvS), the National Biodiversity Strategy, the National Integrated Energy and Climate Plan 2030 (PNIEC), which outlines decarbonization objectives, the long-term Italian strategy for greenhouse gas emissions reduction, and the National Air Pollution Control Program (PNCIA). The National Recovery and Resilience Plan (PNRR) serves as the main tool for achieving the objectives of political strategies. It defines Mission 2—Green Revolution and Ecological Transition, Component 4—Protection of Land and Water Resources, with Investment 3.1 focusing on the protection and enhancement of urban and extra-urban green areas. The PNRR project for Investment 3.1 designates Metropolitan Cities as the implementing bodies responsible for planning and executing interventions under an Urban and Peri-Urban Forestry Plan. The Italian Government, through the Ministry of Ecological Transition (MiTE), has established the National Urban and Peri-Urban Forestry Plan, aligned with the broader objectives of the PNRR [22]. This plan outlines the planting of 6.6 million trees, aiming to develop urban and peri-urban forests across the 14 largest metropolitan areas, restoring degraded habitats, and reconnecting ecological corridors in urbanized contexts. Metropolitan areas, as key centers of human activity, face significant environmental challenges, including high levels of air pollution and the Urban Heat Island effect due to extensive land use and multiple pollution sources (e.g., vehicular traffic, industrial activities, heating). These pollutants negatively impact human health and impose economic costs on society through environmental and health damages. Urban areas also contribute substantially to global warming by emitting GHG. In Italy, 3.3% of the urban population lives in areas exceeding European air quality limits, prompting the European Commission to initiate infringement procedures related to PM2.5, PM10, nitrogen dioxide, and non-compliance with Directive 2016/2284 on national emission reductions [23].

Resources allocation is based on environmental and social criteria, such as air quality and areas subject to infringement procedures. The plan provides a standardized methodology for selecting appropriate tree species suited to local ecological conditions. MiTE established a Steering Committee to ensure effective implementation to oversee operations, provide scientific support, and monitor impacts. The plan emphasizes using local native plants to avoid biodiversity loss while enhancing ecosystem resilience [23]. In terms of cooling effects, peri-urban forests can reduce summer temperatures up to 170 m from their edge, extensive urban forests up to 100 m, and roadside trees up to 30 m [23].

Vegetation types were evaluated based on their ability to absorb CO_2, remove atmospheric particulate matter (PM10), and mitigate the Urban Heat Island (UHI) effect. A study of 10 Italian metropolitan cities found that 1 hectare of urban forest removes approximately 17 kg/year of PM10, valued at EUR 1825, with a total estimated value of EUR 1.532 billion across the cities. Given the European Union Emissions Trading System (EU ETS) carbon price of EUR 100.34 per metric ton as of February 2023, the monetary value of CO_2 absorbed by Italian forests over the first five years can be calculated. Assuming an average absorption rate of 20 to 30 tons per hectare over five years, the estimated value ranges from EUR 2006.80 to EUR 3010.20 per hectare.

Deciduous broadleaf species have demonstrated superior capacities for atmospheric pollutant removal, including CO_2 sequestration (3,953,280 g/m^2/year), O_3 absorption (5677.76 g/m^2/year), and NO_2 uptake (2358.30 g/m^2/year). Additionally, several species with exceptional carbon sequestration capacities have been identified, reaching up to 1025.47 g CO_2/m^2/year in species such as *Celtis australis, Platanus × acerifolia, Ulmus pumila,* and *Quercus rubra* [24]. On a per-tree basis, it is estimated that a single tree can absorb carbon dioxide at an average rate of 21.8 kg/year, underscoring the cumulative impact of widespread urban reforestation. In the Mediterranean region, *Pinus pinea* has been identified as a species with the highest potential for carbon absorption and storage. This naturalized archeophyte species, while not spontaneously occurring, has shown significant promise in urban forestry programs tailored to the local environment [25].

According to the plan, evergreen broad-leaved trees and Mediterranean conifers are the most effective for CO_2 sequestration and PM10 removal (Table 1). While assessments currently show variability, more precise data will be available with advanced planning.

Table 1. Quantitative and monetary evaluation of CO_2 absorption through the planting of different tree species according to Urban and Peri-Urban Forestry Plan. Adapted from: https:/mase.gov.it/ accessed on 16 December 2024.

Functional Type of Vegetation	Average CO_2 Absorption (t/ha/yr)	CO_2 Absorption First 5 Years (t/ha)	Lower Monetary Value of CO_2 Absorbed First 5 Years (EUR/ha)	Higher Monetary Value of CO_2 Absorbed First 5 Years (EUR/ha)	CO_2 Absorption at 20 Years (t/ha)	Lower Monetary Value of CO_2 Absorbed First 20 Years (EUR/ha)	Higher Monetary Value of CO_2 Absorbed First 20 Years (EUR/ha)
Deciduous broadleaf	6.6	21.9	208.2	835.2	121.6	1154.8	4631.5
Evergreen broadleaf	9.5	31.5	299.1	1199.7	174.6	1658.9	6653.0
Mediterranean conifers	9.0	29.7	281.9	1130.5	164.5	1563.2	6269.1
Mountain conifers	5.7	18.7	177.2	710.6	103.4	982.6	3940.6

To maintain structural efficiency and ecosystem functionality, it is required to plant species that are suited to the local environment and consistent with the area's natural vegetation. Effective reforestation necessitates the use of plant material specifically adapted to the microclimatic and soil conditions of each locality. A recent detailed review of nurseries in Sicily and Apulia indicates that these facilities are capable of producing high-quality, locally adapted plants with strong potential to withstand the effects of climate change [26]. This approach is critical for preserving biodiversity, safeguarding endangered species, and preventing genetic pollution. Forest nurseries are integral to this process, as they provide seedlings derived from native seeds, ensuring ecological compatibility and resilience. However, recent Italian regional forestry policies have largely overlooked nursery activities, resulting in a decline in both regional nurseries and skilled personnel. This has constrained the availability of high-quality native plant material. Hence, the success of this ambitious initiative is contingent upon a substantial supply of native trees and shrubs, which are currently underproduced in Italian nurseries. Despite this, the quantity of plant material remains insufficient to meet the increasing demand driven by contemporary green policies [26].

To assist Metropolitan Cities in identifying suitable reforestation areas, summary maps of potential vegetation based on the Vegetation Series of Italy mapping were provided in the Plan. These maps, which outline the mature stage of vegetation for each series, are intended as a general guide to promote more detailed studies that consider local soil, vegetation, and climate conditions. A recommended list of trees and shrubs serves as a basic guide for experts, who may also select additional species that align with the biogeographical and ecological context. The list integrates evergreen and deciduous species to maximize CO_2 storage and particulate matter removal throughout the year. It also includes species tolerant of high temperatures, promoting adaptation to climate change. Overall, a high level of biodiversity is recommended to ensure the resilience and robustness of reforestation efforts.

The following section analyzes the case of the Metropolitan City of Genoa, selected for its proximity, access to data on an ongoing reforestation project, and collaboration with local experts.

3.2. *The Case of Genoa, Italy*

Genoa was selected as a case study due to its combination of Mediterranean climatic conditions, dense urban fabric, and ongoing urban greening initiatives. These factors create an ideal setting for assessing the efficacy and challenges of implementing nature-based solutions (NBS) in mitigating extreme heat effects. Despite spatial constraints and complex topography, Genoa's urban forestry initiatives—including reforestation efforts, green corridors, and revitalization of underutilized spaces—demonstrate the potential of nature-based interventions in enhancing urban climate resilience.

The city is wedged between the Ligurian Sea and the Apennine mountains, resulting in urban districts that climb steep hillsides. The average slope gradient in parts of Genoa reaches about 50%, with peaks up to 70–75% incline on the surrounding hills [27]. This steep terrain poses obvious challenges for afforestation. Planting trees on very steep slopes is technically difficult—access for planting and maintenance is hard, soil tends to be shallow and prone to drying, and gravity can cause newly planted saplings (and soil amendments) to wash downhill during heavy rain.

The Metropolitan City of Genoa spans 183,375 hectares, with 12% designated as agricultural land and 8% as built-up areas. The landscape of Genoa is marked by a clear contrast between the densely built-up coastal and valley areas, which have minimal greenery, and the rest of the municipality, dominated by forests, shrublands, and abandoned fields, though transitional zones emerge along the hills and riverbeds where urban and natural elements intermingle. Forested and semi-natural areas cover 80% of the total municipal territory, the highest percentage of such land cover among Italy's metropolitan cities. These areas are primarily characterized by deciduous broadleaf forests, such as chestnut and beech woods, which play a crucial ecological and cultural role. Advances in remote sensing technologies have significantly improved the ability to study and manage these ecosystems. Metrics such as the Leaf Area Index (LAI) and phenological indicators provide valuable data for ecological research, urban green infrastructure planning, and the refinement of models assessing the ecosystem services offered by urban vegetation. In Genoa, the dominance of forested and semi-natural areas is particularly evident in their contribution to seasonal patterns in LAI. During summer, the dense canopy structure is reflected in high average (4.47) and maximum (6.37) LAI values. These values drop sharply in winter to an average of 0.77 and a maximum of 3.93, underscoring the deciduous nature of the predominant vegetation. This pronounced seasonal variation highlights the need to incorporate temporal dynamics into ecosystem assessments, enhancing our understanding and management of these landscapes.

Despite its high forest cover, air quality suffers from the negative effects of port activities. Recent monitoring data highlight that Genoa's geographical setting as a coastal city with a busy port significantly influences its air quality. According to ARPAL, Urban PM10 concentrations average around 30 μg/m^3—near or slightly above the EU annual limit of 25 μg/m^3—with particularly elevated levels recorded in areas adjacent to the port. Moreover, NO_2 measurements in the port vicinity average approximately 50 μg/m^3 [28], with episodic peaks coinciding with high shipping activity, clearly indicating the substantial contribution of port operations to local air pollution. Elevated levels of black carbon further corroborate the impact of combustion-related emissions from maritime traffic on Genoa's air quality, underscoring the need for targeted mitigation strategies.

Genoa's climate is Mediterranean but highly variable and locally intense—this adds another layer of challenge for urban forestry. On the one hand, the city enjoys mild, wet winters and hot, dry summers typical of its region. On the other hand, the Ligurian coast is infamous for sudden severe storms. Liguria (and Genoa in particular) is characterized by short but intense rainfall events that can drop huge volumes of water in just a few hours [29]. These cloudbursts regularly cause flash floods and severe runoff down the steep streets. For newly planted trees or forested slopes, heavy rain can mean soil erosion around roots, seedlings getting washed out, or even debris flows taking out vegetation. Afforestation projects must therefore include erosion control—for example, using mulch, geotextiles, or terracing to protect young plantings until they are established.

In 2024, temperatures exhibited frequent fluctuations. The coolest values, occasionally dropping below historical minimums for the period, were observed at the end of April, in June, and mid-September. Conversely, the warmest values, nearing historical maximums, occurred between mid-January and March, early July to mid-August, as well as in October and December. The annual average temperature was approximately 16 °C. During July and August, average minimum temperatures ranged between +20.8 °C and +20.9 °C, while average maximum temperatures varied from +27.2 °C to +27.5 °C. High humidity levels, particularly at night, often resulted in a pronounced feeling of mugginess, even when temperatures were not exceptionally high [28].

The Mediterranean region is experiencing a notable intensification of heat waves, with rising nighttime temperatures and extreme daytime peaks (Figure 1). During the summer of 2024, particularly in late July, Genoa endured prolonged extreme heat, with temperatures exceeding 35 °C and locally approaching 40 °C, accompanied by high humidity levels (Figure 2a,b). This follows an alarming trend, with an increasing frequency of "tropical nights"—when nighttime temperatures fail to drop below 28–30 °C. The situation is further underscored by record-breaking heat events in 2023, including Liguria's first-ever coastal 40 °C, recorded at Rapallo (GE) on 21 August 2023, with a peak of 40.6 °C. Inland areas experienced even more extreme conditions, with temperatures reaching 41–42 °C.

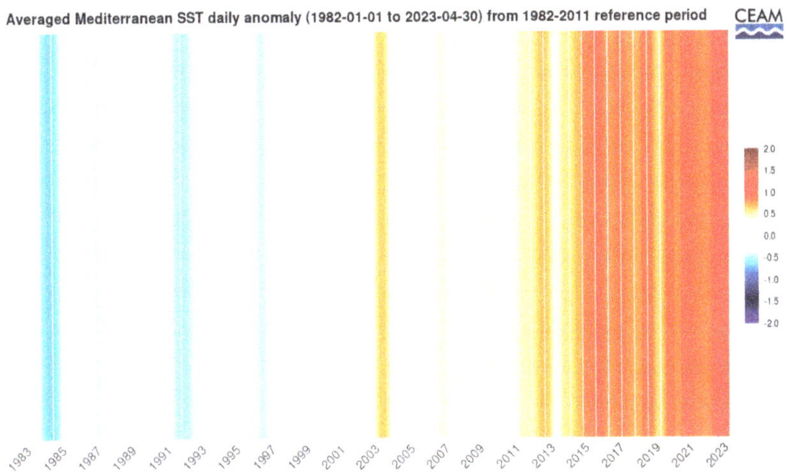

Figure 1. Average Mediterranean SST daily anomaly (1982–2023) from 1982–2011 reference period. Data sources: https://www.ceam.es, accessed on 21 December 2024.

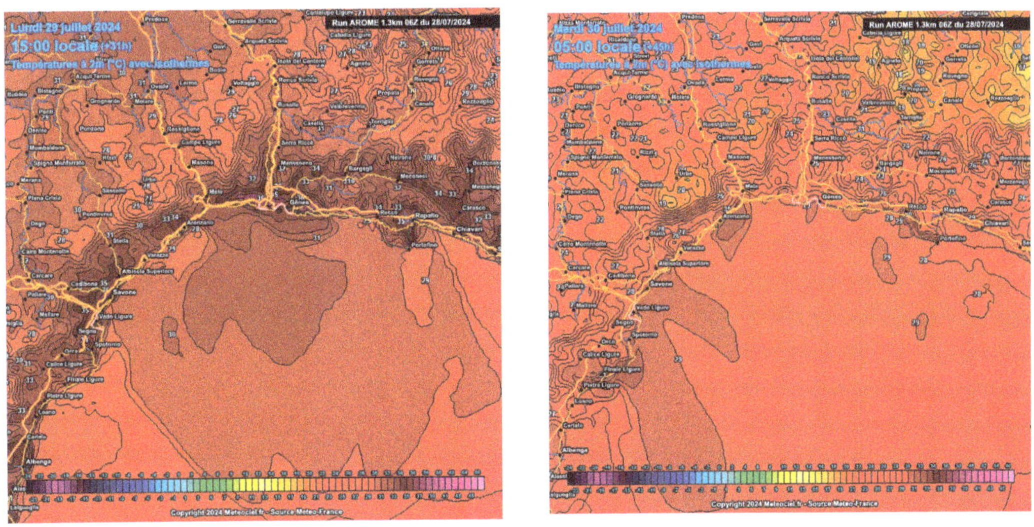

(a) (b)

Figure 2. Forecasted ground temperatures from the AROME model for central Liguria: (**a**) Maximum temperatures for Monday, 29 July 2024, 3:00 p.m. (**b**) Minimum temperatures for Tuesday, 30 July 2024. Temperatures at 2 m (°C) with isotherms. (AROME 1.3 km 06Z, 07/28/2024). Source: https://www.meteociel.fr/modeles/aromezoom.php?mode=41&x=10372.61&y=7505.73&ech=5&zoom=3.4, accessed on 20 December 2024.

Three key factors are driving these temperature anomalies:

- Exceptionally Warm Sea Surface Temperatures (SSTs)—The Ligurian Sea's surface temperatures have reached unprecedented levels, locally approaching 29 °C, reducing the thermal contrast between sea and land. This has diminished the cooling effect of sea breezes, intensifying coastal heat stress.
- Elevated 850 hPa Isotherm Temperatures—Atmospheric temperatures at ~1500 m altitude have been recorded at +20/22 °C, typically corresponding to surface

temperatures of 36–38 °C. While Liguria's complex topography and sea influence historically moderated these effects, the mitigating mechanisms appear to be weakening.

- Föhn Winds and Adiabatic Compression—Under certain meteorological conditions, downslope föhn winds descending from Liguria's mountain ranges accelerate warming through adiabatic compression, causing temperature spikes from early morning hours, compounding extreme heat conditions.

Given these climatic trends, the implementation of urban forestry strategies in Genoa is particularly urgent. Increasing urban vegetation can help counteract rising temperatures by:

- Reducing surface and air temperatures through evapotranspiration and shade provision;
- Mitigating urban heat island effects, particularly in densely built environments;
- Improving thermal comfort by enhancing wind circulation and reducing heat retention in urban materials.

The Municipality received PNRR funding titled "Protection and Enhancement of Urban and Extra-Urban Greenery" in 2023. The grant funding for afforestation projects originates from the national strategy outlined in the Urban and Peri-Urban Forestry Plan [23]. The funded projects that have been executed include three main projects: "Genoa Verde", "Extra-urban Afforestation", and "Genoa and Eastern Genoese". These projects highlight a commitment to the sustainable and resilient management of urban and peri-urban green areas with objectives for environmental improvement and climate change adaptation (Table 2).

Table 2. List of ongoing projects in the Metropolitan City of Genoa. Source: Metropolitan City of Genoa.

Project	Location	Description	Status	Duration	Final Target	Total Cost(EUR)
"Genoa Verde" (code MASE GE01)	Genoa, Scarpino area (where the city's main landfill is located)	Urban and peri-urban afforestation to stabilize soil, increase biodiversity, restore degraded areas and CO_2 sequestration	Intervention completed and under maintenance	Start Date: 8 May 2022 End Date: 31 March 2026	30 hectares 30,000 plants	1,291,156.52
"Extra-urban Afforestation" (code MASE GE02)	Arenzano, Bric Bardella Punta Goetta Cogoleto, Prato Zanino Ronco Scrivia	Planting in riparian zones to improve biodiversity, supporting the ecological resilience of areas not covered by Natura 2000	To be completed by the end of 2023 with planting, work inspection, and start of maintenance	Start Date: 8 May 2022 End Date: 31 March 2026	40 hectares 40,000 plants	1,728,882.20
"Genoa and Eastern Genoese" (code MASE GE01-2023)	Bogliasco, Genoa, Pieve Ligure	The project aims to increase forest cover and improve biodiversity and the ecological functionality of ecosystems in these areas	Finalization of project. To be completed by the end of 2024	Start Date: 31 January 2024 End Date: 30 June 2026	40 hectares 40,000 plants	1,726,267.06

Based on the analyses carried out during the preparation phase of the Plan, resources were allocated specifying the targets (hectares/number of plants) for each Metropolitan City.

Recent studies emphasize the importance of prioritizing tree planting in urban areas and along linear infrastructures, as these locations offer significant opportunities to integrate vegetation into human-dominated landscapes. Furthermore, they argue that afforestation efforts yield the greatest biodiversity benefits when focused on degraded lands. This approach not only supports ecological restoration but also reduces the pressure to afforest wild open areas, thereby helping to conserve their natural ecosystems and maintain biodiversity [30]. In this case, the selected areas include degraded former industrial areas, riparian zones, and open fields upon the hills. The selection of areas for the afforestation projects was made in collaboration with the metropolitan municipalities.

3.3. Afforestation Projects in the Metropolitan City of Genoa: Plant Choice

The initial project sheets approved by the Ministry do not explicitly mention a detailed calculation of CO_2 absorption during the growth of the plants in the afforestation projects. However, they discuss the overall positive impacts on the environment and air quality through the reduction of greenhouse gases and the increase in vegetation cover, particularly using evergreen species that can absorb CO_2 more effectively.

Recent studies demonstrated that a newly planted urban forest in Southern Europe requires 13 years to function as an effective carbon sink; results indicated that 7–8 years are required for canopy development to offset carbon emissions from soil and other sources, with full carbon neutrality achieved after 13 years [31]. The exponential increase in CO_2 removal capacity beyond this point aligns with similar findings [32]. Tree biodiversity was found to play a decisive role in enhancing CO_2 sequestration. The selection of tree species should prioritize a mix of evergreen and deciduous species to ensure year-round CO_2 assimilation. For example, evergreen species like *Cupressus sempervirens* act as carbon sinks during winter, while deciduous species such as those from the *Tilia* and Ulmus genera exhibit higher photosynthetic rates during the growing season. Incorporating CO_2 sequestration potential as a criterion for species selection maximizes the ecological benefits of urban forests. Several abiotic stresses, including drought, nutrient deficiency, and pollution, could constrain CO_2 assimilation in urban environments. For instance, prolonged droughts typical of Mediterranean climates may cause trees to close stomata to conserve water, thereby reducing carbon uptake. Optimal tree density (309 trees/ha), was near the saturation threshold, ensuring efficient carbon sequestration without delays in the offset timeline [33].

The selection of native species for reforestation efforts is critical. However, climate change poses a substantial challenge, as native species and ecotypes may no longer be optimally adapted to current or future climatic conditions. For example, southern ecotypes may already exhibit greater adaptability in northern regions, where heatwaves and prolonged droughts are becoming increasingly frequent and severe. The evolutionary potential of many European tree species could be undermined if genotypes and ecotypes are mixed without careful consideration of the characteristics of local populations [30].

In all three projects, the emphasis on the selection of native species was required to ensure the resilience and environmental integration of the new plantations, as well as to improve the biodiversity and ecological stability of the areas involved (see Table 3). The species were identified according to the criteria established by the Public Notice. Each reforestation intervention must provide for the planting of 1000 plants per hectare, with the presence of shrubs in a percentage ranging between 10 and 30%, chosen according to the successional dynamics of the potential natural vegetation and, when possible, referring to the species indicated for each metropolitan city in the Plan. The forest propagation material must consist of native species, and its procurement by the Metropolitan Cities must be ensured by nurseries that guarantee provenance certification.

Planting was planned for both trees and shrub species. The rules set by the Ministry require that plants, shrubs, or seeds have a well-specified provenance and certificates of origin. It was not required that they come from Ligurian nurseries. In fact, there are no regional forest nurseries in Liguria. For both realized projects, both seeds and plants were used, provided by nurseries identified by the two economic operators who won the work tenders. For the project "Genoa and Eastern Genoese", since the project is in its initial phase, efforts are being made to obtain the entire range of plants needed through the agreements with certified nursery companies.

Table 3. List of Selected Plant Species for the implemented projects in the Metropolitan City of Genoa. Source: Metropolitan City of Genoa.

Project	Tree Species	Shrub Species	Use of Native Plants
Genova Verde	*Quercus ilex* *Quercus pubescent* *Ostrya carpinifolia* *Prunus avium* *Fraxinus ornus* *Alnus glutinosa.*	*Erica arborea* *Prunus spinosa* *Juniperus communis* *Sorbus torminalis* *Cistus salviifolius* *Cytisus scoparius* *Rhamnus alaternus*	The goal was to use native plants to support local resilience and biodiversity.
Extra-urban Afforestation	*Acer campestre* *Alnus glutinosa* *Fraxinus ornus* *Ostrya carpinifolia* *Populus alba* *Populus nigra* *Quercus ilex* *Quercus pubescens* *Salix alba*	*Arbutus unedo* *Cistus salviifolius* *Cornus sanguinea* *Crataegus monogyna* *Myrtus communis* *Pistacia lentiscus* *Salix purpurea,* *Prunus spinosa,* *Rhamnus alaternus* *Rosa sempervirens* *Viburnum lantana* *Viburnum tinus*	Native species only, maintaining a ratio of 70% trees and 30% shrubs consistent with the potential natural vegetation of the area, to preserve and enhance ecological coherence with the local environment.
Genoa and Eastern Genoese	various species including *Quercus ilex, Fraxinus ornus*	various species of *Cistus* and *Thymus*	Use of predominantly native plants to strengthen fire resistance and improve biodiversity in areas prone to clearings and gaps due to previous fires.

The PNRR Afforestation projects "Genoa Verde" and "Extra-Urban Afforestation" are advancing with work execution and on-site testing underway. The planting phase has been completed, and the mandatory five-year monitoring and maintenance plan has started. Meanwhile, the "Genoa and Eastern Genoese" project is in the executive design phase, preparing for tender documentation and subsequent implementation. These projects aim to enhance urban resilience and sustainability by improving air quality, increasing biodiversity with native species, sequestering carbon to mitigate climate change, managing hydrogeological risks, and enhancing overall well-being and quality of life. Despite initial challenges, such as meeting the December 2023 milestone and securing suitable land for afforestation, progress has been consistent. Activities now focus on ongoing maintenance, including plant health monitoring, replacement of dead plants, irrigation, and control of invasive vegetation, aiming to meet the planting targets by November 2024, with completion deadlines set for December 2024 and December 2025. Afforestation efforts extend beyond urban areas, influencing the urban microclimate, although a detailed climatic analysis was not explicitly required by the Ministry. The Metropolitan City is proceeding with its maintenance and monitoring phases for "Genoa Verde" and "Extra-Urban Afforestation", while also moving forward with the planning and execution of the "Genoa and Eastern Genoese" project. These initiatives collectively represent a comprehensive strategy to enhance the resilience and ecological sustainability of the metropolitan area. During planting, forest plants have been marked with orange paint and/or shelters to help monitor their future growth. Monitoring and verification of planting are conducted using a drone for irregular, naturalistic spacing (Figure 3).

Figure 3. Plants relocated to their final habitat (courtesy of Metropolitan City of Genoa).

Key Aspects of Reforestation in Genoa include:

- Plant Density: Reforestation requires 1000 plants per hectare, with 10–30% shrubs based on successional dynamics of natural vegetation and recommended species.
- Using Native Species: The selection of native species emphasized resilience biodiversity, and ecological stability.
- Focus on evergreen broad-leaved trees and Mediterranean conifers, most effective for CO_2 sequestration and PM10 removal.
- Provenance Certification: Forest propagation material must consist of native species with certification of origin from approved nurseries.

The "Genova Verde" project focuses on targeted interventions across various areas of the metropolitan city of Genoa, aiming to enhance forest and environmental resources. In the Sestri Ponente area, located northwest near the Scarpino landfill, the forest parcels are characterized by mixed conifer and broadleaf forests interspersed with vegetation-free zones showing signs of surface erosion. The forest management plan includes measures to diversify and expand shrub and tree cover, employing species with strong biotechnical properties for soil stabilization. Additionally, the plan seeks to enhance biodiversity and increase the area's attractiveness to birdlife.

As for the extra-urban reforestation efforts, two key areas stand out. The Cogoleto-Arenzano zone is marked by Mediterranean shrublands and bushy areas, shallow soils, and rocky terrains with sparse vegetation cover, along with clusters of maritime pine forests. The Ronco Scrivia area, on the other hand, features a riparian zone with the potential for poplar and willow groves, alongside former industrial zones with sandy and deep soils.

Here, the primary goal is to establish a buffer zone to counterbalance urban settlements and infrastructure, while also creating recreational and sports areas along the river.

Finally, the potential natural vegetation communities in these areas include the Ligurian sub-acidophilous holm oak series (*Viburno tini—Querco ilicis sigmetum*) and the Ligurian edaphically indifferent downy oak series (*Rubio—Querco pubescentis sigmetum*), both of which are well-suited to local conditions and contribute to restoring ecological balance.

4. Discussions

Although our methodological framework identified comprehensive evaluation criteria (ecological benefits, socio-economic factors, and spatial constraints), due to the current status of urban forestry initiatives in Genoa, only preliminary qualitative assessments were possible. Specifically, our results confirmed adherence to national guidelines regarding native species selection and vegetation composition. However, quantitative assessments of socio-economic benefits or detailed ecological impacts (such as biodiversity restoration or long-term climate mitigation effects) are not yet available and will require systematic monitoring over time. Spatial constraints were qualitatively discussed, emphasizing the challenges of densely built environments that led to prioritizing peri-urban zones.

The projects, aligned with both national and European Union strategies, aim to enhance citizens' quality of life and well-being through a series of reforestation, afforestation, and densification initiatives. These efforts address critical challenges such as air pollution, climate change mitigation, and biodiversity loss.

In particular, the initiatives adhere to modern environmental and urban planning policies that emphasize afforestation as a tool for promoting urban resilience and sustainability. The Metropolitan Municipality has set general objectives that reflect these priorities. First, the projects aim to improve air quality by leveraging the natural capacity of plants to absorb CO_2 and other air pollutants, thereby enhancing the atmosphere in urban and peri-urban areas. Another key objective is the increase of biodiversity, achieved by introducing a diverse range of native plant species to create and strengthen habitats for local flora and fauna. Additionally, the interventions contribute to climate change mitigation through carbon sequestration by the trees and shrubs planted. Vegetation also plays a critical role in responding to hydrogeological risks by stabilizing soil and managing rainwater runoff, thus mitigating erosion and flood risks. Finally, the creation of green spaces seeks to enhance urban well-being by offering psychological benefits and improving the quality of life. These spaces provide opportunities for rest and recreation, fostering a healthier and more connected local community.

Evergreen broad-leaved trees and Mediterranean conifers yield the highest returns and are also the most efficient in terms of particulate matter removal (see Table 4).

Table 4. Quantitative and monetary evaluation of CO_2 absorption through the planting of different tree species. Adapted from: https://www.mase.gov.it, accessed on 20 Decembere 2024.

Functional Type of Vegetation	Average CO_2 Absorption (t/ha/yr)	CO_2 Absorption in the First 5 Years (t/ha)	Lower Monetary Value of CO_2 Absorbed in the First 5 Years (EUR/ha)	Higher Monetary Value of CO_2 Absorbed in the First 5 Years (EUR/ha)	CO_2 Absorption at 20 Years (t/ha)	Lower Monetary Value of CO_2 Absorbed in the First 20 Years (EUR/ha)	Higher Monetary Value of CO_2 Absorbed in the First 20 Years (EUR/ha)
Deciduous broadleaf	6.6	21.9	208.2	835.2	121.6	1154.8	4631.5
Evergreen broadleaf	9.5	31.5	299.1	1199.7	174.6	1658.9	6653.0

Quantitative CO_2 evaluation aligns with IPCC guidelines and established scientific literature [34,35]. The estimated average CO_2 absorption rates of 6.6 tCO_2/ha/yr for deciduous broadleaf forests and 9.5 tCO_2/ha/yr for evergreen broadleaf forests are consistent with standard carbon density values and turnover rates. Given that the carbon stock for trees is 120 tC/ha, and applying a moderate turnover rate of 2.5% per year, the estimated annual carbon sequestration would be 3 tC/ha, translating to approximately 11 tCO_2/ha/yr (using the IPCC conversion factor of 3.67). These reported values fall within a plausible range, considering that deciduous forests experience seasonal biomass loss, whereas evergreen forests maintain more consistent carbon uptake. While these figures align with general scientific estimates, specific variations depend on forest type, age, and environmental conditions.

To enhance the robustness of the quantitative CO_2 sequestration assessment, it is essential to incorporate confidence intervals for these estimates. This will allow for a more precise evaluation of potential variations arising from environmental conditions, such as soil composition, climatic fluctuations, and species-specific growth dynamics. Additionally, scenario-based analyses should be conducted to account for uncertainties linked to land-use changes, forest management practices, and long-term climate trends.

By calculating the arithmetic average of the minimum ministerial estimates of total CO_2 absorption over 20 years for deciduous broadleaf forests (121.6 tCO_2/ha) and evergreen broadleaf forests (174.6 tCO_2/ha), an average value of 154 tCO_2/ha was obtained. Therefore, the total CO_2 absorption for 110 hectares of urban forestry interventions would amount to 16,940 tons, corresponding to a total monetary value of EUR 643,632 in the first 20 years. However, the economic valuation methodology requires further clarification, particularly regarding the assumptions underlying the monetary calculations.

Official estimates from ministerial sources indicate that the lower and higher monetary values of CO_2 absorbed in the first 20 years range between EUR 1154.8 and 4631.5 per hectare for deciduous broadleaf forests (assuming an absorption rate of 6.6 tCO_2/ha/year) and between EUR 1658.9 and 6653.0 per hectare for evergreen broadleaf forests (assuming an absorption rate of 9.5 tCO_2/ha/year). Based on these values, the implicit carbon price ranges between approximately 8.75 EUR/tCO_2 and 35.09 EUR/tCO_2 for deciduous broadleaf forests and between 8.75 EUR/tCO_2 and 35.02 EUR/tCO_2 for evergreen broadleaf forests. This implicit price range is considerably higher than the 6.11 EUR/tCO_2 reported in some literature sources [36], but remains below recent market prices for carbon credits, which have fluctuated between 58 EUR/tCO_2 and 102 EUR/tCO_2 in the past two years. This suggests that the ministerial estimates fall within a reasonable, albeit somewhat conservative, valuation framework. A more detailed discussion of the economic parameters used in the present analysis—such as carbon pricing mechanisms, inflation-adjusted valuation, and sensitivity to market fluctuations—would enhance the credibility and applicability of the monetary assessment. Given the wide range of values, it is recommended to incorporate updated market prices for carbon credits when assessing the financial benefits of CO_2 sequestration projects. Additionally, periodic recalibrations of economic models should be considered to reflect evolving market conditions.

Forestation has been successfully carried out in peri-urban areas following sustainable reforestation principles, respecting native vegetation series, and enhancing biodiversity. However, the specific impacts of these interventions on the urban environment remain unclear, particularly in relation to adjacent green spaces in densely populated areas. Effects such as the reduction of the urban heat island effect, improvement of air quality through atmospheric pollutant absorption, noise reduction, and the enhancement of residents' psychophysical well-being require further study. Conducting additional research to better quantify and understand these potential benefits will be crucial in guiding future urban forestry policies toward interventions specifically tailored to the urban context. It is also worth noting the mitigation effect exerted by vegetation on the local climate, which is extremely important in an urban context characterized by the heat island effect. Research indicates that urban green spaces, including forests and roadside trees, can significantly mitigate surrounding temperatures, with the extent of this cooling effect varying based on the size and type of vegetation. Regarding urban forests, another study observed that the temperature within urban forests decreased by approximately 1.9 °C over a distance of ~3 km from the traffic island near the city to the forest, indicating a substantial cooling effect extending several kilometers [37].

Urban reforestation initiatives supported by the Italian Recovery and Resilience Plan represent an important opportunity to enhance urban environments and improve quality of life for future generations. However, the realization of these objectives is hindered by several obstacles. Limited spaces for reforestation in densely built urban areas, political inaction, and insufficient nursery stocks of native plants pose significant barriers. Addressing these issues requires enhanced inventory management, expanded production capacities, and greater investment in nursery infrastructure to ensure the successful implementation of reforestation strategies. The decision to focus reforestation efforts on peri-urban areas appears to be influenced by the characteristics of densely populated urban structures, where space for planting is limited. In urban centers, planting opportunities are constrained by small park sizes and partial private ownership of green spaces due to historical factors. Additionally, there has been limited integration of greenery within buildings, a situation exacerbated by stringent landscape protection regulations applied across large portions of urban areas.

Although afforestation offers multiple benefits, its capacity to absorb CO_2 remains limited when compared to the overall emissions produced by urban areas. This highlights the need to complement peri-urban interventions with targeted measures within urban boundaries, aiming to enhance the ecological performance of cities. At present, urban forestry projects are in the process of being completed, and therefore no documentation on their outcomes has yet been made public. However, it is interesting to analyze how different administrative bodies (municipalities, metropolitan cities, and regions) have chosen to communicate the ongoing processes. In most cases, official websites have published documents related to administrative procedures and calls for tenders for reporting the works. This approach aims to provide citizens with useful tools to understand the actions of public administrations, also in compliance with the necessity of transparency. However, the nature of such documentation makes it difficult to read for an audience that is not well-versed in legal-administrative matters. To address this, some institutions have included news articles or press kits on their official channels, designed to be picked up by newspapers and media outlets at both local and national levels. Among these, notable examples include the Municipality of Rome [38] and the Metropolitan City of Genoa [39], which has also produced a video to better inform citizens about the implementation of ongoing projects. Other institutions have included technical-scientific documentation to complement the information provided by the ministry. Among them, the Metropolitan City of Turin [40] has made a detailed list of project sheets accessible on its website, while the Metropolitan City of Bologna has added a series of attachments containing technical-scientific guidelines for metropolitan forestry [41].

In general, however, there is a certain inconsistency in the information and content published, particularly regarding public outreach communication as well as technical-scientific information, which is essential for ensuring greater involvement of the scientific community in the crucial phase of evaluating the objectives achieved and the results obtained.

5. Conclusions

This study evaluates the reforestation initiatives implemented in the metropolitan area of Genoa, focusing on their outcomes and challenges. Reforestation efforts in peri-urban areas have been executed adhering to sustainable principles, respecting native vegetation series, and enhancing local biodiversity. These initiatives are estimated to have an average carbon dioxide absorption capacity of 154 tons per hectare over 20 years. For the 110 hectares targeted by urban forestry interventions, the total anticipated CO_2 absorption amounts to approximately 16,940 tons, with a corresponding monetary valuation of EUR 643,632 for the carbon sequestered over the same period. Given the preliminary nature of this evaluation, detailed economic assessments should be addressed in future studies, incorporating comprehensive cost-benefit analyses based on finalized implementation data and updated market carbon prices.

Despite these positive outcomes, the specific impacts of these initiatives on key urban environmental factors remain uncertain. Afforestation efforts have the potential to influence the urban microclimate beyond the immediate area of intervention. However, detailed climatic analyses have not been required by local authorities. Moreover, forest nurseries have faced significant challenges in maintaining native biodiversity, often resorting to non-native genotypes whose adaptability to local conditions is uncertain. These challenges highlight critical gaps in the planning and implementation of reforestation efforts.

Qualitative considerations highlighted key strengths and limitations, particularly related to Genoa's dense urban fabric and challenging topography. The primary strengths include enhanced ecological connectivity, improved biodiversity through the planting of native species, and potential contributions to mitigating urban heat islands and enhancing air quality. Conversely, significant limitations include limited citizen accessibility due to the peri-urban location of interventions, increased maintenance challenges posed by steep terrain, and insufficient integration of greenery within densely built urban areas. These findings suggest the necessity for complementary urban-scale interventions—such as microforests, pocket parks, and vertical greening—to effectively overcome spatial constraints and maximize ecological and socio-economic benefits.

Integrating greenery into urban areas as Genoa necessitates innovative, scientifically grounded approaches to address spatial constraints while maximizing ecological and environmental benefits. Regulatory constraints—from tight urban planning rules to heritage protections—require creative navigation and policy innovation, while the physical hurdles of steep, unstable terrain call for engineering savvy and adaptive planting techniques. Small-scale afforestation projects, such as microforests or the conversion of derelict urban lots into vegetated pocket parks, can enhance biodiversity and ecological stability. Vertical greening systems, such as modular green walls, could offer a viable solution for enhancing building envelopes, improving air quality, and mitigating the urban heat island effect. These systems can be designed with native or drought-tolerant species, ensuring low water consumption and ecological compatibility. Similarly, rooftop forests present a transformative potential by repurposing underutilized spaces into carbon-sequestering ecosystems. When integrated with photovoltaic systems, rooftop greenery can simultaneously optimize energy efficiency and urban sustainability. To ensure widespread adoption, urban planning regulations must mandate green infrastructure in new developments, supported by spatial analysis to identify optimal intervention sites.

The realization of these interventions depends on addressing key challenges in nursery capacity, native plant availability, and biodiversity conservation through robust policy measures. Expanding the production of high-quality, climate-resilient native plant species requires significant investment in nursery infrastructure and operations. Policy frameworks should establish certification programs to ensure that nurseries adhere to standards of biodiversity preservation and ecological suitability. Furthermore, fostering collaboration between nurseries and research institutions can enhance species selection and propagation techniques, prioritizing genotypes that are resilient to changing climatic conditions. A centralized database for plant stock and demand, integrated with a national monitoring system, could align nursery outputs with regional and municipal afforestation objectives. To maximize carbon sequestration and ecosystem services, reforestation strategies should employ a mix of evergreen and deciduous species, ensuring year-round functionality. Public-private partnerships and economic incentives, such as subsidies for nurseries employing sustainable practices, can bolster these efforts. Enhanced inventory management, coupled with region-specific species recommendations, will be critical for achieving long-term ecological sustainability and meeting national climate targets.

Further research is needed to quantify the impacts of peri-urban afforestation interventions, particularly in terms of their effectiveness in mitigating the urban heat island effect and improving air quality through pollutant absorption. Given the diverse environmental conditions of urban areas, each city must develop tailored solutions that align with its specific territorial characteristics. In this regard, Genoa could stand out as a valuable model for cities situated in hilly landscapes. Future studies should also play a crucial role in shaping urban forestry policies, steering them towards context-specific interventions that consider local soil properties, vegetation dynamics, and climatic conditions.

Author Contributions: Conceptualization, A.M. and P.S.; methodology, A.M. and P.S.; writing—original draft preparation, P.S.; writing—review and editing, A.M. and P.S.; supervision, A.M.; project administration, A.M.; funding acquisition, A.M. All authors have read and agreed to the published version of the manuscript.

Funding: This research was funded by the European Union—Next Generation EU, under Italian Government PRIN (Project of Significant National Interest) "BETTER POLICY" research project, project code D53D23011320006.

Data Availability Statement: The raw data supporting the conclusions of this article will be made available by the authors on request.

Acknowledgments: Laura Papaleo, Unità di Gestione, monitoraggio e rendicontazione progetti PNRR/PNC ed Ufficio Europa, Direzione Risorse Città Metropolitana di Genova, Metropolitan City of Genoa.

Conflicts of Interest: The authors declare no conflicts of interest. The funders had no role in the design of the study; in the collection, analyses, or interpretation of data; in the writing of the manuscript; or in the decision to publish the results.

Abbreviations

The following abbreviations are used in this manuscript:

GI	Green Infrastructure
UF	Urban Forests
LAI	Leaf Area Index
LCA	Life Cycle Assessment
MASE	Ministry of the Environment and Energetic Safety
MiTE	Ministry of Ecological Transition
NBS	Nature-Based Solutions
NETP	National Ecological Transition Plan
PNRR	National Recovery and Resilience Plan

References

1. Duncanson, L.; Liang, M.; Leitold, V.; Armston, J.; Krishna Moorthy, S.M.; Dubayah, R.; Costedoat, S.; Enquist, B.J.; Fatoyinbo, L.; Goetz, S.J.; et al. The effectiveness of global protected areas for climate change mitigation. *Nat. Commun.* **2023**, *14*, 2908. [CrossRef]
2. Food and Agriculture Organization of the United Nations (FAO). Guidelines on Urban and Peri-Urban Forestry (FAO Forestry Papers No. 178). 2016. Available online: https://openknowledge.fao.org/server/api/core/bitstreams/9c27d98b-8071-4ee7-8fc4-e0b430f8a8dc/content/ (accessed on 20 December 2024).
3. Benedict, M.A.; McMahon, E.T. *Green Infrastructure: Linking Landscapes and Communities*; Island Press: Washington, DC, USA, 2012.
4. Food and Agriculture Organization of the United Nations (FAO). *Forests and Sustainable Cities*; Unasylva, 2018; Volume 69, pp. 1–84. Available online: http://www.fao.org/3/I8707EN/i8707en.pdf (accessed on 13 December 2024).
5. Endreny, T.A. Strategically growing the urban forest will improve our world. *Nat. Commun.* **2018**, *9*, 1160.
6. Abbate, S.; Di Paolo, L.; Carapellucci, R.; Cipollone, R. Carbon uptake dynamics associated to the management of unused lands for urban CO2 planning. *Renew. Energy* **2021**, *178*, 946–959.
7. Nicese, F.P.; Colangelo, G.; Comolli, R.; Azzini, L.; Lucchetti, S.; Marziliano, P.A.; Sanesi, G. Estimating CO2 balance through the Life Cycle Assessment prism: A case study in an urban park. *Urban For. Urban Green.* **2021**, *57*, 126869. [CrossRef]
8. World Bank. Advancing Climate Action Through an Urban Lens. 2021. Available online: https://www.worldbank.org/en/topic/urbandevelopment/brief/climate-action-through-an-urban-lens (accessed on 13 December 2024).
9. Taha, H. Development of an urban heat mitigation plan for the Greater Sacramento Valley, California, a Csa Koppen climate type. *Sustainability* **2021**, *13*, 9709. [CrossRef]
10. López, F.P.; Jaramillo, R.S.; de la Torre, R.A. Sustainable urban forest management: A case study in Colombia. In *E3S Web of Conferences*; EDP Sciences: Les Ulis, France, 2021; Volume 88, p. 10002.
11. Penn State Extension. Urban Forestry. 14 October 2023. Available online: https://extension.psu.edu/forests-and-wildlife/urban-forestry (accessed on 12 December 2024).
12. Teo, H.C.; Zeng, Y.; Sarira, T.V.; Fung, T.K.; Zheng, Q.; Song, X.P.; Chong, K.Y.; Koh, L.P. Global urban reforestation can be an important natural climate solution. *Environ. Res. Lett.* **2021**, *16*, 034059. [CrossRef]
13. Moore, G.M. Urban trees and the global greenhouse. In Proceedings of the 7th National Street Tree Symposium, Adelaide, Australia, 7–8 September 2006.
14. Di Sacco, A.; Hardwick, K.A.; Blakesley, D.; Brancalion, P.H.; Breman, E.; Cecilio Rebola, L.; Chomba, S.; Dixon, K.; Elliott, S.; Ruyonga, G.; et al. Ten golden rules for reforestation to optimise carbon sequestration, biodiversity recovery and livelihood benefits. *Glob. Change Biol.* **2021**, *27*, 1328–1348. [CrossRef]
15. Yang, H.; Ciais, P.; Frappart, F.; Li, X.; Brandt, M.; Fensholt, R.; Fan, L.; Saatchi, S.; Besnard, S.; Deng, Z. Global increase in biomass carbon stock dominated by growth of northern young forests over past decade. *Nat. Geosci.* **2023**, *16*, 886–892. [CrossRef]
16. Rodriguez Mendez, Q.; Fuss, S.; Lück, S.; Creutzig, F. Assessing global urban CO2 removal. *Nat. Cities* **2024**, *1*, 413–423. [CrossRef]
17. Forest Research. Forest Research. 14 October 2023. Available online: https://www.forestresearch.gov.uk/ (accessed on 12 December 2024).
18. Frosini, G.; Amato, A.; Mugnai, F.; Cinelli, F. The Impact of Trees on the UHI Effect and Urban Environment Quality: A Case Study of a District in Pisa, Italy. *Atmosphere* **2024**, *15*, 123. [CrossRef]
19. Montaldi, A.; Iamonico, D.; Del Vico, E.; Valeri, S.; Lasinio, G.J.; Capotorti, G. Green infrastructure design for the containment of biological invasions. Insights from a peri-urban case study in Rome, Italy. *J. Environ. Manag.* **2024**, *365*, 121555. [CrossRef]
20. Masiero, M.; Biasin, A.; Amato, G.; Malaggi, F.; Pettenella, D.; Nastasio, P.; Anelli, S. Urban Forests and Green Areas as Nature-Based Solutions for Brownfield Redevelopment: A Case Study from Brescia Municipal Area (Italy). *Forests* **2022**, *13*, 444. [CrossRef]
21. Bartoli, F.; Savo, V.; Caneva, G. Biodiversity of urban street trees in Italian cities: A comparative analysis. *Plant Biosyst.-Int. J. Deal. All Asp. Plant Biol.* **2021**, *156*, 649–662. [CrossRef]
22. MASE Ministero Della Transizione Ecologica. National Ecological Transition Plan. 2022. Available online: https://www.mase.gov.it/sites/default/files/archivio/allegati/PTE/PTE-definitivo.pdf (accessed on 12 December 2024).
23. MiTE-Ministero della Transizione Ecologica. Urban and Peri-Urban Forestry Plan: Part of the National Recovery and Resilience Plan (PNRR). 2021. Available online: https://www.mase.gov.it/sites/default/files/archivio/allegati/PNRR/PNRR_piano_forestazione.pdf (accessed on 13 December 2024).
24. Zappitelli, I.; Conte, A.; Alivernini, A.; Finardi, S.; Fares, S. Species-Specific Contribution to Atmospheric Carbon and Pollutant Removal: Case Studies in Two Italian Municipalities. *Atmosphere* **2023**, *14*, 285. [CrossRef]
25. Giampaoletti, P. Reforestation of Urban Districts in Mediterranean Climate: New Decarbonization Models for Cities. *Athens J. Mediterr. Stud.* **2023**, *9*, 221–236. [CrossRef]
26. Pardi, R.; Venturella, G.; Domina, G.; Di Gristina, E.; Cirlincione, F.; Gargano, M.L. Forest nurseries and the National Recovery and Resilience Plan: The case of Sicily and Apulia (Italy). *Ital. Bot.* **2023**, *15*, 49–63. [CrossRef]

27. Paliaga, G.; Luino, F.; Turconi, L.; Marincioni, F.; Faccini, F. Exposure to Geo-Hydrological Hazards of the Metropolitan Area of Genoa, Italy: A Multi-Temporal Analysis of the Bisagno Stream. *Sustainability* **2020**, *12*, 1114. [CrossRef]
28. ARPAL. Agenzia Regionale per la Protezione dell'Ambiente Ligure, Ente Strumentale di Regione Liguria. 2025. Available online: https://www.arpal.liguria.it/ (accessed on 12 December 2024).
29. Acquaotta, F.; Faccini, F.; Fratianni, S.; Paliaga, G.; Sacchini, A. Rainfall intensity in the Genoa Metropolitan Area: Secular variations and consequences. *Weather* **2018**, *73*, 356–362. [CrossRef]
30. Abeli, T.; Giulio, A.D. Risks of massive tree planting in Europe should be considered by the EU Forestry Strategy 2030. *Restor. Ecol.* **2023**, *31*, e13834. [CrossRef]
31. Manzini, J.; Hoshika, Y.; Sicard, P.; Anav, A.; De Marco, A.; Sorrentino, B.; Trentanovi, G.; Baesso Moura, B.; Ferrini, F.; Azzini, L.; et al. Life Cycle Analysis and species-specific net CO2 assimilation model to assess when a new urban forest becomes a carbon sink in a Mediterranean city. *Sci. Total Environ.* **2025**, *959*, 178267. [CrossRef]
32. Strohbach, M.W.; Arnold, E.; Haase, D. The carbon footprint of urban green space—A life cycle approach. *Landsc. Urban Plan.* **2012**, *104*, 220–229.
33. Drolen, J.; Brandt, L.; Wei, Y.; Dybzinski, R. Modeling management-relevant urban Forest stand characteristics to optimize carbon storage and sequestration. *Forests* **2023**, *14*, 2207. [CrossRef]
34. Scarascia-Mugnozza, G.; Oswald, H.; Piussi, P.; Radoglou, K. Forest of the Mediterranean region: Gaps in knowledge and research needs. *For. Ecol. Manag.* **2000**, *132*, 97–109. [CrossRef]
35. Post, W.M.; Kwon, K. Soil Carbon Sequestration and Land-Use Change: Processes and Potential. *Glob. Chang. Biol.* **2008**, *6*, 317–327. [CrossRef]
36. Kazak, J.; Malczyk, J.; Castro, D.G.; Szewrański, S. Carbon Sequestration in Forest Valuation. *Real Estate Manag. Valuat.* **2016**, *24*, 76–86.
37. Hwang, B.; Sou, D.; Oh, H.; Park, R. Cooling effect of urban forests on the urban heat island in Seoul, South Korea. *PLoS ONE* **2023**, *18*, e0288774. [CrossRef]
38. Comune.Roma. Available online: https://www.comune.roma.it/web/it/notizia/ambiente-avanti-piano-forestazione-mun02.page#:~:text=In%20citt%C3%A0%20stiamo%20concludendo%20le,abbiamo%20un%20bilancio%20ecologico%20positivo (accessed on 28 February 2025).
39. Pnrr.Regione.Liguria. Available online: https://pnrr.regione.liguria.it/progetti/altri-soggetti-attuatori/pnrr-forestazione-2-genova-e-levante-genovese-terreni-da-rimboschire-nei-comuni-di-bogliasco-pieve-ligure-genova-della-citt-metropolitana-di-genova-diverse-aree-da-rimboschire-nei-tre-comuni-fogli-e-mappali-da-progetto-terreni-da-rimboschire-nei-52217.html (accessed on 28 February 2025).
40. Cittametropolitana.Torino. Available online: http://www.cittametropolitana.torino.it/cms/pnrr/missioni-pnrr/missione-2/progetti-forestazione (accessed on 28 February 2025).
41. Cittametropolitana.Bologna. Available online: https://www.cittametropolitana.bo.it/agenda_sviluppo_sostenibile/Home_Page/Archivio_news/Linee_guida_per_la_forestazione_metropolitana (accessed on 28 February 2025).

land

MDPI

Article

Does Multidimensional Urban Morphology Affect Thermal Sensation? Evidence from Shanghai

Haochen Qian, Minqi Wang, Shurui Zheng, Bing Qiu and Fan Zhang *

College of Landscape Architecture, Nanjing Forestry University, Nanjing 210037, China; qhc@njfu.edu.cn (H.Q.); 8241011231@njfu.edu.cn (M.W.); reliyzheng@njfu.edu.cn (S.Z.); qiubing@njfu.edu.cn (B.Q.)
* Correspondence: nlzhangfan@njfu.edu.cn; Tel.: +86-189-1385-5338

Abstract: The inappropriate thermal conditions resulting from increasingly severe climate issues have led to numerous complications for urban residents, decreased urban settlement comfort, and increased average and peak energy demands in built environments. Existing studies have demonstrated the significant influence of urban morphology (UM) on the urban thermal environment (UTE); however, at the meso-scale and macro-scale, UTE is often simplified to land surface temperature (LST) and building surface temperatures. To investigate the impact of UM on UTE, we developed an evaluation framework consisting of thermal sensing feedback (TSF) and LST. We employed the seven-level TSF scale to evaluate TSF data obtained from the Internet, emphasizing individualized thermal perceptions of urban spaces and reorienting UTE research towards a human-centric perspective. Using a regression model, we examined the relationships between two-dimensional and three-dimensional UM variables and UTE at the meso-scale in the central urban area of Shanghai, China, during August and December 2024. The results indicated the following: (1) The normalized difference vegetation index (NDVI), building density (BD), floor area ratio (FAR), impervious surface index (ISI), building height (BH), average building volume (ABV), sky view fraction (SVF), and building shape (BS_{sh}) effectively explained TSF. However, area weighted mean shape index ($SHAPE_{AM}$), aggregation index (AI), edge density (ED), elevation, building spacing (BS_{sp}), and spatial congestion degree (SCD) showed no significant correlation with TSF. (2) Significant variables, including NDVI, FAR, ISI, UM, BD, and BH, exhibited opposite effects on cold perception in winter compared to heat perception in summer, indicating a consistent influence on thermal perception across seasons. (3) In summer, the significant variables SVF, BS_{sh}, and ISI showed opposite effects on TSF and LST, while in winter, FAR demonstrated contrasting impacts on TSF and LST. The results of this study advance understanding of the mechanisms through which UM influences UTE, providing valuable insights for the development of sustainable, thermally comfortable urban environments.

Keywords: urban thermal environment (UTE); urban heat island; thermal sensation feedback (TSF); urban morphology (UM); land surface temperature (LST); sustainable development goals

1. Introduction

In 2024, the global average temperature has been approximately 1.55 °C above pre-industrial levels, marking the first year exceeding the 1.5 °C threshold established by *the Paris Agreement*. This severe climatic situation underscores the necessity for urban planners and designers to create more attractive, thermally comfortable, and sustainable

https://doi.org/10.3390/land14040769

urban environments. Amid intensified global urban expansion, alterations in urban surface properties and structural characteristics have significantly affected urban-scale energy balance and microclimatic conditions [1]. Consequently, this has intensified climate change phenomena, increased the frequency of extreme weather events [2], and exacerbated urban heat islands (UHIs) [3].

For urban residents, the urban thermal environment (UTE) is a critical indicator of urban life quality and well-being and significantly impacts urban activity. Adverse UTE often causes dermatological issues, cardiovascular strain, heart attacks, strokes, respiratory distress, psychological disorders, and increased mortality among vulnerable populations [4]. The human perception of thermal comfort, essentially determined by the equilibrium between body heat production and dissipation rates, generates subjective sensations of cold or heat. Thermal neutrality, characterized by balanced heat exchange, corresponds with comfort.

Historically, UTE characterization has been limited primarily to land surface temperature (LST) measurements or data obtained from fixed and mobile devices, providing straightforward but effective assessments at macroscales [5]. However, individual-level thermal comfort in urban settings cannot be accurately represented by simple temperature indices alone. The first thermal comfort evaluation index, effective temperature (ET), emerged in the 1920s [6]. Subsequent development has produced comprehensive frameworks emphasizing psychological, physiological, and energy dimensions [7] or physiological, psychological, and physical dimensions [8]. Physical parameters typically include air temperature, airflow velocity, solar radiation, and relative humidity. Physiological factors involve skin temperature, perspiration rate, clothing insulation, and activity levels. Psychological factors are particularly complex, encompassing social habits and behavioral patterns [9].

ASHRAE Standard 55 [10] defines thermal comfort as a subjective state of satisfaction with the thermal environment, involving both objective environmental parameters and subjective physiological factors such as health, endurance, and adaptability. Hence, comprehensive assessments combining subjective perceptions and objective data are typically necessary [11]. In 1970, P.O. Fanger introduced the influential Predicted Mean Vote (PMV) index, based on air temperature, radiant temperature, airflow velocity, humidity, and two human-body-related variables [12]. PMV is recognized by ISO 7730 [13] and ASHRAE Standard 55–92 [10], though it is predominantly applied within built environments [14]. Outdoor thermal perception, influenced by spatiotemporal human–climate interactions, encompasses diverse and dynamic thermal experiences and expectations [15], requiring approaches beyond static temperature regulation. Outdoor thermal comfort refers explicitly to an individual's degree of satisfaction or discomfort within urban environments [4]. Common indicators for outdoor thermal comfort include Wet Bulb Globe Temperature (WBGT) [16,17], Universal Thermal Climate Index (UTCI) [17–19], and Physiologically Equivalent Temperature (PET) [20], widely employed in evaluating long-term climate impacts and microclimatic conditions. These indices vary temporally (instantaneous to long-term) and spatially (micro-, meso-, and macro-scale), necessitating climate-zone-specific calibration for practical application.

At the micro-scale, researchers often rely on physiological parameters, such as body temperature and heart rate variability (HRV)—correlated with metabolic rate, oxygen saturation, and blood pressure—to objectively assess individual thermal comfort [21–24]. However, physiological indicators alone do not entirely capture UTE perception; memory, adaptation, and expectation significantly influence comfort assessments [25,26], requiring further characterization through standardized scales [27]. Practically, PMV and Actual Mean Vote (AMV) discrepancies are common [28,29]. Studies suggest that populations tolerate higher temperatures in colder climates and prefer lower temperatures in hot climates [30,31]. Neutral thermal conditions (PMV = 0) are not universally optimal [32], prompting researchers to incorporate thermal preference scales alongside comfort judgments for accuracy [33]. Consequently, the concept of "semantic shift" can be interpreted as a dynamic "neutral point" on the ASHRAE thermal comfort scale, where the optimal thermal comfort value is not fixed at neutrality (zero). Analysis of the ASHRAE scale's underlying dynamics suggests the need to consider not only individuals' perceived thermal state (e.g., feeling warm or cool) but also their desired thermal state [32].

Concerned by the adverse evolution of urban temperatures, researchers utilize quantitative indicators to analyze urban morphology (UM) impacts on UTE from an urban planning perspective [3]. Techniques including regression models, Geodetector analysis, and machine learning (network regression, Random Forest, eXtreme Gradient Boosting, and Convolutional Neural Networks) [34] have clarified the complex relationships between UTE and various indicators. Influencing factors are traditionally categorized as follows: (1) spatial morphology, emphasizing landscape features affecting microclimate regulation and shading [35], urban density generating internal heat, and impermeable surfaces (concrete, asphalt) absorbing solar radiation; (2) climatic conditions [36]; and (3) socio-economic factors (GDP, population, nighttime illumination) [37].

The influence of UM indices on UTE varies across macro-, meso-, local-, and micro-scales. Different scales offer distinct yet interactive insights, ranging from macro-scale remote sensing-based LST measurements [38] to local-scale Local Climate Zones (LCZs) classifications by Stewart and Oke [3,39–41], and micro-scale physiological and individual feedback evaluations. Such scale-specific research effectively elucidates urban microclimatic dynamics and their implications for human thermal comfort.

At present, Pearson, Spearman, and other relevant correlation models have been employed to investigate the impact of UM on TSF. However, these models typically neglect the synergistic relationships among various urban morphology variables, resulting in insufficient integration between micro-scale and macro-, meso-, and local-scale analyses. Previous studies have demonstrated the feasibility of utilizing Internet-based big data to establish multi-scale correlations linking individual perceptions, such as noise perception, with urban morphology [42]; however, equivalent research specifically addressing TSF remains unexplored. To address this research gap, this study examines the influence of UM on the UTE in the central urban area of Shanghai by comprehensively analyzing both two-dimensional land use and three-dimensional architectural forms, emphasizing correlations with TSF. Specifically, the study addresses three key questions: (1) How do UM variables independently and jointly influence LST and TSF? (2) What mechanisms account for the differential impacts of UM variables on TSF and LST? (3) How can UM shaping effectively regulate UTE at both LST and TSF levels?

2. Study Area and Data Sources

2.1. Study Area

The study was conducted in the central urban area of Shanghai, China (see Figure 1). Shanghai is located in the alluvial plain of the Yangtze River Delta, with a flat topography and subtropical monsoon climate. The average annual temperature is 17.7 °C, and the average yearly precipitation is 1086.8 mm. The central urban area of Shanghai is delineated as the region within the Shanghai Outer Ring Expressway, with an area of about 664 square kilometers and a planned resident population of about 11 million. The surface spatial pattern dataset of the study area is comprehensive and openly accessible; the residential and living population is dense, the young and middle-aged groups are active, and the Internet socializing activities are rich and varied.

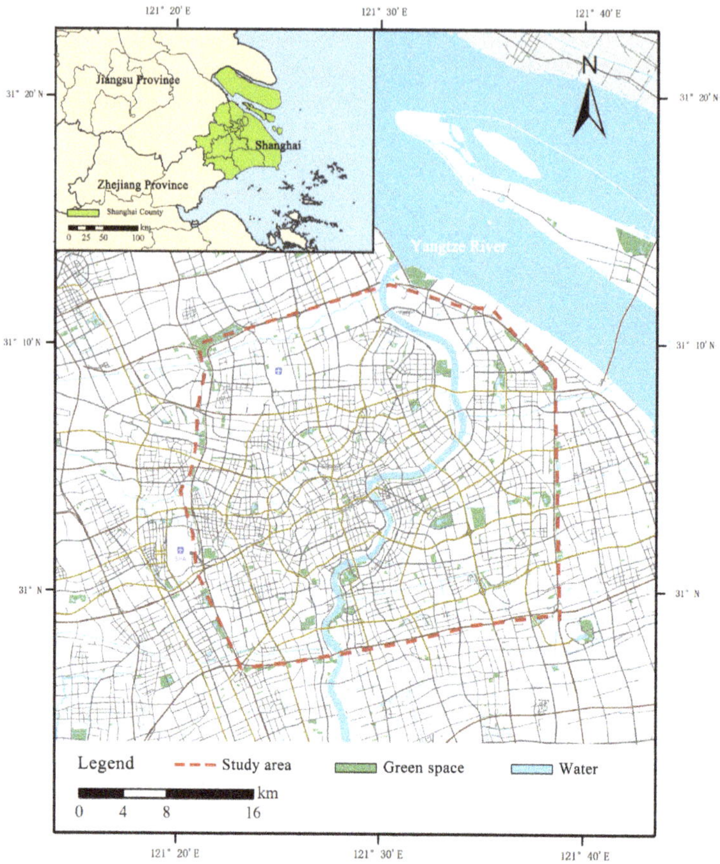

Figure 1. Study area.

To facilitate this study, the research area was divided into grid cells. Typically, grid resolutions ranging from 100 to 500 m are selected for urban form research [43]. Aggregation units at varying scales yield different insights into spatial characteristics [44], and differing grid scales influence experimental outcomes [45]. A grid resolution of 300 m × 300 m was selected, as it satisfies the requirements for statistical analysis of spatial form indicators while accurately capturing surface temperature inversion results. Grid units were generated using the fishnet tool in ArcGIS 10.8.1, and all spatial data utilized in this study were reprojected to match the spatial reference system of the grid.

2.2. Data Resources

2.2.1. Acquisition and Calibration of Remote Sensing Images of Urban Surface Space

Landsat-8 imagery (Band 10, originally at 100 m spatial resolution and resampled to 30 m) was collected to retrieve LST using the radiative transfer equation method [46]. Remote sensing allows detailed observation of surface characteristics, providing significant advantages for surface thermal environment studies [47]. High-precision infrared remote sensing data further enable detailed measurements of urban surface temperatures [2].

In China, spring typically spans from March to May, summer from June to August, autumn from September to November, and winter from December to February. This study specifically addresses thermal conditions in the hot summer and the cold winter. In 2024, Shanghai experienced significant warming, recording 52 high-temperature days

(≥35 °C)—the second-highest in historical records—with 23 days reaching extreme heat levels (≥37 °C). Notably, a 12-day heatwave occurred between July 31 and August 11, breaking the 152-year record for consecutive high-temperature days. These climatic events are closely linked to global warming trends. In the winter of 2024, the central urban area of Shanghai faced notable cold extremes. Specifically, the maximum temperature on February 23–24 was just 2.2 °C, marking the coldest period recorded in late February over the past 43 years. All available images from January 1 to February 29, June 1 to August 31, and December 1 to 31, 2024, were screened, with only imagery from August 11 and December 17 meeting cloud-free conditions in the study area. The weather on these selected dates and preceding days was characterized by calm winds and no rainfall (https://rp5.ru (accessed on 25 February 2025)), resulting in high-quality data. Multi-spectral bands used in the study had a spatial resolution of 30 m. Image preprocessing included geometric corrections, radiometric calibration, atmospheric correction, and cropping.

2.2.2. Acquisition and Management of Urban Surface Spatial Morphology Data in Central Urban Area of Shanghai

The building vector data required for the calculation of urban surface spatial morphology indicators in the central area of Shanghai were obtained from the Resource and Environmental Science Data Platform (https://www.resdc.cn/ (accessed on 28 February 2025)) of the Institute of Geographic Sciences and Resources of the Chinese Academy of Sciences (IGSR). The Digital Elevation Model (DEM) data of the central urban area of Shanghai were obtained from the Geospatial Data Cloud (http://www.gscloud.cn/ (accessed on 28 February 2025)) (see Figure 2).

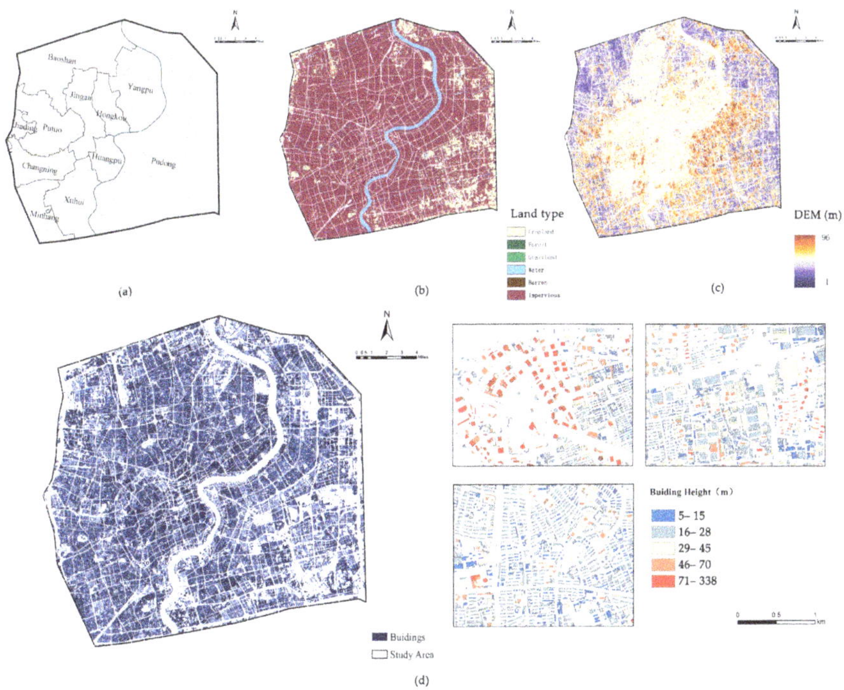

Figure 2. Spatial information data of central urban area of Shanghai: (**a**) administrative division; (**b**) land use type; (**c**) DEM data; (**d**) building vector data (building distribution and local building height).

2.2.3. Acquisition of TSF Social Media Data

Social media data provide valuable insights into human activities and emotional states within urban areas, enabling researchers to investigate potential directions in urban planning, management, and development [48]. Comments on social media platforms are typically unstructured; hence, converting these data into analyzable information requires advanced text mining and analysis techniques [49]. Since 2009, China has introduced various social media platforms such as Sina Weibo and Dianping, where users can share text, images, and other information. Using GPS-enabled smart devices, users can also share their real-time geographic locations. The act of publicly sharing one's environment, physical sensations, and emotional states on social media is referred to as "checking in". Location-Based Social Network Services (LBSNS) have progressively become significant channels through which individuals share daily experiences and express real-time emotions. According to the Weibo User Consumption Trend Report published in December 2024, Weibo has reached 587 million monthly active users and 257 million daily active users.

In selecting the research period, priority was given to continuous intervals exhibiting distinctly cold or hot climates accompanied by favorable weather conditions. Consequently, August and December 2024 were selected as representative intervals for summer and winter analysis, respectively. Remote sensing data collected for this study were also derived from these two periods. Geo-tagged, weather-related comments posted on Sina Weibo during these periods were collected, specifically filtering for texts containing explicit descriptions of perceived heat or cold (e.g., "The sun is melting me today" or "The wind is biting"). User-generated spatiotemporal data were compiled into an Excel database and subsequently integrated with geographic location data using ArcGIS 10.8.1, establishing a foundational geospatial database for further analysis.

3. Methodology

3.1. Selection and Treatment of UM Indicators

As multi-layered and highly interactive systems, urban morphology is closely related to two-dimensional landscape patterns and three-dimensional building patterns [42,50]. We selected 14 UM variables related to TSF and LST, including 8 two-dimensional variables and 6 three-dimensional variables. Two-dimensional variables include normalized difference vegetation index (NDVI), building density (BD), floor area ratio (FAR), building spacing (BS_{sp}), impervious surface index (ISI), area weighted mean shape index ($SHAPE_{AM}$), aggregation index (AI), and edge density (ED). The three-dimensional variables include elevation, building height (BH) and average building volume (ABV), sky view fraction (SVF), building shape (BS_{sh}), and spatial congestion degree (SCD) (see Table 1).

Table 1. The summary of 2D and 3D metrics in this study.

Norm	Formula	Description	References
2D Indicators			
$NDVI$	$NDVI = \frac{\rho_{nin} - \rho_{red}}{\rho_{nin} + \rho_{red}}$	Show the extent of the green space.	[51]
BD	$BD = \sum_{i=1}^{n} \frac{A_i}{S} \times 100\%$	Reflects the extent to which the ground is covered by buildings.	[51]
FAR	$FAR = \frac{\sum_{i=1}^{n} f_i \times A_i}{S}$	Characterize the intensity of land development.	[42,51]
BS_{sp}	$BS_{sp} = \frac{1}{n}\sum_{i=1}^{n} d_i$	Average building-to-building distance between buildings in each grid pixel.	[42]
ISI	$ISI = \left(\frac{A_{impervious}}{A}\right)$	The proportion of impervious surface coverage per unit area.	[52]

Table 1. *Cont.*

Norm	Formula	Description	References
$SHAPE_{AM}$	$SHAPE_{AM} = \frac{\sum_{j=1}^{n} \frac{0.25P_{ij}}{\sqrt{a_{ij}}} a_{ij}}{\sum_{j}^{n} a_{ij}}$	Reflecting the complexity of specific types of landscape patch boundaries.	[34]
AI	$AI = \sum_{i=1}^{n} \left(\frac{g_{ii}}{max}\right) P_i$	The aggregation characteristics reflecting the distribution of patches.	[34,53]
ED	$ED = \frac{\sum_{j=1}^{n} e_{ij}}{A}$	Measure the degree of landscape fragmentation.	[34,53]
3D Indicators			
Elevation	Directly from DEM data	Indicates the elevation status of the urban space.	-
BH	Directly from the building vector dataset	Indicates the vertical height of the building.	[50]
ABV	$ABV = \frac{\sum_{i}^{n} V_i}{n}$	Reflecting the degree of urban intensification.	[34]
SVF	$SVF = 1 - \frac{\sum_{i=1}^{n} \sin \gamma_i}{m}$	Reflects spatial openness.	[20,54,55]
BS_{sh}	$BS_{sh} = \frac{P_i - \sqrt{P_i^2 - 16SA_i}}{4}$	The average of the shape indices of all buildings in each grid pixel.	[42]
SCD	$SCD = \frac{\sum_{i=1}^{n} V_i}{\sum_{i=1}^{n} H_i}$	Reflects the density of the distribution of elements within the spatial unit.	[56]

Note: ρ_{nin} and ρ_{red} are the reflectance values of the near infrared and red channels, respectively. A_i, f_i, P_i, V_i, and H_i are the area, the number of floors, the perimeter, the volume, and the height, respectively, occupied by the building i or patch i. n denotes the number of buildings, d_i is the distance between neighboring buildings, i is the total logarithm between buildings, and A is the area of the unit. γ denotes the elevation angle of the topographic horizon, and m denotes the number of directions used to estimate γ. e_{ij} denotes the length of the edge segment j in the landscape patch of category i. g_{ii} denotes the number of adjacencies between the basic spatial units of the landscape patches of category i, and P_i denotes the proportion of the landscape patches of category i in the landscape area.

3.2. LST Retrieval

LST retrieval was conducted using Landsat-8 imagery, obtained from the Thermal Infrared Sensor (TIRS) onboard the Landsat-8 satellite and accessed via the United States Geological Survey (https://www.usgs.gov/landsat-missions/landsat-8 (accessed on 25 February 2025)). The primary thermal infrared bands employed for surface temperature retrieval are bands 10 and 11, with band 10 exhibiting higher atmospheric transmittance, making it particularly suitable for accurate surface temperature inversion.

Surface temperature was derived using the radiative transfer equation (RTE), where thermal infrared energy measured by satellite sensors is corrected for atmospheric effects. The formula is presented below:

$$LST = \frac{K_2}{\ln \frac{K_1}{B_{LST}} + 1} - 273.15 \quad (1)$$

$$P_v = \frac{NDVI - NDVI_{soil}}{NDVI_{veg} - NDVI_{soil}} \quad (2)$$

$$\varepsilon = 0.004P_v + 0.986 \quad (3)$$

$$L_\lambda = [\varepsilon B_{LST} + (1-\varepsilon)L_{down}]\tau + L_{up} \quad (4)$$

$$B_{LST} = \frac{[L_\lambda - L_{up} - \tau(1-\varepsilon)L_{down}]}{\tau\varepsilon} \quad (5)$$

where P_v represents the vegetation coverage fraction. $NDVI_{veg} = 0.70$ denotes the vegetation coverage in vegetated areas, while $NDVI_{soil} = 0.05$ corresponds to the vegetation coverage in bare soil regions. ε is the surface emissivity. P_v refers to the overall vegetation coverage fraction within a mixed pixel. L_λ stands for the infrared radiation heat value. B_{LST} represents the combined radiation brightness. τ is the atmospheric transmittance in the thermal infrared band. L_{up} and L_{down} are the upward and downward atmospheric radiation brightness, respectively. We use the NASA EARTHDATA (https://earthdata.nasa.gov/ (accessed on 25 February 2025)) for atmospheric parameters. K_1 and K_2 are constants, with $K_1 = 774.8853\ \mathrm{W \cdot m^{-2} \cdot sr^{-1} \cdot \mu m^{-1}}$ and $K_2 = 1321.0789K$.

3.3. TSF Data Processing, Induction, and Quantification

As previously mentioned, due to the complex interplay of physical, physiological, and psychological factors influencing cold and heat perception, no single factor can comprehensively explain human thermal sensations. To simplify the experimental procedure and enhance feasibility, this study referenced the Thermal Sensation Vote (TSV) scale proposed by the American Society of Heating, Refrigerating, and Air-Conditioning Engineers (ASHRAE) and developed a 7-level Thermal Sensation Factor (TSF) scale (see Table 2). This scale is based on the laboratory-derived Predicted Mean Vote (PMV) model formulated by P.O. Fanger, who established correlations between thermal sensation and thermal balance through extensive laboratory observations [57]. By employing this scale, researchers can effectively capture spatial users' immediate thermal sensations related to UTE. However, it should be noted that the TSF scale reflects immediate spatial experiences and does not comprehensively represent long-term thermal adaptability.

Table 2. Seven-level TSF scale.

Scale	Ruler	Human Perception and Physiological Response
−3	Cold	Skin tingling, shortness of breath, high risk of core temperature decline.
−2	Cool	The limbs are stiff and trembling, and may be frostbitten after exposure for 30 min.
−1	Slightly cool	Hands and feet are slightly cold and can be adjusted independently.
0	Neutral	No active regulation behavior, dry skin, stable heart rate.
1	Slightly warm	Slight sweating and increased thirst.
2	Warm	Continuous sweating, increased heart rate, and decreased attention.
3	Hot	Risk of heat cramps, and failure of thermoregulation.

We preprocessed a total of 234,248 social media comment datasets to exclude advertisements, duplicate content, comments related to indoor environments, comments unrelated to TSF, and comments regarding specific outdoor spaces such as Shanghai Ice World. Ultimately, 3867 comments involving temperature perception feedback were selected, comprising 2276 in summer and 1591 in winter. A team of seven expert raters was assembled, and the seven-level TSF scale was employed to score the review text data item by item. Based on Fleiss's Kappa test for inter-rater consistency, the differences among the seven groups of data were found to correlate with grade distance, and Linear Weights were applied to calculate Kappa values of 0.663 (>0.6), indicating moderate agreement and fulfilling the coefficient requirements for subjective tasks in social science research [58]. Additionally, $p < 0.05$ confirmed that the consistency was statistically significant. The final score was determined as the rounded average of the experts' scores. The Thermal Feedback Orientation Index (*TFOI*) can be computed by integrating the degree of individual cold and hot perception with the frequency of each degree within the unit.

$$TFOI = \sum (S_i \cdot F_i) \tag{6}$$

where the sentiment temperature score (S_i) for each comment is categorized and F_i denotes the frequency of occurrence for each scale i, with $i \in \{-3, -2, -1, 0, 1, 2, 3\}$. $TFOI$ stands for the overall intensity of thermal perception feedback within the unit, where a more positive value indicates a warmer perception, while a more negative value signifies a colder perception.

3.4. Model Selection

To assess how various variables influence the UTE, a regression analysis is required [59]. Common methods include Ordinary Least Squares (OLS) regression [36,60] and Geographically Weighted Regression (GWR) [61]. The traditional OLS regression model, widely employed in urban planning research [62], particularly for identifying relationships between urban thermal environments and their influencing factors [42], is considered the standard global regression approach [63]. Conversely, the GWR model is a local regression technique that requires data to exhibit spatial autocorrelation and demands a more rigorous sample size compared to OLS.

Spatial autocorrelation analysis was performed to characterize the spatial relationships and attribute similarities between each observation and adjacent statistical entities [64]. To assess spatial clustering, the Global Moran's I statistic was computed individually for each variable through ArcGIS 10.8.1. The formula used in the calculations is as follows:

$$I = \frac{n\sum_{i=1}^{n}\sum_{j=1}^{n} w_{ij}(x_i - \overline{x})(x_j - \overline{x})}{\left(\sum_{i=1}^{n}\sum_{j=1}^{n} w_{ij}\right)\sum_{i=1}^{n}(x_i - \overline{x})^2} \tag{7}$$

where n is the total number of spatial units in the study area; x_i and x_j are the attribute values of the i and jth spatial units, respectively. $\overline{x}$ is the average of all spatial unit attribute values; w_{ij} is the element in the spatial weight matrix that represents the spatial relationship between unit i and unit j. Values range from -1 to 1, where a positive value indicates a positive correlation, a negative value indicates a negative correlation, and a zero value indicates that there is no spatial autocorrelation [63]. In addition, the z-score needs to be calculated to determine whether the statistic significantly deviates from the expectation of the null hypothesis [61]. The calculation formula is as follows:

$$Z = \frac{I - E(I)}{\sqrt{var(I)}} \tag{8}$$

If $|Z| > 1.96$, the null hypothesis is rejected, which indicates that Moran's I is statistically significant at the 0.05 level.

After calculation, except AI, the Moran's I values of UM variable data are all greater than 0.3, and $|Z| > 1.96$, indicating that there is a significant positive spatial correlation and a spatial clustering pattern. AI's Moran's I = 0.15, $z = 22.7$, and $p = 0.000$, indicating that there is a significant weak positive spatial autocorrelation in the data, that is, a non-random low-intensity clustering pattern in space. The Moran I values of the dependent variables *TFOI* in this study are all close to 0, and $|Z| < 1.96$, indicating that the distribution is random in space and there is no obvious global spatial autocorrelation. Therefore, OLS should be used as the research model in this study.

This study adopts the OLS model and takes it as the basic model of the research. OLS is expressed as follows:

$$Y = \beta_0 + \beta_1 X_1 + \beta_2 X_2 + \ldots + \beta_n X_n + \xi \tag{9}$$

where Y is the dependent variable; X_1, $X_2 \ldots X_n$ is the explanatory variable; β is the estimated coefficient; and ξ is the random error.

4. Results

4.1. Spatial Distribution Characteristics of TSF and LST in the Central Urban Area of Shanghai

The calculated TFOI data were integrated with the geographical location data of comments and visualized in ArcGIS 10.8.1 to generate the TSF index map based on microblog sign-in data (as shown in Figure 3).

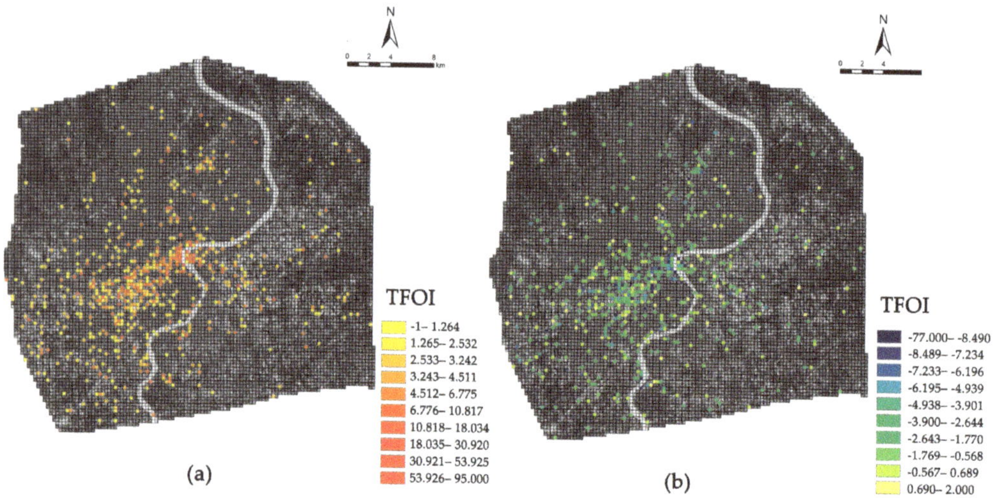

Figure 3. Description of the TFOI index generated based on Weibo check-in data represents (**a**) summer data and (**b**) winter data.

The calculated LST data were imported into ArcGIS 10.8.1 to generate the surface temperature map of the central urban area of Shanghai. As depicted in Figure 4, summer temperatures in the central urban area of Shanghai are generally elevated, with high-temperature zones predominantly concentrated in the Baoshan District, Yangpu District, Hongkou District, Jing'an District, and Pudong New Area, indicating pronounced urban heat island effects. The spatial distribution of temperature in winter exhibits a pattern similar to that observed in summer.

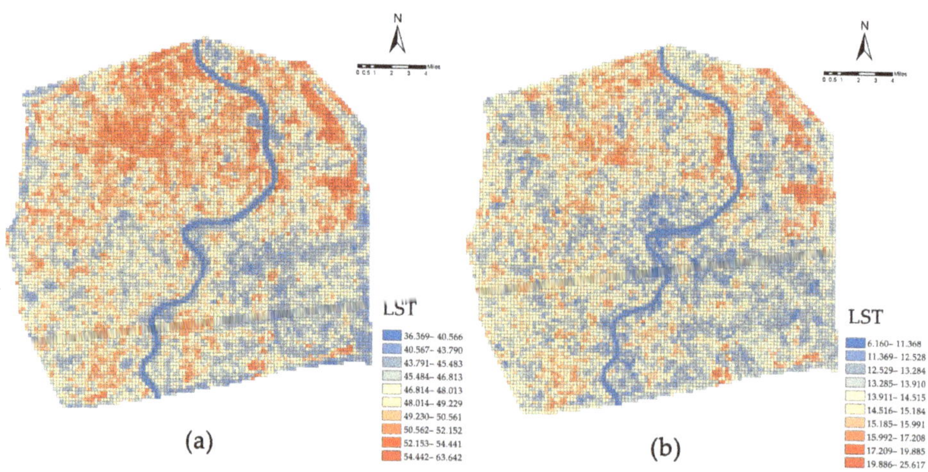

Figure 4. (**a**) Summer LST and (**b**) Winter LST of the central urban area of Shanghai derived from Landsat-8 image.

4.2. The Surface Spatial Form of the Central Urban Area of Shanghai

The spatial data for the central urban area of Shanghai were imported into ArcGIS 10.8.1 to calculate the UM index. As illustrated in Figure 5, UM in the central urban area of Shanghai exhibits complexity. The built environment in Huangpu District, Jing'an District, and Putuo District is characterized by high density and congestion. Urbanization diminishes gradually in a radial pattern from the central urban core outward. The peripheral wedge-shaped green belts demonstrate a higher rate of natural coverage. Additionally, the overall development intensity of the Pudong New Area is comparatively lower than that of the area north of the Huangpu River.

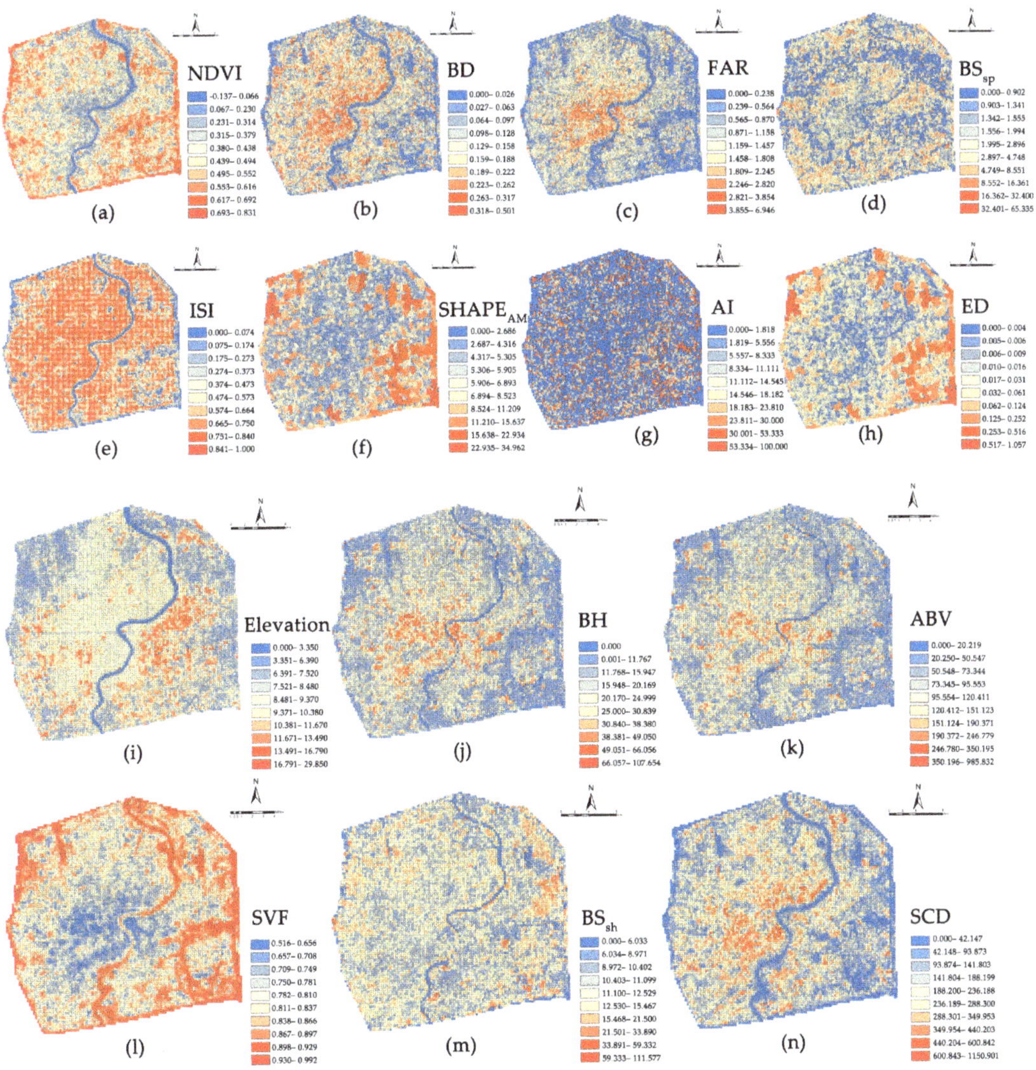

Figure 5. Central urban area of Shanghai UM: (**a**) NDVI; (**b**) BD; (**c**) FAR; (**d**) BS_{sp}; (**e**) ISI; (**f**) $SHAPE_{AM}$; (**g**) AI; (**h**) ED; (**i**) elevation; (**j**) BH; (**k**) ABV; (**l**) SVF; (**m**) BS_{sh}; (**n**) SCD.

Considering that multicollinearity can lead to bias in the introduced OLS model and increase the standard error [65], we diagnosed the covariance in the linear regression of the UM data with the help of SPSS 26 before proceeding with the regression analysis of the OLS model. The VIF of all UM variables was tested to be less than 10. Therefore, all 14 indicators were ready for the next stage of analysis.

4.3. The Relationship Between the UM Variable and UTE

4.3.1. Results of the UM Variables with the TSF OLS Model

We incorporated UM variables and the TSF dataset of the central urban area of Shanghai into the OLS model. The OLS results are presented in Table 3. BD, BH, and TSF exhibit significant positive effects during summer, suggesting that higher building density and height intensify residents' heat perception. Conversely, NDVI, FAR, SVF, ABV, and TSF demonstrate significant negative effects, implying that expanding urban green spaces and maintaining a more open urban environment can mitigate heat perception. These findings align with theoretical studies on urban ventilation, greening, and cooling during summer. Additionally, BS_{sh} and TSF show significant negative correlations. The negative association of BS_{sh} suggests that as the shape index of buildings increases, residents' thermal perception decreases. A higher shape index typically reflects more complex, multi-boundary, or irregular building profiles, which may enhance local ventilation, increase shadowed areas, or alter near-surface heat exchange characteristics, thereby alleviating heat perception. Similarly, ISI and TSF also exhibit a significant negative correlation, which contradicts conventional studies.

In winter, FAR, NDVI, and ISI show a significant positive relationship with TSF, indicating that higher building density, increased vegetation cover, and a greater proportion of hard surfaces contribute to a warmer perception among the population. In contrast, BH, BD, and TSF display a significant negative correlation, suggesting that high-rise buildings and densely packed structures may reduce outdoor warmth, potentially by obstructing sunlight. This phenomenon can likely be attributed to the venturi effect, which occurs when air flows through constricted spaces between tall buildings, causing accelerated airflow and significantly increased wind speeds. These conditions subsequently reduce pedestrians' perception of warmth and thermal comfort outdoors [66].

Table 3. Results of OLS model on UM and TSF in the central urban area of Shanghai.

Variables	Coef.	Std. Err.	T	*p*
		Summer (August)		
NDVI	−12.501	3.154	−3.963	0.013 **
BD	22.578	11.323	1.994	0.064 *
FAR	−8.878	2.496	−3.557	0.006 ***
BS_{sp}	0.057	0.060	0.963	0.307
ISI	−11.925	2.214	−5.387	0.001 ***
$SHAPE_{AM}$	−0.237	0.164	−1.438	0.259
AI	−0.220	0.211	−1.044	0.207
ED	11.662	10.763	1.084	0.334
Elevation	0.218	0.175	1.246	0.425
BH	4.377	1.241	3.527	0.028 **
ABV	−0.012	0.007	−1.739	0.089 *
SVF	−18.137	8.134	−2.230	0.039 **
BS_{sh}	−0.179	0.137	−1.303	0.086 *
SCD	−0.031	0.300	−0.103	0.906
R-squared		0.119		

Table 3. Cont.

Variables	Coef.	Std. Err.	T	p
		Winter (December)		
NDVI	6.61	1.799	3.675	0.041 **
BD	−7.921	5.235	−1.513	0.072 *
FAR	3.374	1.345	2.508	0.002 ***
BS_{sp}	0.058	0.062	0.940	0.137
ISI	4.938	1.280	3.857	0.014 **
$SHAPE_{AM}$	0.027	0.095	0.289	0.695
AI	0.146	0.118	1.236	0.147
ED	−4.567	6.359	−0.718	0.404
Elevation	0.061	0.099	0.622	0.491
BH	−1.692	0.640	−2.645	0.069 *
ABV	−0.001	0.004	−0.191	0.872
SVF	6.838	4.678	1.462	0.256
BS_{sh}	0.007	0.031	0.241	0.733
SCD	−0.104	0.161	−0.646	0.351
R-squared		0.081		

Note: *** $p < 0.01$, ** $p < 0.05$, * $p < 0.1$.

4.3.2. Results of the UM Variables with the LST OLS Model

We investigated the relationship between UM form and LST in the central urban area of Shanghai using an OLS model (see Table 4).

During summer, BD, BH, SVF, BS_{sh}, and ISI exhibit significant positive effects on surface temperature. This suggests that a dense, high-rise built environment and a high proportion of impervious surfaces exacerbate the urban thermal environment. In contrast, FAR, NDVI, ABV, AI, and $SHAPE_{AM}$ show significant negative correlations with LST, implying that higher plot ratios, greater green coverage, and higher land surface aggregation can mitigate urban heat to some extent. However, the correlation between LST and NDVI in summer is not statistically significant, which contradicts findings from most existing studies.

Table 4. Results of OLS model on UM and LST in the central urban area of Shanghai.

Variables	Coef.	Std. Err.	t	p
		Summer (August)		
NDVI	−1.288	0.193	−6.672	0.000 ***
BD	19.777	0.591	33.477	0.000 ***
FAR	−2.335	0.199	−11.744	0.000 ***
BS_{sp}	−0.004	0.008	−0.523	0.646
ISI	2.704	0.150	18.049	0.000 ***
$SHAPE_{AM}$	−0.025	0.011	−2.35	0.019 **
AI	−0.043	0.016	−2.779	0.006 ***
ED	0.069	0.592	0.116	0.907
Elevation	0.011	0.012	0.886	0.370
BH	1.298	0.051	25.653	0.000 ***
ABV	−0.004	0.001	−6.706	0.000 ***
SVF	3.755	0.729	5.149	0.000 ***
BS_{sh}	0.058	0.005	11.765	0.000 ***
SCD	−0.004	0.010	−0.453	0.716
R-squared		0.314		

Table 4. *Cont.*

Variables	Coef.	Std. Err.	t	p
		Winter (December)		
NDVI	0.928	0.097	9.546	0.000 ***
BD	8.012	0.297	26.934	0.000 ***
FAR	−1.196	0.100	−11.945	0.000 ***
BS_{sp}	0.001	0.004	0.153	0.893
ISI	0.650	0.075	8.612	0.000 ***
$SHAPE_{AM}$	−0.029	0.005	−5.332	0.000 ***
AI	−0.040	0.008	−5.071	0.000 ***
ED	0.851	0.298	2.857	0.002 ***
Elevation	0.036	0.006	5.751	0.000 ***
BH	0.389	0.025	15.264	0.000 ***
ABV	−0.003	0.000	−9.166	0.000 ***
SVF	3.049	0.367	8.303	0.000 ***
BS_{sh}	0.030	0.002	12.348	0.000 ***
SCD	−0.012	0.005	−2.318	0.029
R-squared		0.174		

Note: *** $p < 0.01$, ** $p < 0.05$.

In winter, BD, BH, SVF, ISI, ED, and elevation demonstrate significant positive effects on LST, indicating that denser and taller buildings, along with increased surface roughness, contribute to higher winter temperatures. Conversely, FAR, ABV, $SHAPE_{AM}$, and AI exhibit significant negative correlations with winter LST, suggesting that higher plot ratios, greater building volumes, and more compact urban forms help moderate cold-season surface temperatures.

5. Discussion

5.1. The Effect of UM Variables on UTE

Using an OLS model, we analyzed the diverse correlations between UM variables, TSF, and LST (see Figure 6). Regarding TSF, in summer, multiple UM variables significantly influence TSF. Specifically, NDVI, FAR, SVF, ABV, BS_{sh}, and ISI exhibit significant negative correlations with TSF, indicating that increases in these factors reduce residents' heat perception. Conversely, BD and BH show significant positive effects, suggesting that higher building density and height intensify heat perception. The negative correlation between ISI and TSF transcends conventional understanding. Given that the correlation between NDVI and LST in this study is also insignificant, it suggests that deeper microclimatic and surface material factors may be influencing the pronounced UTE in the central urban area of Shanghai. The relationship between SVF and TSF deviates from traditional assumptions, as low SVF is generally believed to provide shading effects [67]. However, spaces with high SVF tend to experience faster air circulation but also stronger solar radiation, leading to an opposite effect on TSF [68]. This suggests the existence of a threshold in the relationship between SVF and TSF. Previous studies have demonstrated that the impact of SVF on LST is often determined by a threshold, where beyond a certain value, the effect becomes negative [69]. For instance, while SVF generally exhibits a positive correlation with LST, when its value ranges between 0.75 and 1.00, the relationship follows a linear trend. However, for values below 0.75, the correlation weakens significantly [70]. A similar threshold effect is observed in the relationship between BD, BH, SCD, FAR, and LST [69,71,72]. Whether such thresholds exist in the correlation between UM variables and TSF remains uncertain. In winter, TSF is influenced by several UM variables, primarily in the form of cold perception (negative TSF values in this study). FAR, NDVI, ISI, and TSF demonstrate significant positive correlations, whereas BH and BD show significant negative correlations. According to traditional studies, a higher NDVI is generally associated with a more pronounced cooling effect and light obstruction during winter. However, this contradicts the findings of the present study. This discrepancy may be attributed to the fact that in warmer winter months, individuals are more likely to engage in outdoor activities on grassy areas and express positive sentiments regarding the mild climate. This indicates that NDVI, FAR, and ISI improve thermal comfort within appropriate limits, while higher BD and BH decrease comfort. The phenomenon manifests as the urban canyon effect, impeding heat dissipation, while in winter, it shifts to the Venturi effect, amplifying surface wind speed.

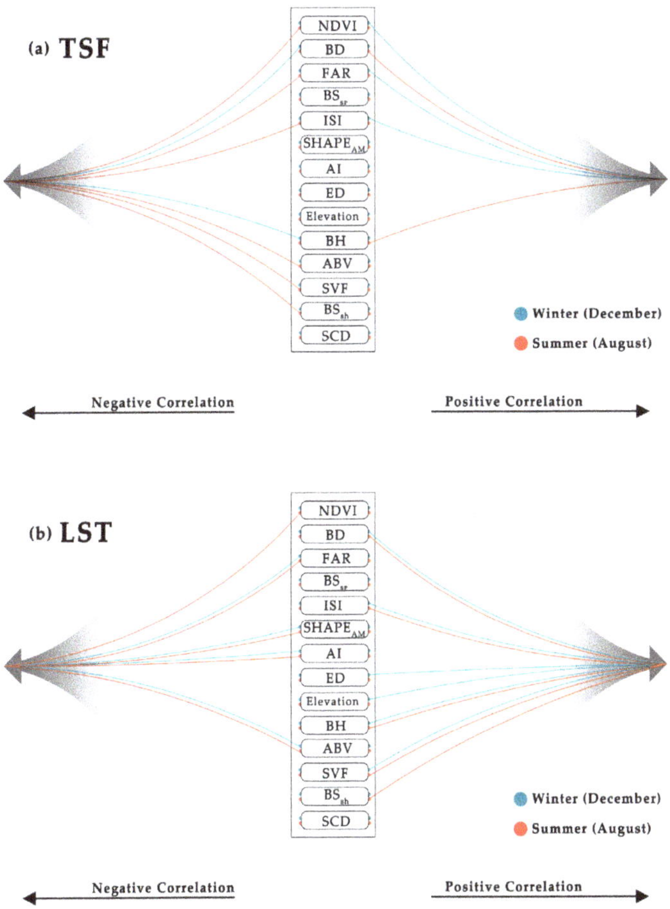

Figure 6. Correlation between the UM variable and TSF and LST. (**a**) TSF; (**b**) LST.

Regarding LST, our findings largely align with existing literature, with the primary discrepancy being the non-significant correlation between NDVI and LST in summer. Generally, vegetation is expected to reduce LST in summer but exhibits a positive correlation in winter [36,68,73], which is consistent with its effect on TSF. This discrepancy may stem from factors such as urban microclimate variations and seasonal changes in vegetation cover in the central urban area of Shanghai. In summary, the summer thermal environment is primarily influenced by high-density buildings and impervious surfaces, which exacerbate the UHI effect by increasing solar radiation absorption and heat retention, thereby raising LST and leading to extreme TSF occurrences. Vegetation cover effectively mitigates UTE through transpiration and shading mechanisms. In winter, a dual mechanism is observed: high building density, building height, and sky view limitation restrict long-wave radiation diffusion [70], while moderate plot ratios and landscape complexity enhance surface turbulent exchange, forming thermal diffusion channels that promote winter warm perception [74].

Existing studies on the correlation between UM variables and UTE have integrated both two-dimensional and three-dimensional indicators. Common 2D indicators encompass urban land use patterns, surface attributes, and landscape ecology metrics. In addition to NDVI, BD, FAR, BS_{sh}, ISI, $SHAPE_{AM}$, AI, and ED examined in this study, frequently used variables include Mean Architecture Projection Area (MAPA), Patch Density (PD), Largest Patch Index (LPI), Shannon's Diversity Index (SHDI), and Open Space Ratio (OSR) in city-scale studies [34,51,53,75]. These 2D UM variables influence energy exchange efficiency, thereby affecting surface heat flux and significantly altering UTE [76]. The 3D indicators primarily describe architectural morphology. In addition to elevation, BH, ABV, SVF, BS_{sh}, and SCD analyzed in this study, relevant research has incorporated the Mean Compactness Factor (MCF), Average Building Height Standard Deviation (AHSD), Facade Area Index (FAI), and Urban Canyon Ratio (UCR) [34,45,77,78]. Some studies suggest that LST variations in summer are primarily driven by 3D building indices, while winter LST is more strongly associated with 2D land use indices [79]. However, other studies report conflicting conclusions [55], though generally, 3D UM variables exert a greater thermal effect [77]. Moreover, UM variables are closely linked to urban microclimate conditions. Specifically, four key climate indicators—actual temperature, relative humidity, solar radiation, and wind environment—are directly influenced by UM variables, ultimately shaping UTE patterns [80,81]. Given the complexity of urban spatial morphology's influence on LST, comprehensively evaluating the impact of UM variables on UTE remains a challenging and intricate task.

5.2. Mechanism of Different Effects of UM on TSF and LST

Our research suggests that effective UTE management requires a dual focus on both physical cooling strategies and human-centered design to enhance outdoor thermal comfort [27]. LST is governed by the surface energy balance, whereas human thermal comfort is influenced by atmospheric and physiological processes. LST reflects the temperature of surfaces, including buildings, and is primarily determined by solar radiation, surface materials, and geometric structures. In contrast, TSF represents subjective human thermal perception, which depends on environmental factors such as air temperature, humidity, wind speed, and mean radiant temperature. From a causal perspective, UM affects human thermal perception primarily by altering radiation exposure and wind flow, an indirect influence mediated by atmospheric conditions. When evaluating UTE through both TSF and LST, the following scenarios emerge: High-rise buildings may reduce LST while simultaneously decreasing street-level wind speed. In low-wind environments, even if surface temperatures are relatively cool, individuals may still experience heat stress due to limited evaporative cooling. This phenomenon explains why some UM variables in our study exhibited weaker correlations with TSF compared to LST. Moreover, human thermal perception is influenced by additional factors, including wind patterns, which depend on street orientation and building layout—elements that play a minimal role in LST but are critical for TSF. Consequently, remote sensing of surface temperatures alone is insufficient for assessing livable thermal comfort in urban streetscapes. Conversely, subjective thermal perception data, such as human-body questionnaires, should be supplemented with objective heat distribution measurements for a comprehensive assessment [67]. Comparing these two levels of analysis allows us to infer distinct mechanisms underlying UTE regulation. In summary, UM serves as a fundamental lever in UTE management, facilitating holistic urban microclimate optimization by influencing both LST and TSF. Ultimately, this integrated approach enhances urban climate livability and promotes sustainable urban development.

5.3. Limitations and Shortcomings

This study explores the use of text-based comments on TSF from microblog check-in data as a novel approach to characterizing the UTE. While this method holds significant research value in both process and outcome, it also presents several limitations. First, the comment data primarily capture surface-level semantics in the publisher's text, whereas deeper semantic interpretation often requires consideration of contextual factors such as environment, clothing, and emotions, which cannot be fully inferred from simple discourse alone. Second, as a major tourist destination, Shanghai is expected to receive 120 million visitors in 2024. Foreign tourists often exhibit varying sensitivities, tolerances, and perception thresholds regarding extreme temperatures, as they may not be fully acclimated to Shanghai's climate. Consequently, their TSF expressions on social media are strongly influenced by individual subjective experiences. Additionally, tourists' thermal perception often differs significantly from that of residents [26]. Finally, the findings indicate that the OLS model has limitations in explaining the relationship between UM, LST, and TSF, suggesting that more suitable methodologies should be explored in future research. Moving forward, we aim to further quantify the dynamic relationship between morphological indicators and individual thermal perception, as well as investigate cross-scale collaborative optimization mechanisms.

6. Conclusions

In this study, the OLS model was employed to examine the relationship between LST, TSF, and UTE representation modes, alongside UM variables. The findings provide preliminary evidence that UM regulates UTE by influencing both LST and TSF. From a physical perspective, UM determines how the urban underlying surface absorbs, stores, and releases heat. From a human perception standpoint, UM shapes microclimatic conditions and pedestrian comfort. The study yields three key conclusions: (1) NDVI, BD, FAR, ISI, BH, ABV, SVF, and BS_{sh} significantly explain TSF, whereas $SHAPE_{AM}$, AI, ED, elevation, BS_{sp}, and SCD show no correlation with TSF. (2) Significant variables, including NDVI, FAR, ISI, UM, BD, and BH, exhibited opposite effects on cold perception in winter compared to heat perception in summer, indicating a consistent influence on thermal perception across seasons. (3) In summer, the significant variables SVF, BS_{sh}, and ISI exhibit opposing effects on TSF and LST. In contrast, during winter, the significant variable FAR demonstrates opposite influences on TSF and LST. These findings suggest that higher building density and height intensify residents' heat perception and contribute to increased LST. Enhancing urban green spaces and maintaining more open urban environments are effective in mitigating heat perception. The study's in-depth examination of TSF highlights the importance of human-centered urban planning. Specifically, cooling strategies should not focus solely on reducing LST while overlooking pedestrian activity patterns, as this may lead to an overestimation of actual thermal comfort improvements. Instead, urban planning and behavioral science must jointly account for individuals' dynamic spatial adaptations. This perspective clarifies the discrepancies between TSF and LST, guiding more comprehensive and effective urban green infrastructure strategies to enhance the UTE.

Author Contributions: Conceptualization, H.Q., F.Z., and B.Q.; methodology, H.Q., M.W., and S.Z.; software, H.Q., M.W., and S.Z.; validation, H.Q., M.W., S.Z., F.Z., and B.Q.; formal analysis, S.Z.; investigation, H.Q., S.Z., and M.W.; resources, F.Z. and B.Q.; data curation, H.Q. and S.Z.; writing—original draft preparation, H.Q., S.Z., M.W., F.Z., and B.Q.; writing—review and editing, F.Z., B.Q., H.Q., S.Z., and M.W.; visualization, M.W.; supervision, F.Z. and B.Q.; project administration, F.Z. and B.Q.; funding acquisition, F.Z. and B.Q. All authors have read and agreed to the published version of the manuscript.

Funding: Sponsored by the Seventh Jiangsu 333 High-level Talent Programme Phase Third-tier Cultivation Candidates Project (2024), the National Natural Science Foundation of China (NSFC) General Project (31971721) and the Priority Academic Program Development of Jiangsu Higher Education Institutions (PAPD).

Data Availability Statement: The research data are available upon request from the corresponding author.

Acknowledgments: We express our gratitude to the United States Geological Survey (USGS) for providing free access to Landsat data. We also acknowledge the Resource and Environmental Science Data Platform of the Institute of Geographic Sciences and Natural Resources Research, Chinese Academy of Sciences, for offering free vector data on buildings in central Shanghai. Additionally, we thank Geospatial Data Cloud (http://www.gscloud.cn/ (accessed on 28 February 2025)) for providing free online Digital Elevation Model (DEM) data for central Shanghai.

Conflicts of Interest: The authors declare no conflicts of interest.

References

1. Wang, Y.; Chen, L.; Kubota, J. The Relationship between Urbanization, Energy Use and Carbon Emissions: Evidence from a Panel of Association of Southeast Asian Nations (ASEAN) Countries. *J. Clean. Prod.* **2016**, *112*, 1368–1374. [CrossRef]
2. Coutts, A.M.; Harris, R.J.; Phan, T.; Livesley, S.J.; Williams, N.S.G.; Tapper, N.J. Thermal Infrared Remote Sensing of Urban Heat: Hotspots, Vegetation, and an Assessment of Techniques for Use in Urban Planning. *Remote Sens. Environ.* **2016**, *186*, 637–651. [CrossRef]
3. Zhao, J.; Guo, F.; Zhang, H.; Dong, J. Mechanisms of Non-Stationary Influence of Urban Form on the Diurnal Thermal Environment Based on Machine Learning and MGWR Analysis. *Sustain. Cities Soc.* **2024**, *101*, 105194. [CrossRef]
4. Patel, S.; Indraganti, M.; Jawarneh, R.N. A Comprehensive Systematic Review: Impact of Land Use/Land Cover (LULC) on Land Surface Temperatures (LST) and Outdoor Thermal Comfort. *Build. Environ.* **2024**, *249*, 111130. [CrossRef]
5. Mirzaei, P.A. Recent Challenges in Modeling of Urban Heat Island. *Sustain. Cities Soc.* **2015**, *19*, 200–206. [CrossRef]
6. Aynsley, R.; Spruill, M. Thermal Comfort Models for Outdoor Thermal Comfort in Warm Humid Climates and Probabilities of Low Wind Speeds. *J. Wind Eng. Ind. Aerodyn.* **1990**, *36*, 481–488. [CrossRef]
7. Höppe, P. Different Aspects of Assessing Indoor and Outdoor Thermal Comfort. *Energy Build.* **2002**, *34*, 661–665. [CrossRef]
8. Nikolopoulou, M.; Steemers, K. Thermal Comfort and Psychological Adaptation as a Guide for Designing Urban Spaces. *Energy Build.* **2003**, *35*, 95–101. [CrossRef]
9. Ng, E.; Cheng, V. Urban Human Thermal Comfort in Hot and Humid Hong Kong. *Energy Build.* **2012**, *55*, 51–65. [CrossRef]
10. *ASHRAE 55-1992*; Thermal Environmental Conditions for Human Occupancy. American Society of Heating, Refrigerating and Air-conditioning Engineers Inc.: Atlanta, GA, USA, 1992.
11. Alghamdi, S.; Tang, W.; Kanjanabootra, S.; Alterman, D. Field Investigations on Thermal Comfort in University Classrooms in New South Wales, Australia. *Energy Rep.* **2023**, *9*, 63–71. [CrossRef]
12. Kim, J.; Zhou, Y.; Schiavon, S.; Raftery, P.; Brager, G. Personal Comfort Models: Predicting Individuals' Thermal Preference Using Occupant Heating and Cooling Behavior and Machine Learning. *Build. Environ.* **2018**, *129*, 96–106. [CrossRef]
13. *ISO 7730*; Moderate Thermal Environment, Determination of PMV and PPD Indices and Specification of the Condition for Thermal Comfort. International Organization for Standardization: Geneva, Switzerland, 1994.
14. Kumar, P.; Sharma, A. Study on Importance, Procedure, and Scope of Outdoor Thermal Comfort—A Review. *Sustain. Cities Soc.* **2020**, *61*, 102297. [CrossRef]
15. Kántor, N.; Égerházi, L.; Unger, J. Subjective Estimation of Thermal Environment in Recreational Urban Spaces—Part 1: Investigations in Szeged, Hungary. *Int. J. Biometeorol.* **2012**, *56*, 1075–1088. [CrossRef] [PubMed]
16. Mirzabeigi, S.; Khalili Nasr, B.; Mainini, A.G.; Blanco Cadena, J.D.; Lobaccaro, G. Tailored WBGT as a Heat Stress Index to Assess the Direct Solar Radiation Effect on Indoor Thermal Comfort. *Energy Build.* **2021**, *242*, 110974. [CrossRef]

17. Kumar, P.; Sharma, A. Assessing Outdoor Thermal Comfort Conditions at an Urban Park during Summer in the Hot Semi-Arid Region of India. *Mater. Today Proc.* **2022**, *61*, 356–369. [CrossRef]
18. Bröde, P.; Krüger, E.L.; Rossi, F.A.; Fiala, D. Predicting Urban Outdoor Thermal Comfort by the Universal Thermal Climate Index UTCI—A Case Study in Southern Brazil. *Int. J. Biometeorol.* **2012**, *56*, 471–480. [CrossRef]
19. Mahdavinejad, M.; Shaeri, J.; Nezami, A.; Goharian, A. Comparing Universal Thermal Climate Index (UTCI) with Selected Thermal Indices to Evaluate Outdoor Thermal Comfort in Traditional Courtyards with BWh Climate. *Urban Clim.* **2024**, *54*, 101839. [CrossRef]
20. Sharmin, T.; Steemers, K.; Humphreys, M. Outdoor Thermal Comfort and Summer PET Range: A Field Study in Tropical City Dhaka. *Energy Build.* **2019**, *198*, 149–159. [CrossRef]
21. Ugursal, A.; Culp, C. Gender Differences of Thermal Comfort Perception UnderTransient Environmental and Metabolic Conditions—ProQuest. *ASHRAE Trans.* **2013**, *119*, 52–62.
22. Xiong, J.; Lian, Z.; Zhou, X.; You, J.; Lin, Y. Effects of Temperature Steps on Human Health and Thermal Comfort. *Build. Environ.* **2015**, *94*, 144–154. [CrossRef]
23. Nkurikiyeyezu, K.N.; Suzuki, Y.; Lopez, G.F. Heart Rate Variability as a Predictive Biomarker of Thermal Comfort. *J. Ambient. Intell. Hum. Comput.* **2018**, *9*, 1465–1477. [CrossRef]
24. Xiong, J.; Lian, Z.; Zhang, H. Physiological Response to Typical Temperature Step-Changes in Winter of China. *Energy Build.* **2017**, *138*, 687–694. [CrossRef]
25. Nikolopoulou, M.; Baker, N.; Steemers, K. Thermal Comfort in Outdoor Urban Spaces: Understanding the Human Parameter. *Sol. Energy* **2001**, *70*, 227–235. [CrossRef]
26. Lindner-Cendrowska, K.; Błażejczyk, K. Impact of Selected Personal Factors on Seasonal Variability of Recreationist Weather Perceptions and Preferences in Warsaw (Poland). *Int. J. Biometeorol.* **2018**, *62*, 113–125. [CrossRef]
27. Villadiego, K.; Velay-Dabat, M.A. Outdoor Thermal Comfort in a Hot and Humid Climate of Colombia: A Field Study in Barranquilla. *Build. Environ.* **2014**, *75*, 142–152. [CrossRef]
28. Ge, J.; Wu, J.; Chen, S.; Wu, J. Energy Efficiency Optimization Strategies for University Research Buildings with Hot Summer and Cold Winter Climate of China Based on the Adaptive Thermal Comfort. *J. Build. Eng.* **2018**, *18*, 321–330. [CrossRef]
29. Aghniaey, S.; Lawrence, T.M.; Sharpton, T.N.; Douglass, S.P.; Oliver, T.; Sutter, M. Thermal Comfort Evaluation in Campus Classrooms during Room Temperature Adjustment Corresponding to Demand Response. *Build. Environ.* **2019**, *148*, 488–497. [CrossRef]
30. Arens, E.; Humphreys, M.A.; de Dear, R.; Zhang, H. Are 'Class A' Temperature Requirements Realistic or Desirable? *Build. Environ.* **2010**, *45*, 4–10. [CrossRef]
31. Hoyt, T.; Lee, K.H.; Zhang, H.; Arens, E.; Webster, T. *Energy Savings from Extended Air Temperature Setpoints and Reductions in Room Air Mixing*; UC Berkeley: Berkeley, CA, USA, 2005.
32. Humphreys, M.A.; Hancock, M. Do People like to Feel 'Neutral'?: Exploring the Variation of the Desired Thermal Sensation on the ASHRAE Scale. *Energy Build.* **2007**, *39*, 867–874. [CrossRef]
33. Fox, R.H.; Woodward, P.M.; Exton-Smith, A.N.; Green, M.F.; Donnison, D.V.; Wicks, M.H. Body Temperatures in the Elderly: A National Study of Physiological, Social and Environmental Conditions. *Br. Med. J.* **1973**, *1*, 200–206. [CrossRef]
34. Yuan, B.; Zhou, L.; Hu, F.; Wei, C. Effects of 2D/3D Urban Morphology on Land Surface Temperature: Contribution, Response, and Interaction. *Urban Clim.* **2024**, *53*, 101791. [CrossRef]
35. Liu, Y.; Peng, J.; Wang, Y. Efficiency of Landscape Metrics Characterizing Urban Land Surface Temperature. *Landsc. Urban Plan.* **2018**, *180*, 36–53. [CrossRef]
36. Peng, J.; Jia, J.; Liu, Y.; Li, H.; Wu, J. Seasonal Contrast of the Dominant Factors for Spatial Distribution of Land Surface Temperature in Urban Areas. *Remote Sens. Environ.* **2018**, *215*, 255–267. [CrossRef]
37. Li, J.; Wang, F.; Fu, Y.; Guo, B.; Zhao, Y.; Yu, H. A Novel SUHI Referenced Estimation Method for Multicenters Urban Agglomeration Using DMSP/OLS Nighttime Light Data. *IEEE J. Sel. Top. Appl. Earth Obs. Remote Sens.* **2020**, *13*, 1416–1425. [CrossRef]
38. Liu, D.; Chen, N.; Zhang, X.; Wang, C.; Du, W. Annual Large-Scale Urban Land Mapping Based on Landsat Time Series in Google Earth Engine and OpenStreetMap Data: A Case Study in the Middle Yangtze River Basin. *ISPRS J. Photogramm. Remote Sens.* **2020**, *159*, 337–351. [CrossRef]
39. Geletič, J.; Lehnert, M.; Savić, S.; Milošević, D. Inter-/Intra-Zonal Seasonal Variability of the Surface Urban Heat Island Based on Local Climate Zones in Three Central European Cities. *Build. Environ.* **2019**, *156*, 21–32. [CrossRef]
40. Stewart, I.D.; Oke, T.R. Local Climate Zones for Urban Temperature Studies. *Bull. Am. Meteorol. Soc.* **2012**, *93*, 1879–1900. [CrossRef]
41. Oke, T.R. Towards Better Scientific Communication in Urban Climate. *Theor. Appl. Climatol.* **2006**, *84*, 179–190. [CrossRef]
42. Chen, S.; He, P.; Yu, B.; Wei, D.; Chen, Y. The Challenge of Noise Pollution in High-Density Urban Areas: Relationship between 2D/3D Urban Morphology and Noise Perception. *Build. Environ.* **2024**, *253*, 111313. [CrossRef]

43. Guo, F.; Schlink, U.; Wu, W.; Hu, D.; Sun, J. Scale-Dependent and Season-Dependent Impacts of 2D/3D Building Morphology on Land Surface Temperature. *Sustain. Cities Soc.* **2023**, *97*, 104788. [CrossRef]
44. Fotheringham, A.S.; Wong, D.W.S. The Modifiable Areal Unit Problem in Multivariate Statistical Analysis. *Environ. Plan A* **1991**, *23*, 1025–1044. [CrossRef]
45. Han, W. Analyzing the Scale Dependent Effect of Urban Building Morphology on Land Surface Temperature Using Random Forest Algorithm. *Sci. Rep.* **2023**, *13*, 19312. [CrossRef]
46. Weng, Q.; Lu, D.; Schubring, J. Estimation of Land Surface Temperature–Vegetation Abundance Relationship for Urban Heat Island Studies. *Remote Sens. Environ.* **2004**, *89*, 467–483. [CrossRef]
47. Li, X.; Zhou, W.; Ouyang, Z. Relationship between Land Surface Temperature and Spatial Pattern of Greenspace: What Are the Effects of Spatial Resolution? *Landsc. Urban Plan.* **2013**, *114*, 1–8. [CrossRef]
48. Niu, H.; Silva, E.A. Crowdsourced Data Mining for Urban Activity: Review of Data Sources, Applications, and Methods. *J. Urban Plan. Dev.* **2020**, *146*, 04020007. [CrossRef]
49. Chen, Y.; Niu, H.; Silva, E.A. The Road to Recovery: Sensing Public Opinion towards Reopening Measures with Social Media Data in Post-Lockdown Cities. *Cities* **2023**, *132*, 104054. [CrossRef]
50. Wang, Z.; Zhou, R.; Yu, Y. The Impact of Urban Morphology on Land Surface Temperature under Seasonal and Diurnal Variations: Marginal and Interaction Effects. *Build. Environ.* **2025**, *272*, 112673. [CrossRef]
51. Yelixiati, H.; Tong, L.; Luo, S.; Chen, Z. Spatiotemporal Heterogeneity of the Relationship between Urban Morphology and Land Surface Temperature at a Block Scale. *Sustain. Cities Soc.* **2024**, *113*, 105711. [CrossRef]
52. Wen, D.; Wang, L.; Cao, Q.; Hong, M.; Wang, H.; Bian, G. A Comparative Study of the Effects of Urban Morphology on Land Surface Temperature in Chengdu and Chongqing, China. *Sci. Rep.* **2024**, *14*, 25130. [CrossRef]
53. Zhou, L.; Hu, F.; Wang, B.; Wei, C.; Sun, D.; Wang, S. Relationship between Urban Landscape Structure and Land Surface Temperature: Spatial Hierarchy and Interaction Effects. *Sustain. Cities Soc.* **2022**, *80*, 103795. [CrossRef]
54. Yin, C.; Yuan, M.; Lu, Y.; Huang, Y.; Liu, Y. Effects of Urban Form on the Urban Heat Island Effect Based on Spatial Regression Model. *Sci. Total Environ.* **2018**, *634*, 696–704. [CrossRef] [PubMed]
55. Hu, Y.; Dai, Z.; Guldmann, J.-M. Modeling the Impact of 2D/3D Urban Indicators on the Urban Heat Island over Different Seasons: A Boosted Regression Tree Approach. *J. Environ. Manag.* **2020**, *266*, 110424. [CrossRef] [PubMed]
56. Wu, Y.; Che, Y.; Liao, W.; Liu, X. The Impact of Urban Morphology on Land Surface Temperature across Urban-Rural Gradients in the Pearl River Delta, China. *Build. Environ.* **2025**, *267*, 112215. [CrossRef]
57. Fanger, P.O. *Thermal Comfort*; Danish Technical Press: Copenhagen, Denmark, 1970.
58. Fleiss, J.L.; Cohen, J. The Equivalence of Weighted Kappa and the Intraclass Correlation Coefficient as Measures of Reliability. *Educ. Psychol. Meas.* **1973**, *33*, 613–619. [CrossRef]
59. Morabito, M.; Crisci, A.; Messeri, A.; Orlandini, S.; Raschi, A.; Maracchi, G.; Munafò, M. The Impact of Built-up Surfaces on Land Surface Temperatures in Italian Urban Areas. *Sci. Total Environ.* **2016**, *551–552*, 317–326. [CrossRef]
60. Deilami, K.; Kamruzzaman, M.; Liu, Y. Urban Heat Island Effect: A Systematic Review of Spatio-Temporal Factors, Data, Methods, and Mitigation Measures. *Int. J. Appl. Earth Obs. Geoinf.* **2018**, *67*, 30–42. [CrossRef]
61. Gao, Y.; Zhao, J.; Han, L. Exploring the Spatial Heterogeneity of Urban Heat Island Effect and Its Relationship to Block Morphology with the Geographically Weighted Regression Model. *Sustain. Cities Soc.* **2022**, *76*, 103431. [CrossRef]
62. Li, L.; Zha, Y.; Zhang, J. Spatially Non-Stationary Effect of Underlying Driving Factors on Surface Urban Heat Islands in Global Major Cities. *Int. J. Appl. Earth Obs. Geoinf.* **2020**, *90*, 102131. [CrossRef]
63. Yu, H.; Peng, Z.-R. Exploring the Spatial Variation of Ridesourcing Demand and Its Relationship to Built Environment and Socioeconomic Factors with the Geographically Weighted Poisson Regression. *J. Transp. Geogr.* **2019**, *75*, 147–163. [CrossRef]
64. Anselin, L. Local Indicators of Spatial Association—LISA. *Geogr. Anal.* **1995**, *27*, 93–115. [CrossRef]
65. Wheeler, D.; Tiefelsdorf, M. Multicollinearity and Correlation among Local Regression Coefficients in Geographically Weighted Regression. *J. Geogr. Syst.* **2005**, *7*, 161–187. [CrossRef]
66. Nugroho, N.Y.; Triyadi, S.; Wonorahardjo, S. Effect of High-Rise Buildings on the Surrounding Thermal Environment. *Build. Environ.* **2022**, *207*, 108393. [CrossRef]
67. He, X.; Miao, S.; Shen, S.; Li, J.; Zhang, B.; Zhang, Z.; Chen, X. Influence of Sky View Factor on Outdoor Thermal Environment and Physiological Equivalent Temperature. *Int. J. Biometeorol.* **2015**, *59*, 285–297. [CrossRef] [PubMed]
68. Xiang, Y.; Huang, C.; Huang, X.; Zhou, Z.; Wang, X. Seasonal Variations of the Dominant Factors for Spatial Heterogeneity and Time Inconsistency of Land Surface Temperature in an Urban Agglomeration of Central China. *Sustain. Cities Soc.* **2021**, *75*, 103285. [CrossRef]
69. Han, D.; An, H.; Wang, F.; Xu, X.; Qiao, Z.; Wang, M.; Sui, X.; Liang, S.; Hou, X.; Cai, H.; et al. Understanding Seasonal Contributions of Urban Morphology to Thermal Environment Based on Boosted Regression Tree Approach. *Build. Environ.* **2022**, *226*, 109770. [CrossRef]

70. Scarano, M.; Sobrino, J.A. On the Relationship between the Sky View Factor and the Land Surface Temperature Derived by Landsat-8 Images in Bari, Italy. *Int. J. Remote Sens.* **2015**, *36*, 4820–4835. [CrossRef]
71. Wang, J.; Meng, F.; Lu, H.; Lv, Y.; Jing, T. Individual and Combined Effects of 3D Buildings and Green Spaces on the Urban Thermal Environment: A Case Study in Jinan, China. *Atmosphere* **2023**, *14*, 908. [CrossRef]
72. Li, J.; Song, C.; Cao, L.; Zhu, F.; Meng, X.; Wu, J. Impacts of Landscape Structure on Surface Urban Heat Islands: A Case Study of Shanghai, China. *Remote Sens. Environ.* **2011**, *115*, 3249–3263. [CrossRef]
73. Sun, D.; Kafatos, M. Note on the NDVI-LST Relationship and the Use of Temperature-Related Drought Indices over North America. *Geophys. Res. Lett.* **2007**, *34*. [CrossRef]
74. Wu, Z.; Tong, Z.; Wang, M.; Long, Q. Assessing the Impact of Urban Morphological Parameters on Land Surface Temperature in the Heat Aggregation Areas with Spatial Heterogeneity: A Case Study of Nanjing. *Build. Environ.* **2023**, *235*, 110232. [CrossRef]
75. Nasar-u-Minallah, M.; Haase, D.; Qureshi, S. Evaluating the Impact of Landscape Configuration, Patterns and Composition on Land Surface Temperature: An Urban Heat Island Study in the Megacity Lahore, Pakistan. *Environ. Monit. Assess.* **2024**, *196*, 627. [CrossRef] [PubMed]
76. Liu, H.; Weng, Q. Seasonal Variations in the Relationship between Landscape Pattern and Land Surface Temperature in Indianapolis, USA. *Environ. Monit. Assess.* **2008**, *144*, 199–219. [CrossRef] [PubMed]
77. Li, H.; Li, Y.; Wang, T.; Wang, Z.; Gao, M.; Shen, H. Quantifying 3D Building Form Effects on Urban Land Surface Temperature and Modeling Seasonal Correlation Patterns. *Build. Environ.* **2021**, *204*, 108132. [CrossRef]
78. Coseo, P.; Larsen, L. How Factors of Land Use/Land Cover, Building Configuration, and Adjacent Heat Sources and Sinks Explain Urban Heat Islands in Chicago. *Landsc. Urban Plan.* **2014**, *125*, 117–129. [CrossRef]
79. Yan, J.; Yin, C.; An, Z.; Mu, B.; Wen, Q.; Li, Y.; Zhang, Y.; Chen, W.; Wang, L.; Song, Y. The Influence of Urban Form on Land Surface Temperature: A Comprehensive Investigation from 2D Urban Land Use and 3D Buildings. *Land* **2023**, *12*, 1802. [CrossRef]
80. Cao, J.; Zhou, W.; Zheng, Z.; Ren, T.; Wang, W. Within-City Spatial and Temporal Heterogeneity of Air Temperature and Its Relationship with Land Surface Temperature. *Landsc. Urban Plan.* **2021**, *206*, 103979. [CrossRef]
81. Goldblatt, R.; Addas, A.; Crull, D.; Maghrabi, A.; Levin, G.G.; Rubinyi, S. Remotely Sensed Derived Land Surface Temperature (LST) as a Proxy for Air Temperature and Thermal Comfort at a Small Geographical Scale. *Land* **2021**, *10*, 410. [CrossRef]

land

MDPI

Systematic Review

Nature-Based Solutions in Urban Green Infrastructure: A Systematic Review of Success Factors and Implementation Challenges

Melika Zarei [1] and Sina Shahab [2,*]

[1] Department of Urban Planning and Design, School of Art and Architecture, Shiraz University, Shiraz 71877335, Iran; melikazarei76@hafez.shirazu.ac.ir
[2] School of Geography and Planning, Cardiff University, Cardiff CF10 3WT, UK
* Correspondence: shahabs@cardiff.ac.uk

Abstract: Nature-Based Solutions (NBSs) have gained prominence in urban planning as integrative strategies that utilize natural processes to address complex environmental and societal challenges while advancing green infrastructure development. Despite growing academic interest, the practical integration of NBSs into urban green infrastructure remains hindered by fragmented methodologies and limited understanding of context-specific implementation dynamics. This study contributes to addressing these gaps through a systematic review and bibliometric analysis of 90 peer-reviewed articles published between 2014 and 2024. It examines the range of NBSs employed in urban green infrastructure, the factors shaping their successful implementation, and the barriers, financial, technical, social, and political, that constrain their adoption. The analysis also explores the roles of key stakeholders, including local governments, private actors, and communities, in the planning, execution, and maintenance of NBS projects. The findings reveal both conceptual convergence and contextual variation in how NBSs are deployed and evaluated, highlighting critical enablers, such as spatial justice, governance integration, financial viability, and technical capacity. By clarifying the conditions under which NBSs function effectively, this review offers insights for researchers and policymakers seeking to embed these approaches within sustainable urban development frameworks.

Keywords: nature-based solutions; urban green infrastructure; environmental sustainability; implementation challenges; systematic review

1. Introduction

Urbanization and extensive land consumption have significantly transformed cityscapes, disrupting the balance between built environments and natural ecosystems, and increasing the vulnerability of ecological and life-support systems [1–4]. These systems provide essential services to urban areas, attracting the attention of planners and policymakers seeking to address the growing pressures of urban growth [5]. In response, scholars have revisited conventional definitions of infrastructure, highlighting the structural and functional limitations of post-industrial systems and advocating for frameworks more attuned to contemporary urban complexities [6–8]. Among the most influential concepts to emerge from this rethinking is green infrastructure, introduced by Benedict and McMahon [9]. This concept underscores the critical distinction between natural and gray infrastructure (e.g., roads, water, and electricity systems), highlighting their unique characteristics and the differences in their respective planning and development approaches. Green infrastructure refers to the spatial abundance and distribution of natural elements within the landscape, such as forests, wetlands, and other ecological features [10]. Just as gray infrastructure is essential for modern societies; green infrastructure also provides ecosystem services that are equally vital for the health of living organisms. The loss of these natural systems incurs significant societal and environmental costs [3,11,12].

Although cities continue to expand their gray infrastructure—through highways, water systems, and other urban services—they are simultaneously confronted by mounting environmental challenges, including air and water pollution, stormwater runoff, biodiversity loss, rapid population growth, and the urban heat island effect [13,14]. These pressures underscore one of the central challenges of twenty-first-century urban governance: addressing environmental degradation, inefficient resource use, and fragmented green infrastructure networks [15]. Within this context, green infrastructure emerges as a strategic means of integrating natural and built environments, redefining urban development trajectories [16]. At the urban level, green infrastructure serves as a network of strategic interventions, offering a wide range of ecosystem services to urban residents and holding significant environmental importance [17,18].

Nature-Based Solutions (NBSs) provide a framework for planning urban green infrastructure effectively. NBSs refer to interventions inspired and supported by nature that address societal challenges while simultaneously delivering environmental, social, and economic benefits [19,20]. When integrated into urban green infrastructure, they help manage material and energy flux dynamics, enhance biodiversity, and improve ecosystem services within urban spatial patterns [21]. By promoting multifunctional landscapes, NBSs contribute to climate resilience, flood mitigation, and the reduction in the urban heat island effect [22]. They also support ecological connectivity, enabling the movement of species across fragmented urban environments. Additionally, NBSs enhance social well-being by creating accessible green spaces, improving air and water quality, and fostering community engagement in environmental stewardship [23]. These combined benefits make NBSs crucial tools for achieving sustainable and resilient urban development.

Existing research highlights significant conceptual and semantic variations in how theoretical and spatial models define and interpret NBSs in urban green infrastructure projects [24,25]. Despite significant progress in the study of NBS, critical research gaps persist, hindering a comprehensive understanding and broader application of these strategies. One major gap is the absence of a comprehensive and standardized research framework. A unified framework that integrates key dimensions, such as geographic focus, research topics, methodologies, and driving forces, has yet to be developed. Without such a framework, systematic comparison and synthesis of findings across studies remain challenging, underscoring the need for a standardized approach to advance the field cohesively.

Another critical gap lies in the underexplored economic, social, and technical dimensions of NBS. Existing literature primarily focuses on governance and environmental aspects [26], while the economic, social, and technical factors remain relatively neglected despite their importance in understanding the broader impacts of NBS implementation. Holistic analyses encompassing these multifaceted dimensions are crucial for informing effective solutions and best practices. Although the environmental benefits and implementation barriers of NBSs are well-documented [16,27], there is a notable lack of rigorous, quantitative evaluations of their contributions to these global challenges. Addressing this deficiency requires robust empirical studies that quantify the effectiveness of NBSs in mitigating climate change and enhancing urban resilience. Furthermore, an overreliance on case studies (e.g., [26,28–30]) has constrained theoretical advancements in NBS research. While localized comparative analyses have provided valuable insights into practical implementation, this approach has limited the development of spatially explicit models for NBS design and scaling. Expanding research to incorporate broader theoretical frameworks and modeling approaches is essential to enhance the generalizability of findings and advance the field beyond context-specific case studies.

This study contributes to addressing these critical research gaps by systematically reviewing how NBSs have been used to develop urban green infrastructure over the past decade. Along with a systematic investigation of prevailing research trends, thematic developments, and prospective trajectories for future exploration, this paper explores three interrelated questions: (1) What are the main NBSs employed in urban green infrastructure projects? (2) What factors shape their successful implementation across diverse urban contexts? And (3) what barriers, e.g., financial, technical, social, or political, limit their uptake and effectiveness? To answer these questions, this study applies a dual-method approach combining bibliometric mapping with in-depth qualitative content analysis. This enables both a high-level view of intellectual and geographic trends in the field and a grounded examination of the thematic priorities, stakeholder roles, and implementation dynamics reported in empirical studies. In doing so, this paper identifies not only recurring patterns in how NBSs are conceptualized and applied but also critical contextual factors—such as governance arrangements, spatial justice considerations, and knowledge capacity—that influence their performance. The findings contribute to ongoing debates on how NBSs can be more systematically integrated into urban planning and climate resilience strategies, offering insights for researchers, practitioners, and policymakers seeking to operationalize these solutions in varied urban environments.

2. Methodology

This study employs a hybrid methodological approach to achieve a comprehensive understanding of the key themes explored in NBSs within urban green infrastructure projects. By combining a bibliometric analysis using VOSviewer 1.6.20 [31] and a content analysis utilizing JetBrains WebStorm 2023.3 [32], this integrated approach leverages the strengths of both methods to interpret explicit and latent content within the literature. The bibliometric analysis plays a crucial role in evaluating scholarly outputs across diverse domains, including articles, authors, keywords, journals, institutions, and contributing countries [33]. It enables the systematic examination of a research domain's intellectual, social, and conceptual structures over time, analyzing relationships and interactions among these components [34]. Consequently, the content analysis techniques provide a systematic and structured approach for summarizing, evaluating, documenting, synthesizing, and interpreting scientific findings [35]. It adheres to structured and goal-oriented procedures, guided by predefined and established criteria for data selection, summarization, analysis, and conclusion drawing. Content analysis facilitates the identification of gaps between research and practice while offering a documented synthesis of existing knowledge [36].

The data search for this study began with the selection of the Scopus database, a comprehensive repository of scientific literature managed by Elsevier [37]. The primary methodological steps for analyzing research on NBSs in urban green infrastructure projects were systematically organized into four stages: (1) data collection; (2) document evaluation; (3) data extraction, organization, and coding; and (4) data reporting [33]. The structured sequence of these stages aims to minimize bias, enhance validity, and ensure study reproducibility. In the first stage (data collection), a systematic search was conducted within the Scopus database to retrieve peer-reviewed articles published in the last 10 years, i.e., between 2014 and 2024. To search articles focused on NBSs in urban green infrastructure, the following search query was applied: TITLE-ABS-KEY-AUTH (("Nature-Based Solutions") AND ("Green Infrastructure") AND (City) OR (Cities)) AND PUBYEAR > 2013 AND PUBYEAR < 2025 AND (LIMIT-

TO (DOCTYPE, "ar")) AND (LIMIT-TO (SRCTYPE, "j")) AND (LIMIT-TO (LANGUAGE, "English")). The search was conducted on 31 December 2024.

Following data collection, the second stage (document evaluation) was carried out in five steps (Figure 1). The initial dataset comprised 324 articles extracted from the Scopus database. After removing duplicate entries, the dataset was refined to 318 articles. Titles were then reviewed for relevance and clarity, reducing the dataset to 294 articles. Further evaluation of abstracts narrowed the dataset to 131 articles. To ensure the completeness and comprehensiveness of the dataset, manual searches were performed by examining the reference lists of selected studies. Ultimately, 90 studies were deemed relevant to the research objectives and included in subsequent analyses.

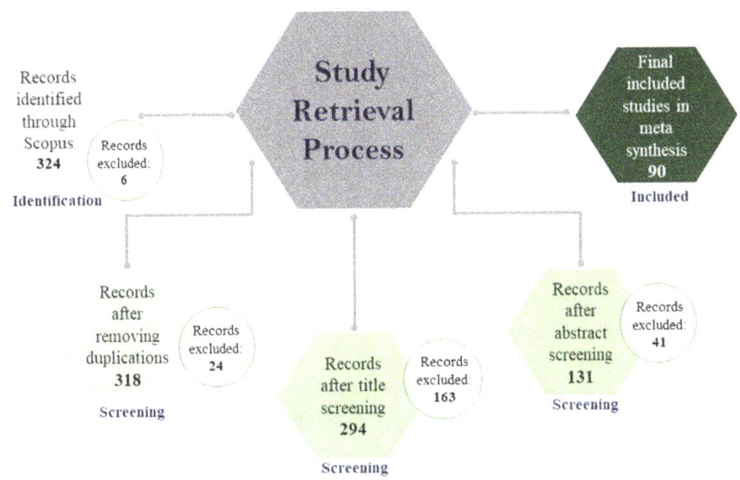

Figure 1. The five-step search and screening process used in the systematic review.

In the third stage (data extraction, organization, and coding), following the identification of relevant research studies, open-coding techniques were applied to textual data (abstracts, discussions, and conclusions) using semantic units (phrases, sentences, or paragraphs) [38]. The coding process continued iteratively until thematic saturation was achieved. Key themes were identified, and their semantic similarities and differences were analyzed [39]. Axial coding was subsequently used to examine overlaps and relationships among the emerging themes and concepts, leading to the development of categories and their associated sub-themes. In the fourth stage (reporting findings), the extracted findings were synthesized into a cohesive analytical framework comprising concepts, indicators, and thematic components for investigating NBSs in urban green infrastructure projects. This structured and multi-stage methodological approach aims to reduce bias, enhance validity, and ensure the reproducibility of the study's findings.

3. Results and Discussions

3.1. Geographical Distribution of Research on NBSs in Urban Green Infrastructure

To understand which countries and regions contribute to research on NBSs in green infrastructure, we conducted a citation analysis at the country level. As shown in Table 1, the United Kingdom and other European countries are the primary contributors to this field, with 49 academic studies on NBSs in urban green infrastructure. The EU plays a leading role in global environmental policy by supporting NBSs through initiatives such as Horizon Europe and the European Green Deal.

Table 1. Ten leading countries in NBS research for urban green infrastructure based on citation metrics.

	Country	No. of Publications	No. of Citations	Average No. of Citation per Paper	Total Link Strength
1	United Kingdom	12	1013	84.4	68
2	The Netherlands	12	994	82.8	64
3	Germany	11	957	87	57
4	Italy	13	916	70.4	73
5	Sweden	8	855	106.8	53
6	Romania	3	726	242	40
7	Belgium	3	508	169.3	24
8	United States	15	450	30	16
9	Denmark	6	269	44.8	19
10	Australia	11	230	20.9	42

East and Southeast Asia follow, with 22 studies examining the ecological and socioeconomic impacts of projects like Singapore's Gardens by the Bay, Beijing's Green Belt, and Malaysia's River of Life. In North America, 12 studies focus on urban resilience initiatives, including New York's High Line and Toronto's wetland system. South America is underrepresented, with only two studies on São Paulo and San José, while Africa has five, including Accra's Green Accra Initiative. Oceania has four studies, highlighting Melbourne's Urban Forest Strategy and New Zealand's native forest restoration, which integrate technology and indigenous knowledge.

Globally, 11 comparative studies explore the adaptability and scalability of NBSs across different contexts. While NBS research and implementation are concentrated in Europe and East Asia, Africa and South America remain underrepresented. Addressing these disparities through collaboration, capacity-building, and knowledge transfer is essential to maximizing the global impact of NBSs on sustainable urban development and climate resilience (see Figure 2, created using ESRI ArcGIS Pro 3.4.2).

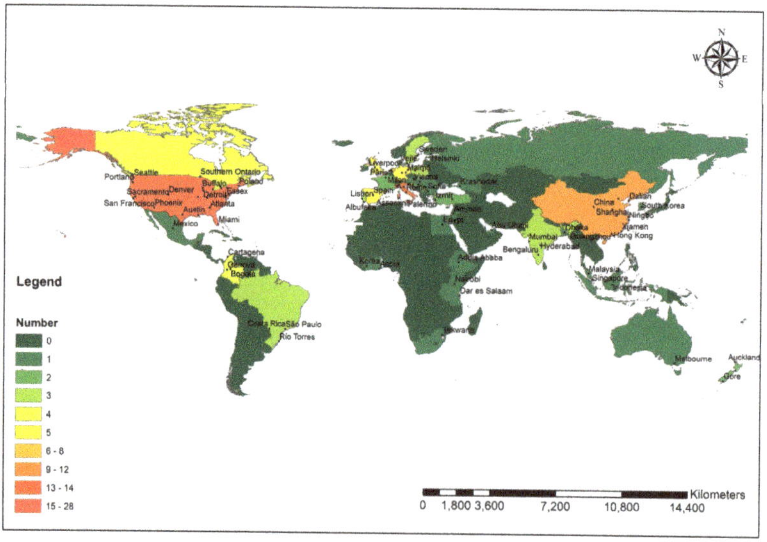

Figure 2. Geographical distribution of research on NBSs in urban green infrastructure.

3.2. How Do Various Analytical Approaches and Methodologies Contribute to the Assessment of NBSs in Urban Green Infrastructure?

An assessment of the selected articles revealed that 53% (48 studies) predominantly employed qualitative methodologies. These studies were primarily concerned with the theoretical elaboration and development of indicators pertinent to NBSs in urban green infrastructure. In contrast, 31% (28 studies) adopted mixed-methods approaches, while only 16% (14 studies) utilized quantitative methods. The notable underrepresentation of quantitative approaches highlights a research gap, particularly in light of their potential to generate measurable and scalable insights that could advance NBS objectives.

The application of analytical methods varied across the studies, depending on the nature of the data and the research design, whether qualitative, quantitative, or mixed methods. A categorization of the analytical techniques employed in the reviewed studies was undertaken to inform future research efforts. Content analysis emerged as the predominant method among qualitative studies, supplemented by methods such as observational and situational analysis, although these were employed less frequently. Quantitative studies, by contrast, primarily relied on descriptive statistical models, reflecting the survey-driven nature of many of these research designs. However, this often limited the analytical capacity of these studies, underscoring the need for more sophisticated statistical frameworks and methodologies to effectively address the multifaceted nature of NBS challenges. Table 2 summarizes the primary methodologies and analysis methods utilized in the reviewed papers.

Table 2. Pioneering methods and their uses in NBS research.

Methodology	Analysis Method		References
	Multi-Criteria Decision Analysis	Spatial Multi-Criteria Analysis	[40,41]
		I-Tree Hydro Plus Model	[42]
	Hydrological Modeling	HEC-HMS Model	[43]
		HEC-RAS Model	[43]
	Descriptive Statistics	Tables and Charts	[23,44,45]
		Analysis of Variance	[45,46]
		Analysis of Covariance	[46]
		Mann–Whitney U Test	[46]
Quantitative		Pearson Correlation Coefficient	[47]
		Multiple Linear Regressions	[48]
	Forecast Model	Adoption and Diffusion Outcome Prediction Tool	[49]
		Ordinary Least Squares Regression	[47]
	Documentary Method	Heat Flow Meter (HFM) Method	[50]
	Economic Analysis Methods	Cost-Effectiveness Analysis	[51]
		Adoption and Diffusion Outcome Prediction Tool	[49]
	Assessment Method	SCS-CN Method	[47,52]
		Pollution Flux Method	[47]
	Situational Analysis	Case Study Method	[53–56]
		Geographic Comparative Analysis	[57–61]
	Critical Analysis	Literature Review	[16,26–30,55,62–65]
		Critically Analyzing	[66]
		Semi-Structured Interviews	[3,67–73]
Qualitative		Depth Interviews	[74]
	Conversation Analysis	Structured Face to Face Interviews	[19,75]
		Interviews	[25,29,55,63]
		Survey	[25,63]
	Content Analysis	Transcribed and Coded	[24,29,70,75–85]
	Observational Analysis	Mapping Analysis	[86–90]
Mixed Method	Mixed Technique	Mixed Quantitative and Qualitative Methods	[1,2,10,18,21,91–113]

3.3. *What Are the Key NBSs That Have Been Used to Develop Urban Green Infrastructure?*

Concerning the key NBSs that have been used to develop urban green infrastructure, we identified five key themes and thirty sub-themes. The key themes include the creation and expansion of urban green ecosystems, storm water management, the protection, sustainable management, and restoration of natural ecosystems; planning and designing cities in harmony with nature; and the development of urban agriculture. In Figure 3, these themes are systematically structured, with recurring codes consistently identified across the analyzed literature.

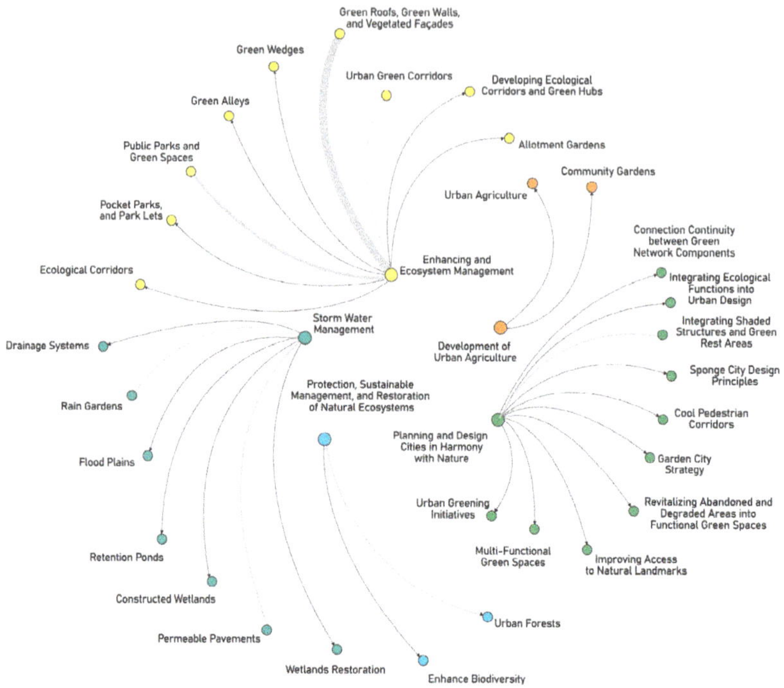

Figure 3. Main themes of NBSs in developing urban green infrastructure.

The first main theme, i.e., the creation and expansion of urban green ecosystems, emerged in its initial form through the development of public parks and green spaces in the 18th and 19th centuries [3,18,45,47,52,53,63,77,86,90,92]. The expanding urban population drove the demand for escapes from the pollution and unhealthy conditions of industrial cities. Public parks came to be viewed as essential 'lungs' of the metropolis, while the growing awareness of the need for accessible urban greenery contributed to the emergence of ecological corridors and green hubs [79]. Frederick Law Olmsted, in collaboration with Calvert Vaux, designed and executed Central Park in New York, the first modern park in the United States. Such initiatives evolved into the design of urban green corridors [50,66,75,77,92,93,100], ecological corridors [2,103], green alleys [96], green wedges [54], allotment gardens [114], pocket parks, and park lets [57]. Additionally, green roofs, green walls, and vegetated façades [18,21,27,29,30,43,44,48–50,52,56,57,63,77,91,96,98,100,101,106,113] have emerged as modern solutions to urban challenges, offering various benefits, such as reducing heating and cooling loads, improving air quality, managing wastewater, mitigating noise pollution, and reducing energy consumption.

The second main theme, i.e., storm water management, represents an integrated approach aimed at justifying the environmental impact of urbanization on watersheds. This approach involves utilizing natural systems to facilitate infiltration, evaporation, or storm water reuse at its source. These strategies focus on controlling and reducing runoff at its origin, restoring natural hydrological flows to their pre-disturbance state. Techniques such as rain gardens [18,42,48,50,51,77,96,97], floodplains [100], retention ponds [100], permeable pavements [18,48,51,52,57,62,77,78,80,104], wetland restoration [30], and constructed wetlands [21,30,41,76] are crucial in mitigating urban storm water runoff's quantitative and qualitative adverse effects. Furthermore, these techniques enhance the performance of conventional urban drainage systems [50,63] and, in some cases, can even replace them, enabling more sustainable and environmentally friendly hydrological management.

The third main theme concerns the protection, sustainable management, and restoration of natural ecosystems. According to the International Union for Conservation of Nature (IUCN), NBSs safeguard, sustainably manage, and reestablish natural or modified ecosystems to address societal challenges [115] effectively and adaptively. These solutions significantly benefit human well-being and enhance biodiversity [10,84], mainly by developing urban forests [63,85,94,97,98,100] that contribute to ecosystem restoration and environmental resilience.

The fourth main theme relates to the planning and design of cities in harmony with nature. NBSs in urban green infrastructure provide innovative solutions by integrating shaded structures and green rest areas [43,57,66,95], contributing to cool pedestrian corridors [97] that mitigate urban heat island effects and support broader urban greening initiatives [60]. The scope of such green infrastructure extends across entire cities, and when combined with multifunctional green spaces [68,81], it delivers several benefits by integrating ecological functions into urban design [61]. One of the primary advantages of green infrastructure is the construction of a unified urban landscape, which strengthens the connection continuity between green network components [45,110], ensures physical and spatial integration, and aligns with other urban landscape layers. This integration is achieved through revitalizing abandoned and degraded areas into functional green spaces [48], improving access to natural landmarks [48], implementing garden city strategies [54], and adopting sponge city design principles [1].

The fifth main theme involves the development of urban agriculture, which has emerged as a novel urban green space development model. This approach integrates ecosystem-based food production with cultural, recreational, and economic functions. From an ecological and financial standpoint, urban agriculture [18,63,75] promotes environmentally compatible food production, shortens food supply chains, and enhances food security within cities. Moreover, agricultural parks, community and allotment gardens [24,25,63,77,95], and agroforestry systems represent key components of urban agriculture, where agrarian production coexists with rural landscapes, esthetic values, and cultural identity. This approach is considered a fundamental pillar of NBSs in developing urban green infrastructure. Across all five key themes, these innovations and strategies emerged in response to deteriorating urban conditions, notably pollution and environmental degradation. Historically, the concept of ecological networks in Europe and greenways in the United States first gained traction in the early 20th century, intending to link urban green systems to surrounding natural and forested areas. This period also marked the global emergence of sustainable urban design, emphasizing the necessity of integrating ecology-driven approaches into contemporary urban planning.

3.4. *Which Factors Contribute to the Successful Implementation of NBSs in Developing Urban Green Infrastructure?*

Our content analysis identified six key themes, ten categories, and forty-eight subthemes. The key themes include the perspective of spatial justice in the implementation of NBS projects, enhancing and ecosystem management, integrated governance strategies, financial resources and economic benefits, technical knowledge and expertise, and optimizing stakeholder engagement in NBS implementation. Figure 4 systematically organizes these themes, highlighting recurring codes observed across the analyzed literature.

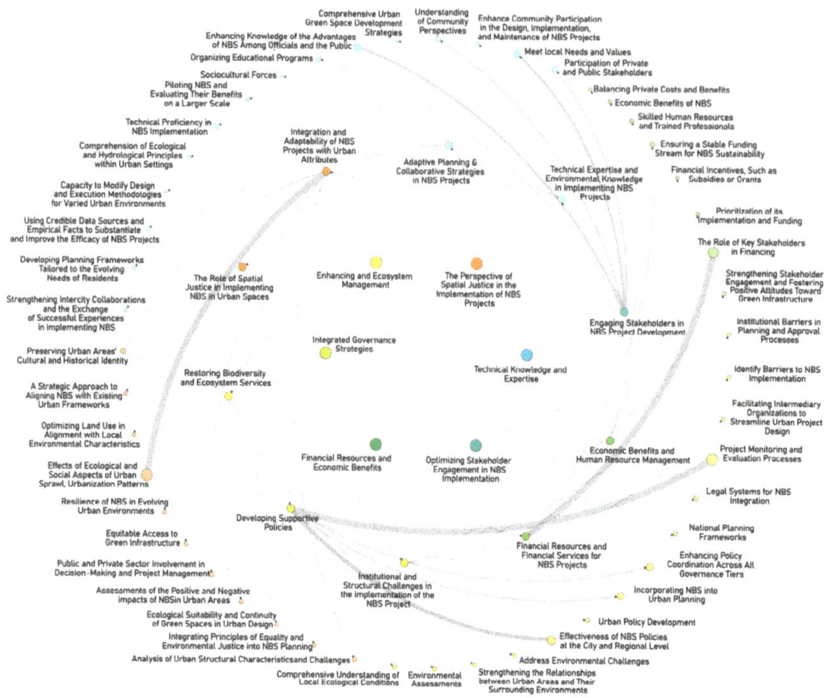

Figure 4. Key factors shaping the successful implementation of NBSs in urban green infrastructure.

3.4.1. The Perspective of Spatial Justice in the Implementation of NBS Projects

- Integration and Adaptability of NBS Projects with Urban Attributes

According to Moorish & Brown [116], green infrastructure should become a foundation for humanity's existence, identity, presence, and collective history. Beyond its profit-driven functions, it can encompass broader cultural, social, and ecological applications. Essentially, rather than being merely profit-oriented, green infrastructure could fulfill objectives such as enriching the sense of place, linking public benefit and interest, and preserving urban areas' cultural and historical identity [1,28,61,105]. In its primary function, green infrastructure is a strategic approach to aligning NBSs with existing urban frameworks [75,110] to protect land. It enables the identification and prioritization of conservation opportunities for future development planning, optimizing land use in alignment with local environmental characteristics [2,10,18,23,24,27,45,51,70,72,74,76,87,91,93,94,96–101,106,107,112,113]. Additionally, the resilience of NBSs in evolving urban environments [103] allows for implementing innovative conservation strategies, which contrasts with the concept of smart growth that emerged in the U.S. to curb uncontrolled urban expansion [117]. This approach affects the ecological and social aspects of urban sprawl, patterns of urbanization [79], land consumption, and fragmentation of open spaces.

- The Role of Spatial Justice in Implementing NBSs in Urban Spaces

Spatial justice is conceptualized through distributive patterns that, by nature, are possibly just or unjust within geographical domains [118]. The incorporation of spatial justice in urban planning requires identifying relevant criteria. Based on humanistic principles, five key criteria for spatial justice are as follows: (1) equality of opportunities, ensuring equitable access to green infrastructure [95,96] for all individuals based on their abilities and merits; (2) participation, emphasizing active citizen engagement, along with public and private sector involvement in decision-making and project management [86,95–97], to ensure spatial justice in NBS initiatives; (3) outcomes, requiring comprehensive assessments of the positive and negative impacts of NBSs in urban areas [24]; (4) Merit, recognizing individuals' capabilities in the social sphere while ensuring the ecological suitability and continuity of green spaces in urban design [112,119,120]; (5) public benefit, integrating principles of equality and environmental justice into NBS planning [87] through the analysis of urban structural characteristics and challenges [50,52,74].

3.4.2. Enhancing and Ecosystem Management

- Restoring Biodiversity and Ecosystem Services

The European Commission [121] defines green infrastructure as a strategically planned network of natural and semi-natural areas designed and managed to provide a wide range of ecosystem services, including water purification, air quality improvement, recreational spaces, pollution reduction, and climate adaptation. Thus, green infrastructure enables multiple ecosystem services through a comprehensive understanding of local ecological conditions [103] and environmental assessments [108]. In this regard, it restores biodiversity, enhances ecosystem services, and fosters ecological relationships between urban areas and their surrounding environments [19,69,122]. This approach also helps address environmental challenges [113], such as air and water pollution, loss of vegetation cover, urban heat effects, and land-use changes to improve urban quality of life.

3.4.3. Integrated Governance Strategies

- Developing Supportive Policies

In regions where land value and economic development drive policymaking, such as China [30,52], engaging politicians and developers in discussions about the added value of green infrastructure is particularly challenging. Increasing green infrastructure visibility through evidence-sharing, regulatory interventions, and policy development is essential to address this. The effectiveness of NBS policies at the city and regional level [16,23,29,45,49,52,64,72,75,78,79,88,95,101,102] varies, as legislators, developers, and planners contribute variably to the accumulating evidence advocating for investments in green infrastructure. Critical considerations for advancing green infrastructure in urban policy development [110] include incorporating NBSs into urban planning [27,42,48,61–63,94,99,104,106], acknowledging green infrastructure's economic, social, and ecological advantages, and enhancing policy coordination across all governance tiers [1,10,21,26,29,30,43,50,91,113]. This entails developing regulatory frameworks, including corporate, public, and environmental partners, and synchronizing policy objectives with enduring urban sustainability aims. Nonetheless, the governance of this process is challenging, especially when national planning frameworks [23], legal systems for NBS integration [24,61,94], and project-monitoring and -evaluation processes [1,2,27,29,30,47–49,52,55,59–61,64,67,70,79,80,83,84,86,91–93,102,110] are inconsistently applied or overly tailored to individual interests or localized agendas.

- Institutional and Structural Challenges in the Implementation of the NBS Project

With the growing focus on localized green infrastructure projects, urban-scale strategies that extend across broader spatial domains have become more common. Regulations in cities such as New York [87] and Philadelphia [110] have sought to institutionalize green infrastructure principles by improving connectivity, environmental access, and sustainability through strategic investment programs. However, achieving consensus at this scale requires long-term efforts, as multiple challenges, including stakeholder participation, land ownership, funding, and competing urban development objectives, must be addressed.

Therefore, facilitating intermediary organizations to streamline urban project design [25], enhance community cooperation, and identify barriers to NBS implementation is crucial [29]. Nevertheless, institutional barriers in planning and approval processes [60], real estate market dynamics, and speculative landholding practices continue to pose challenges. Strengthening stakeholder engagement and fostering positive attitudes toward NBSs in green infrastructure [113] as equivalent to traditional infrastructure is essential. Historical trends, such as Liverpool's prioritization of other urban investments over green spaces, illustrate these difficulties [67]. However, growing recognition of climate adaptation benefits, flood management, and public health are shifting the perspectives of policymakers.

3.4.4. Financial Resources and Economic Benefits

- Financial Resources and Financial Services for NBS Projects

Key stakeholders, including local governments, developers, communities, and businesses, play central roles in funding [1,10,23,27,42,45,48,52,55,60,62,64,71,75,77–79,83,94,99, 103,106,110] green infrastructure. However, their relationships with green infrastructure investments have undergone significant transformation. The increasing awareness of the economic, social, and ecological value of green infrastructure has led to more prioritization of its implementation and funding [97]. Financial incentives, such as subsidies or grants [24], are expected to support urban greening efforts. Although green infrastructure may not yield immediate financial returns comparable to traditional infrastructure, it has the potential for higher long-term economic benefits. Ensuring a stable funding stream for NBS sustainability [2,91,101] can ultimately reduce overall city costs. In the UK, some developers have integrated green infrastructure to enhance the value of residential and commercial properties. Initiatives such as green walls, sustainable drainage systems, and public space greening have been employed to increase property desirability and value [123].

- Economic Benefits and Human Resource Management

Numerous municipal governments and private investors are committing resources to facilitate the development of NBSs in urban green infrastructure. This encompasses renovating structures with green roofs and walls and cultivating public–private partnerships, exemplified by the Atlanta BeltLine in the United States [124]. These partnerships align strategic local government needs with commercial interests, supported by skilled human resources and trained professionals [83,85]. However, private sector entities often express concerns regarding the economic benefits of NBSs [30], balancing private costs and benefits [49], and ensuring transparency in agreements, particularly when land use or access to green infrastructure changes. Transparent policies are essential to ensure that private economic gains do not overshadow green space benefits for the public.

3.4.5. Technical Knowledge and Expertise

- Technical Expertise and Environmental Knowledge in Implementing NBS Projects

Aligning stakeholders' interests, which often focus primarily on outcomes, is not always straightforward. Consequently, many green infrastructure strategies are being developed at city and regional scales. While this expansion provides opportunities for green infrastructure advocates to invest in landscape quality enhancement projects, inconsistencies between policies, strategies, or regulations can lead to conflicts. When formulating plans and policies at the local level, the focus might be on various regions or neighborhoods. This approach facilitates a more detailed analysis of local contexts and the potential value of green infrastructure in these settings. It allows for piloting NBSs and evaluating their benefits on a larger scale [49]. Locally focused programs may, however, overlook broader value propositions, functions, and investment implications of green infrastructure, including ongoing maintenance costs, alignment with other resources, or the effects of investment on urban environmental systems. However, the potential lack of strategic thinking regarding ensuring broader benefits and aligning interventions with city-scale developments remains a fundamental challenge. Effective integration of local policies with broader strategic frameworks necessitates technical proficiency in NBS implementation [27,48,49,55,84], comprehension of ecological and hydrological principles within urban settings [52], and the capacity to modify design and execution methodologies for varied urban environments [112]. Moreover, using credible data sources and empirical facts to substantiate and improve the efficacy of NBS projects [23] may promote sustainable development and the coherence of green infrastructure initiatives.

- Adaptive Planning and Collaborative Strategies in NBS Projects

When residents participate directly in NBS projects, they not only contribute to implementation but also engage more deeply with their surroundings through social interaction, learning, and the enjoyment of nature. This participatory engagement empowers local communities and strengthens their sensory and emotional connection with the natural environment, ultimately enhancing their quality of life. A key challenge in this context lies in overcoming the limitations of contemporary urban landscapes by creating innovative networks that integrate nature more effectively into the built environment. Accordingly, developing planning frameworks that respond to the evolving needs of residents is crucial [55,86]. These frameworks not only support the attainment of long-term sustainability objectives but also promote biodiversity conservation and urban resilience. Moreover, strengthening inter-city collaboration and facilitating the exchange of successful practices in NBS implementation [23] can help scale up adoption. Through shared strategies, cities can collectively move towards more sustainable, inclusive, and nature-adaptive urban futures.

3.4.6. Optimizing Stakeholder Engagement in NBS Implementation

- Engaging Stakeholders in NBS Project Development

While the participation of private and public stakeholders plays a central role in creating and managing green infrastructure [2,27,84,89,92,105,110], there remains significant scope for social groups and environmental organizations to engage as key actors in governance and service delivery, particularly in addressing local needs and values [21,25,77]. To empower these groups and enable the integration of local knowledge, it is essential to promote community participation in the design, implementation, and maintenance of NBS projects [60]. This requires, where possible, the allocation of spaces specifically designated for the establishment and delivery of green infrastructure. The Community Forest Partnership in the UK [125], notably Mersey Forest [126] and initiatives in Manchester [127], offers illustrative examples of successful collaborations between community groups and local authorities. These partnerships have mobilized concerns around climate change, public health and well-being, recreation, and environmental management to strengthen community engagement while generating revenue to reinvest in green infrastructure.

Additionally, debates persist regarding whether communities should assume responsibility for managing and financing green infrastructure. This issue requires careful consideration, as not all governmental and private entities possess the experience, knowledge, or appreciation of community perspectives and the importance of active resident participation in planning processes [1,2,10,50,51,53,61,62,93,96,97]. Even when the transfer of community assets is proposed, these groups may struggle to perform effectively over time without continued support from professionals experienced in social development or landscape management. Moreover, such efforts are likely to remain incomplete or unsustainable unless they are embedded within comprehensive urban green space development strategies [93], especially given the tendency of some institutions to resist high levels of social participation.

Consequently, initiatives must prioritize enhancing awareness of the benefits of NBSs among both public officials and the wider community [3,30,45,52,56,57,65,80,82,88,106,107,113]. This can be achieved through targeted educational programs [2,64,70,88] and through NBS projects that clearly specify the types of green infrastructure to be developed, their optimal locations, and potential funding mechanisms. Such efforts are particularly important in contexts where sociocultural factors [30] shape local perceptions of nature and the environment, often alongside a diverse array of stakeholders engaged in negotiations over investment. In response, many municipal governments and private investors are allocating resources to public spaces to support the expansion of urban green infrastructure.

3.5. What Are the Key Barriers That Hinder the Implementation of NBSs in Urban Green Infrastructure?

Our content analysis identified five key themes, six categories, and twenty-seven sub-themes. The key themes include physical–spatial, economic, technical, social, and political. The findings, presented in Figure 5, highlight the primary barriers to implementing NBSs in urban green infrastructure. These themes are systematically categorized, with recurring codes identified across the analyzed literature.

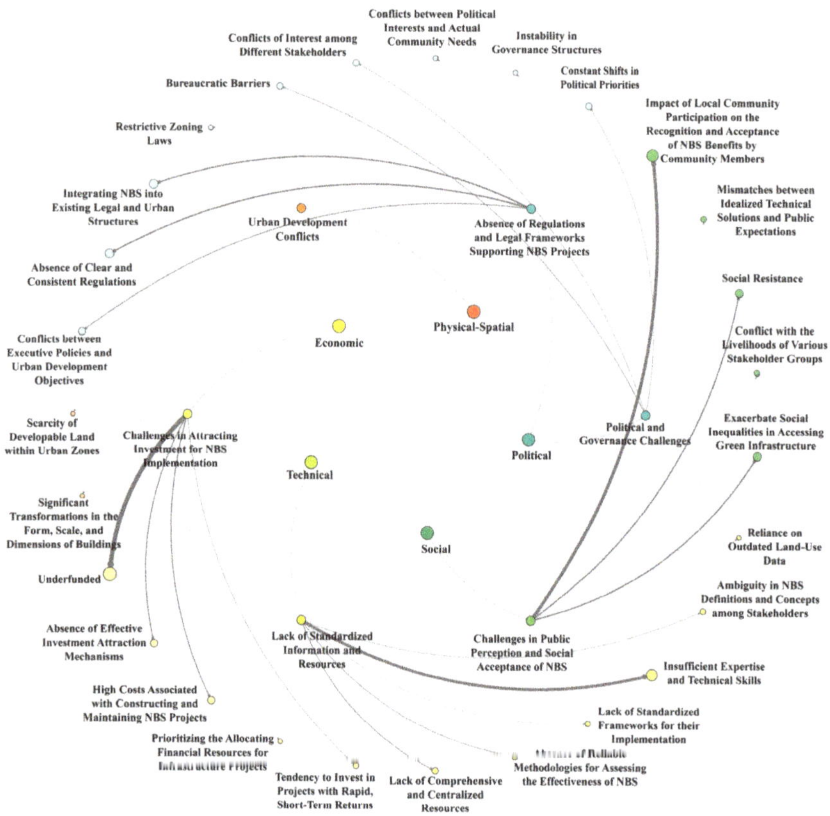

Figure 5. Barriers to implementing NBSs in urban green infrastructure.

3.5.1. Physical–Spatial

- Competing Claims on Urban Land

In many metropolitan areas, the scarcity of developable land [51] often leads to the marginalization of green infrastructure within urban planning agendas. The economic dynamics of land value further compound this challenge; for instance, the escalation of housing prices in proximity to high-quality green spaces tends to reduce the financial incentives for expanding such areas. Properties situated near scenic landscapes typically exhibit higher rental values and improved commercial performance, which, while economically beneficial, can inadvertently disincentivize the allocation of land for NBSs in green infrastructure. A notable example is Maggie Daley Park in downtown Chicago [128], where the development significantly increased local revenues, stimulated business activity, and attracted private investment. However, such projects often require substantial alterations to the built environment, affecting the form, scale, and configuration of buildings [109], as well as associated engineering systems, which can heighten tensions between infrastructure development and the ecological requirements of urban green spaces. More fundamentally, the lack of a coherent and integrated urban green network results in spatial fragmentation, a decline in environmental quality, and reduced urban livability. The situation is further exacerbated by the rapid and often unregulated expansion of urban areas, which has led to the loss of significant green spaces and peri-urban agricultural land, thereby deepening the disconnection between natural ecosystems and urbanized zones.

3.5.2. Economic

- Challenges in Attracting Investment for NBS Implementation

Implementing NBS projects is often hindered by financial constraints [1,2,10,24,27,40,42,48,51,56,60,62,65–67,70,71,74,75,77,78,82,86,93,96,99,107,112] and limited institutional support. Cities require investment strategies that are tailored to their specific economic conditions, yet the absence of effective mechanisms for attracting investment [49,79,89,91,99,100,113] continues to discourage private sector participation. Furthermore, the high costs associated with the construction and maintenance of NBS projects [10,24,43,49,91,100,106] have led many municipalities to abandon initiatives such as sustainable drainage systems, extensive tree planting, and the optimization of urban green land use. For example, in Madrid, large-scale projects such as Madrid Río, dedicated to the development of green space, have incurred substantial costs, thereby limiting the availability of financial resources for other urban areas [129]. Moreover, prioritizing the allocation of financial resources for infrastructure projects [95] sometimes obstructs the advancement of green infrastructure. In Rome, for instance, urban budgets are largely allocated to transportation and heritage restoration, relegating green initiatives such as urban parks and rainwater-harvesting systems to the margins of policy agendas [130]. A further constraint is the prevalent focus on projects promising rapid, short-term returns, while green infrastructure, despite its long-term environmental and social benefits—continues to face chronic underinvestment [2,10,87,95]. In many European cities, commercial and residential developments attract greater investor interest, whereas the expansion of green space requires a long-term vision and sustained financial commitment.

3.5.3. Technical

- Lack of Standardized Information and Resources

NBSs should be tailored to local environmental conditions and socio-spatial challenges, while also ensuring resilience to climate variability and environmental change. Given their site-specific nature, the integration of indigenous and local knowledge can significantly enhance the relevance and effectiveness of NBS interventions. Traditional knowledge, shaped over centuries of community-based resource management and environmental adaptation, contributes meaningfully to the physical, functional, and semantic dimensions of these solutions. Nevertheless, the effective implementation of NBSs continues to encounter substantial barriers. The absence of comprehensive and centralized repositories of knowledge and resources constrains the ability of planners and policymakers to learn from previous experiences [1,2,10,87,93,111]. In parallel, the lack of robust methodologies for evaluating NBS performance [23,78,84,100], along with the absence of standardized implementation frameworks [53,89], impedes the scaling-up of such initiatives. Moreover, limited technical expertise and insufficient capacity in the relevant domains present critical challenges during the execution phase [10,21,27,40,42,44,59,62,65,72,75,82,89,91,95,96,113]. Further complications arise from conceptual ambiguities and inconsistencies in the understanding of NBSs among stakeholders [16,53,70], which hinder coordination and collaborative decision-making. Lastly, reliance on outdated or incomplete land-use data [95] undermines the accuracy and reliability of planning and implementation processes.

3.5.4. Social

- Challenges in Public Perception and Social Acceptance of NBS

Public acceptance of NBS, along with the motivational factors shaping perceptions in participatory processes, remains insufficiently examined. Many NBS initiatives presuppose that the priorities and perspectives of local communities and other stakeholders are aligned with project goals and policy agendas. However, such assumptions can inadvertently reinforce social inequalities in access to green infrastructure [1,2,23,24,62,88,91,97,99,131]. In some cases, these projects may conflict with the livelihoods and interests of particular stakeholder groups [72], leading to adverse consequences for ecosystem integrity and human well-being. In such scenarios, these projects may lack the necessary realism and face social resistance [2,10,18,21,42,48,51,67,84] due to mismatches between idealized technical solutions and public expectations [40], ultimately diminishing their feasibility. Achieving broad-based consensus is vital for strengthening social resilience and supporting the long-term development of green infrastructure. In this context, intermediaries play a crucial role in facilitating dialog, identifying shared priorities, developing co-produced solutions, and cultivating collective knowledge and understanding among stakeholders. Furthermore, the active engagement of local communities is central to enhancing public recognition of, and support for, the benefits of NBSs [1,24,27,44,47–49,54,59,62,63,65,67,71,82,89,91,99,104,105,113]. Addressing these challenges requires moving beyond theoretical discourse towards actionable, inclusive practices that ensure the successful implementation and sustainability of NBS initiatives.

3.5.5. Political

- Political and Governance Challenges

Implementing NBSs in urban green infrastructure projects face fundamental political and managerial challenges affecting their development and conservation. Frequent shifts in political priorities [87,89,101], instability in governance structures [40], and misalignments between political agendas and actual community needs [96] undermine the implementation of sustainable development policies. These challenges weaken long-term programs and complicate coordination among executive institutions. Furthermore, conflicts of interest among different stakeholders [10,101,106], lack of transparency in governmental decisions, and bureaucratic barriers [1,59,60,99], such as those encountered in the Paris Rainwater Plan, can erode public trust and intensify internal resistance within organizations, ultimately slowing or halting environmental projects [132]. Local governments must devise strategies to integrate green infrastructure into local development plans within the framework of national policies without relying on mandatory legislative changes [123]. The Green Belt policy provides a pertinent example: while it has succeeded in preserving a landscape reminiscent of the rural character of the 1950s, it has also faced criticism for lacking contemporary political support. This underscores the importance of stronger engagement by environmental organizations in advocating for green infrastructure. Such organizations play a critical role in promoting awareness of the environmental, economic, and social benefits of investment in green infrastructure, thereby supporting the broader goals of sustainable development. In the UK, institutions such as Community Forests, Natural England, the National Trust, and the Environment Agency have been particularly influential in advancing policies aimed at protecting and enhancing urban green infrastructure [133,134].

- Regulatory and Legal Barriers to NBS Implementation

One of the central questions in green infrastructure discourse is whether the absence of coherent regulatory and legal frameworks significantly constrains investment in landscape enhancement. While developing new policies, regulations, and strategies can foster positive changes, restrictive zoning laws [18] and conventional urban planning approaches continue to pose significant challenges. Furthermore, integrating NBSs into existing legal and urban structures [2,24,75,77,79,83,84,90,100,103] remains a complex and often protracted process, further complicating efforts to mainstream these approaches. Nonetheless, there is growing recognition of these challenges within urban policymaking circles, and in some regions, more robust support mechanisms have begun to emerge. Over the past two decades, green infrastructure has gained increasing prominence in urban planning, becoming a core element of sustainable urban development strategies. Despite this growing recognition, the implementation of NBSs has not been uniformly successful. In several national contexts, the absence of clear and consistent regulatory frameworks [21,24,40,51,64,70,78,89,106], coupled with inadequate oversight mechanisms and a governmental emphasis on short-term economic development goals, has limited progress. Furthermore, in some regions, conflicts between executive-level policy directives and urban development objectives [18,29,43,75,78,113] have obstructed the institutionalization and expansion of NBS. These persistent regulatory misalignments underscore the need for integrated governance approaches that reconcile environmental priorities with broader urban development agendas.

3.6. What Roles Do Stakeholders Play in Planning and Delivering NBSs in Urban Green Infrastructure?

The content analysis identified three main themes, four categories, seventeen sub-themes related to the roles of local governments, private sector actors, and communities. Figure 6 presents the core findings, emphasizing the crucial roles of stakeholders in the planning, implementation, and maintenance of NBSs in urban green infrastructure, with themes systematically categorized and recurring codes identified across the reviewed literature.

3.6.1. Local Governments

- Policy-Making and Regulatory Frameworks

Each NBS project, owing to its specific functions, highlights different aspects of green infrastructure planning. These dimensions are critical for enhancing the capacity of responsible institutions to integrate NBSs into broader urban planning frameworks [42,51,53,56,75,112]. Such integration often necessitates the reallocation of resources across various organizations and planning sectors. While this raises important questions about which actors should oversee the targeted distribution of these resources [21,25,64,88], it also underscores the challenges faced by stakeholders who may lack the traditional expertise required to assume these responsibilities effectively.

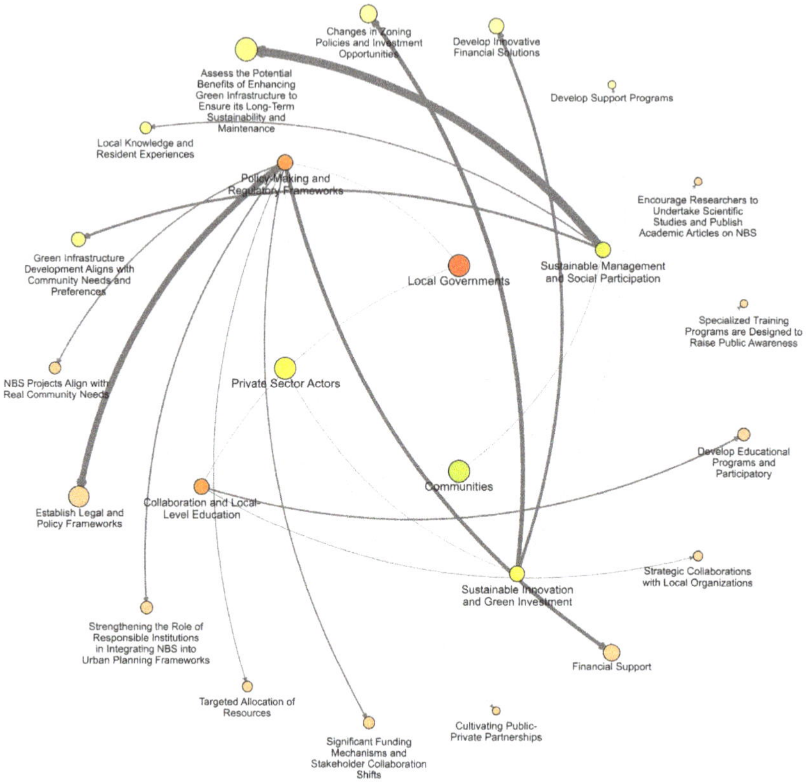

Figure 6. Key stakeholder roles in the planning, implementation, and maintenance of NBSs in urban green infrastructure.

There remains an urgent need to establish legal and policy frameworks that adequately reflect the resource dynamics of green infrastructure and clarify the capacity of different stakeholders to manage it [1,2,18,24,25,27,42,44,46,51,53,61,62,64,70,75,78,87,91–93,95,96,100]. This issue is particularly salient in England, where community asset transfers have raised concerns regarding the long-term governance of green spaces [135]. Even when community groups obtain the necessary legal, managerial, and operational rights, they may lack the knowledge, skills, or time required to govern parks or natural areas effectively.

This situation highlights the need for more robust support mechanisms to ensure that NBS projects are aligned with the actual needs of communities [21,43,62,86,101,111]. These may include direct payments, financial incentives [1,21,24,27,42,44,49,53,61,73,91,95,96,100,107], or public–private partnerships [3,21,24,27,29,44,48,52,54,55,59,65,68,74,77,79,87,89–91,94, 95,100,104,105]. Significant shifts are currently underway in both funding mechanisms and stakeholder collaboration models [17,74,79,91,93,96], opening the door for innovative and adaptive financial strategies. These emerging approaches have the potential to ensure long-term investment and enhance the sustainability of NBS initiatives.

- Collaboration and Local-Level Education

NBSs in urban green infrastructure present distinct political advantages for policymakers. When environmental resources are managed efficiently through strategic collaborations with local organizations [2,62,99], they gain public, business, and political value. Completing the Cheonggyecheon urban renewal project in Seoul is a clear example of positive public and business responses to increased urban usage, prolonged visitor presence in city centers, and economic growth driven by rising expenditures [136]. In this sense, environmental stewardship can translate into electoral gains. If urban governments seek to protect their natural surroundings, they must develop educational programs and participatory [21,24,27,29,44,92,100] workshops involving government agencies, private sectors, and local communities. Specialized training programs are designed to raise public awareness [62] and encourage researchers to undertake scientific studies and publish academic articles on NBSs [62]. The adoption of such collaborative and educational approaches can also provide cities with strategic competitive advantages. For example, urban indices focusing on quality of life and place-making, such as the Green City Index developed by Siemens AG in 2011, have influenced business investment decisions. Cities like Singapore [137] and Vancouver [138] have successfully incorporated NBSs and green infrastructure into their urban branding strategies, enhancing their global competitiveness.

3.6.2. Private Sector Actors

- Sustainable Innovation and Green Investment

A range of new financial models for green infrastructure development has emerged in recent years, providing diverse stakeholders with opportunities to more effectively manage and allocate resources for landscape investment. The involvement of actors from the private, public, voluntary, and social sectors raises important questions about which groups are best positioned to oversee environmental resource management. This issue has become particularly salient in many countries, where diminishing public funds for landscape conservation and management have compelled both governments and environmental professionals to pursue innovative financial strategies [3,42,48,49,52–55,68,89,96,104]. In England, for instance, there is broad consensus that central government funding can no longer be relied upon, prompting a growing interest in alternative mechanisms, such as the sale, transfer, or corporate sponsorship of green infrastructure, particularly by businesses and the real estate development sector [139]. This transition aligns with an increasing awareness among private sector actors that investment in sustainable, nature-based development models may yield higher returns and lower long-term costs compared to conventional engineered solutions. Policy tools, such as zoning reforms and new investment frameworks [1,21,25,27,42,46,51,56,61,72,77,78,87,91,92,95,96], could be used to facilitate funding for green infrastructure, particularly if private actors take proactive steps to develop support programs [56] that encourage public engagement and align urban greening objectives with broader development agendas. In this context, green infrastructure is becoming a strategic business consideration, as corporations increasingly recognize the value of sustainability in enhancing brand identity and promoting high-productivity work environments. Several successful models reflect this emerging shift from state-centric investment and governance towards private-sector-led development (PSD) with a strong environmental orientation. Notable examples include the Parks Trust model in Milton Keynes [140], London's Business Improvement Districts (BIDs) [141], and Paris' participatory budgeting program [142]. These cases illustrate how the private sector can play a leading role in financing and governing urban green infrastructure in ways that are both economically viable and environmentally sustainable.

3.6.3. Communities

- Sustainable Management and Social Participation

Community interactions with green infrastructure are diverse, shaping how individuals perceive, use, and value their surrounding landscapes. These interactions not only influence the immediate functionality of green spaces but also play a critical role in determining their long-term sustainability. As such, global perspectives on the development and stewardship of green infrastructure are highly relevant. Importantly, the variation in community engagement is not inherently problematic; rather, it offers multiple pathways for enhancing landscape quality through the integration of local knowledge and lived experience [45,46,51,65,95,99]. Different communities bring distinct motivations and expectations to their use of green spaces, highlighting the importance of prioritizing public understanding in planning and decision-making processes. Ensuring that green infrastructure development aligns with community needs and preferences [3,48,54,68,84,89,104] is essential for creating urban spaces that are both functional and contextually appropriate. This requires a continuous and iterative process involving a wide range of stakeholders, who must regularly assess their expectations and evaluate the potential benefits of green infrastructure. In doing so, they contribute to its long-term maintenance and sustainability [1,19,21,42,43,48,49,62,63,88,92,94–96].

4. Conclusions

This systematic review has sought to advance the understanding of how NBSs have been conceptualized, implemented, and evaluated in the context of urban green infrastructure over the past decade. By analyzing 90 peer-reviewed articles published between 2014 and 2024, this study has mapped the intellectual contours of the field, identified emerging research trends, and critically examined the conditions that shape the success or failure of NBS initiatives across diverse urban contexts.

The findings suggest that NBSs have increasingly been recognized as a cornerstone of sustainable urban development, particularly in the Global North, where cities such as Barcelona, Milan, Amsterdam, and Paris have embedded NBSs into broader urban greening and climate adaptation strategies. This proliferation is accompanied by growing methodological sophistication, with researchers employing advanced hydrological models, cost-effectiveness tools, and spatial analyses to support evidence-based planning. However, this technical advancement has not been evenly distributed. The geographical distribution of scholarship reveals a concentration of research in Europe and parts of East Asia, with significant gaps in empirical knowledge from the Global South. Addressing these disparities is crucial if NBSs are to fulfill their promise as globally relevant strategies for inclusive and resilient urbanism.

This review has shown that NBS interventions commonly address five interrelated domains: the expansion of urban green ecosystems, stormwater management, the restoration and conservation of natural habitats, nature-integrated urban design, and urban agriculture. These domains reflect the multifunctionality of NBSs, which can simultaneously support biodiversity, enhance climate resilience, improve public health, and foster social cohesion. However, the implementation of such solutions is rarely straightforward. Their success depends not only on ecological suitability or design excellence but also on how effectively they are embedded within governance structures, policy frameworks, and community dynamics.

This study identifies six cross-cutting factors that underpin successful NBS implementation. These include spatial justice, which emphasizes the fair distribution of green infrastructure and equal access to its benefits; effective ecosystem management, which supports ecological integrity and service delivery; integrated governance, which ensures alignment across institutional levels and sectors; sustainable financing, which is necessary for both initial implementation and long-term maintenance; technical capacity, which includes design, engineering, and ecological expertise; and finally, meaningful stakeholder engagement, which builds legitimacy and ensures alignment with local needs and aspirations.

Critically, the analysis has illuminated several persistent barriers that inhibit the wider adoption of NBS. Spatial constraints, such as land scarcity and fragmented green networks, continue to undermine the ecological coherence of urban landscapes. Financial limitations remain acute, especially in contexts where green infrastructure competes with other urban priorities. Technical barriers, such as the lack of standardized data, replicable frameworks, and skilled personnel, further hinder progress. Socially, NBS projects sometimes suffer from low public visibility or acceptance, particularly when designed without adequate community input. Politically, shifting priorities, weak regulatory mechanisms, and institutional silos complicate efforts to integrate NBSs into urban planning in a coherent and sustained manner.

These challenges suggest that the realization of NBS benefits is as much a question of governance and institutional capacity as it is of ecological design. The roles of local governments, private sector actors, and communities are therefore pivotal. Local governments are uniquely positioned to develop enabling policies, coordinate planning processes, and channel public investment towards green infrastructure. At the same time, the private sector can play a complementary role by investing in NBSs as part of broader sustainability strategies, particularly when economic incentives and zoning regulations are aligned. Community actors, in turn, are essential not only as beneficiaries but also as co-creators and stewards of green spaces, contributing valuable local knowledge and fostering long-term care.

Importantly, this study argues that the future of urban green infrastructure planning is poised to diverge from conventional, technocratic approaches. As cities grapple with the compound pressures of climate change, urban inequality, and environmental degradation, there is a growing need to reimagine planning as a more participatory, context-sensitive, and adaptive practice. This includes not only a reconfiguration of how NBSs in green infrastructure are financed and managed but also a shift in how their value is understood, beyond economic returns, to include social well-being, ecological regeneration, and cultural meaning.

Moving forward, several avenues for future research are evident. There is a need to explore the internal dynamics of the thematic clusters identified in this review, examining how technical, institutional, and social dimensions interact to shape project outcomes. Greater attention should also be given to the interconnections among these clusters, particularly in understanding how multi-scalar governance, cross-sectoral partnerships, and temporal factors influence implementation processes. Moreover, comparative research that spans diverse urban contexts can illuminate how local specificities mediate the translation of NBS principles into practice.

In sum, the findings of this review highlight the importance of embracing a more integrated and systemic approach to the planning, implementation, and governance of NBSs in urban green infrastructure. As cities continue to evolve in response to environmental, economic, and social transformations, NBSs offer a promising yet complex pathway toward more resilient and equitable urban futures. Their success, however, will depend on our collective capacity to move beyond narrow sectoral thinking, forge new forms of collaboration, and embed ecological thinking into the very fabric of urban governance.

Author Contributions: Conceptualization, M.Z. and S.S.; methodology M.Z. and S.S.; software, M.Z.; validation, M.Z.; formal analysis, M.Z.; investigation, M.Z. and S.S.; resources, M.Z. and S.S.; data curation, M.Z.; writing—original draft preparation, M.Z. and S.S.; writing—review and editing, S.S.; visualization, M.Z.; supervision, S.S.; funding acquisition, S.S. All authors have read and agreed to the published version of the manuscript.

Funding: This research received no external funding.

Data Availability Statement: Data are open-source and can be extracted from the Scopus database.

Acknowledgments: We thank the anonymous reviewers for their valuable comments and suggestions on our manuscript.

Conflicts of Interest: The authors declare no conflicts of interest.

References

1. Dong, X.; Yang, R.; Ye, Y.; Yi, S.; Haase, D.; Lausch, A. Planning for green infrastructure by integrating multi-driver: Ranking priority based on accessibility equity. *Sustain. Cities Soc.* **2024**, *114*, 105767. [CrossRef]
2. Zhang, H.; Zhou, Q.; Yang, J.; Xiang, H. Change and driving factors of eco-environmental quality in Beijing green belts: From the perspective of Nature-based Solutions. *Ecol. Indic.* **2024**, *166*, 112581. [CrossRef]
3. Sultana, R.; Birtchnell, T.; Gill, N. Grassroots Innovation for Urban Greening within a Governance Vacuum by Slum Dwellers in Dhaka. *Sustainability* **2022**, *14*, 11631. [CrossRef]
4. Nardella, L.; Sebastiani, A.; Stafoggia, M.; Buonocore, E.; Paolo Franzese, P.; Manes, F. Modeling regulating ecosystem services along the urban–rural gradient: A comprehensive analysis in seven Italian coastal cities. *Ecol. Indic.* **2024**, *165*, 112161. [CrossRef]
5. Krull, W.; Berry, P.; Bauduceau, N.; Elmqvist, T.; Fernandez, M.; Hartig, T.; Mayerhofer, E.; Naumann, S.; Noring, L.; Raskin, K.; et al. *Towards an EU Research and Innovation Policy Agenda for Nature-Based Solutions & Re-Naturing Cities. Final Report of the Horizon 2020 Expert Group on 'Nature-Based Solutions and Re-Naturing Cities'*; European Commission: Brussels, Belgium, 2015.
6. Scott, M.; Lennon, M.; Haase, D.; Kazmierczak, A.; Clabby, G.; Beatley, T. Nature-based solutions for the contemporary city/Re-naturing the city/Reflections on urban landscapes, ecosystems services and nature-based solutions in cities/Multifunctional green infrastructure and climate change adaptation: Brownfield greening as an adaptation strategy for vulnerable communities?/Delivering green infrastructure through planning: Insights from practice in Fingal, Ireland/Planning for biophilic cities: From theory to practice. *Plan. Theory Pract.* **2016**, *17*, 267–300. [CrossRef]
7. Kullmann, K. Green-Networks: Integrating Alternative Circulation Systems into Post-industrial Cities. *J. Urban Des.* **2013**, *18*, 36–58. [CrossRef]
8. Mazza, L.; Bennett, G.; De Nocker, L.; Gantioler, S.; Losarcos, L.; Margerison, C.; Kaphengst, T.; McConville, A.; Rayment, M.; ten Brink, P.; et al. *Green Infrastructure Implementation and Efficiency*; Institute for European Environmental Policy: London, UK, 2011.
9. Benedict, M.; MacMahon, E. Green Infrastructure: Smart Conservation for the 21st Century. *Renew. Resour. J.* **2002**, *20*, 12–17.
10. Lyu, L.; Sho, K.; Zhao, H.; Song, Y.; Uchiyama, Y.; Kim, J.; Sakai, T. Construction, assessment, and protection of green infrastructure networks from a dynamic perspective: A case study of Dalian City, Liaoning Province, China. *Urban For. Urban Greening* **2024**, *101*, 128545. [CrossRef]
11. Croeser, T.; Bekessy, S.A.; Garrard, G.E.; Kirk, H. Nature-based solutions for urban biodiversity: Spatial targeting of retrofits can multiply ecological connectivity benefits. *Landsc. Urban Plan.* **2024**, *251*, 105169. [CrossRef]
12. Shahab, S. Transaction Costs in Planning Literature: A Systematic Review. *J. Plan. Lit.* **2021**, *37*, 403–414. [CrossRef]
13. Herreros-Cantis, P.; Hoffman, L.; Kennedy, C.; Kim, Y.; Charles, J.; Gillet, V.; Getzin, A.; Littlefield, D.; Zielinski, A.; Bernstein, J.; et al. Co-producing research and data visualization for environmental justice advocacy in climate change adaptation: The Milwaukee Flood-Health Vulnerability Assessment. *Cities* **2024**, *155*, 105474. [CrossRef]
14. Atri, M.; Nedae-Tousi, S.; Shahab, S.; Solgi, E. The Effects of Thermal-Spatial Behaviours of Land Covers on Urban Heat Islands in Semi-Arid Climates. *Sustainability* **2021**, *13*, 13824. [CrossRef]
15. Grunwald, A. Diverging pathways to overcoming the environmental crisis: A critique of eco-modernism from a technology assessment perspective. *J. Clean. Prod.* **2018**, *197*, 1854–1862. [CrossRef]
16. Morpurgo, J.; Remme, R.P.; Van Bodegom, P.M. CUGIC: The Consolidated Urban Green Infrastructure Classification for assessing ecosystem services and biodiversity. *Landsc. Urban Plann.* **2023**, *234*, 104726. [CrossRef]
17. Bastian, O.; Haase, D.; Grunewald, K. Ecosystem properties, potentials and services-The EPPS conceptual framework and an urban application example. *Ecol. Indic.* **2012**, *21*, 7–16. [CrossRef]
18. Sturiale, L.; Scuderi, A.; Timpanaro, G. Citizens' perception of the role of urban nature-based solutions and green infrastructures towards climate change in Italy. *Front. Environ. Sci.* **2023**, *11*, 1105446. [CrossRef]
19. Faivre, N.; Fritz, M.; Freitas, T.; de Boissezon, B.; Vandewoestijne, S. Nature-Based Solutions in the EU: Innovating with nature to address social, economic and environmental challenges. *Environ. Res.* **2017**, *159*, 509–518. [CrossRef]
20. Dick, J.; Miller, J.D.; Carruthers-Jones, J.; Dobel, A.J.; Carver, S.; Garbutt, A.; Hester, A.; Hails, R.; Magreehan, V.; Quinn, M. How are nature based solutions contributing to priority societal challenges surrounding human well-being in the United Kingdom: A systematic map protocol. *Environ. Evid.* **2019**, *8*, 37. [CrossRef]

21. Marino, R.; Kuller, M.; Zurita, B.; Carrasco, L.; Gosciniak, P.; Kustosik, K.; Riveros, A. Assessment of public space potential for climate change adaptation in three Colombian cities: Nature-based solutions and urban design guidelines. *Water Pract. Technol.* **2024**, *19*, 1696–1709. [CrossRef]
22. Lafortezza, R.; Sanesi, G. Nature-based solutions: Settling the issue of sustainable urbanization. *Environ. Res.* **2019**, *172*, 394–398. [CrossRef]
23. Liu, H.; Jay, M.; Chen, X. The role of nature-based solutions for improving environmental quality, health and well-being. *Sustainability* **2021**, *13*, 10950. [CrossRef]
24. Štrbac, S.; Kašanin-Grubin, M.; Pezo, L.; Stojić, N.; Lončar, B.; Ćurčić, L.; Pucarević, M. Green Infrastructure Designed through Nature-Based Solutions for Sustainable Urban Development. *Int. J. Environ. Res. Public Health* **2023**, *20*, 1102. [CrossRef] [PubMed]
25. Diep, L.; Mulligan, J.; Oloo, M.A.; Guthmann, L.; Raido, M.; Ndezi, T. Co-building trust in urban nature: Learning from participatory design and construction of Nature-Based Solutions in informal settlements in East Africa. *Front. Sustain. Cities.* **2022**, *4*, 927723. [CrossRef]
26. Tiwary, A.; Vilhar, U.; Zhiyanski, M.; Stojanovski, V.; Dinca, L. Management of nature-based goods and services provisioning from the urban common: A pan-European perspective. *Urban Ecosyst.* **2020**, *23*, 645–657. [CrossRef]
27. Sommese, F. Nature-Based Solutions to Enhance Urban Resilience in the Climate Change and Post-Pandemic Era: A Taxonomy for the Built Environment. *Buildings* **2024**, *14*, 2190. [CrossRef]
28. Mosler, S.; Hobson, P. Close-to-nature heuristic design principles for future urban green infrastructure. *Urban Plan.* **2021**, *6*, 67–79. [CrossRef]
29. Adams, C.; Frantzeskaki, N.; Moglia, M. Actors mainstreaming nature-based solutions in cities: A case study of Melbourne's change agents and pathways for urban sustainability transformations. *Environ. Sci. Policy* **2024**, *155*, 103723. [CrossRef]
30. Lechner, A.M.; Gomes, R.L.; Rodrigues, L.; Ashfold, M.J.; Selvam, S.B.; Wong, E.P.; Raymond, C.M.; Zieritz, A.; Sing, K.W.; Moug, P.; et al. Challenges and considerations of applying nature-based solutions in low- And middle-income countries in Southeast and East Asia. *BG. Syst.* **2020**, *2*, 331–351. [CrossRef]
31. Donthu, N.; Kumar, S.; Mukherjee, D.; Pandey, N.; Lim, W.M. How to conduct a bibliometric analysis: An overview and guidelines. *J. Bus. Res.* **2021**, *133*, 285–296. [CrossRef]
32. Bengtsson, M. How to plan and perform a qualitative study using content analysis. *NursingPlus Open* **2016**, *2*, 8–14. [CrossRef]
33. Allan, A.; Soltani, A.; Abdi, M.H.; Zarei, M. Driving Forces behind Land Use and Land Cover Change: A Systematic and Bibliometric Review. *Land* **2022**, *11*, 1222. [CrossRef]
34. Öztürk, O.; Kocaman, R.; Kanbach, D.K. How to design bibliometric research: An overview and a framework proposal. *Rev. Manag. Sci.* **2024**, *18*, 3333–3361. [CrossRef]
35. Kleinheksel, A.J.; Rockich-Winston, N.; Tawfik, H.; Wyatt, T.R. Demystifying Content Analysis. *Am. J. Pharm. Educ.* **2020**, *84*, 7113. [CrossRef] [PubMed]
36. Oleinik, A.; Popova, I.; Kirdina, S.; Shatalova, T. On the choice of measures of reliability and validity in the content-analysis of texts. *Qual. Quant.* **2014**, *48*, 2703–2718. [CrossRef]
37. Aghaei Chadegani, A.; Salehi, H.; Yunus, M.; Farhadi, H.; Fooladi, M.; Farhadi, M.; Ale Ebrahim, N. A Comparison between Two Main Academic Literature Collections: Web of Science and Scopus Databases. *Asian Soc. Sci.* **2024**, *9*, 18. [CrossRef]
38. Dadashpoor, H.; Malekzadeh, N. Driving factors of formation, development, and change of spatial structure in metropolitan areas: A systematic review. *J. Urban Manag.* **2020**, *9*, 286–297. [CrossRef]
39. Zarei, M.; Arasteh, M.; Shahab, S. Exploring Port–City Relationships: A Bibliometric and Content Analysis. *Sustainability* **2024**, *16*, 4341. [CrossRef]
40. Alves, A.; van Opstal, C.; Keijzer, N.; Sutton, N.; Chen, W.S. Planning the multifunctionality of nature-based solutions in urban spaces. *Cities* **2024**, *146*, 104751. [CrossRef]
41. Rizzo, A.; Conte, G.; Masi, F. Adjusted unit value transfer as a tool for raising awareness on ecosystem services provided by constructed wetlands for water pollution control: An italian case study. *Int. J. Environ. Res. Public Health* **2021**, *18*, 1531. [CrossRef]
42. Espinal-Giron, A.; Benegas Negri, L.; Brenes, C.; Birkel, C.; Prins, C. Assessing Potential Effects of Nature-Based Solutions (NBS) on Water Ecosystem Service in the Interurban Micro-Watershed Río Torres, Costa Rica. *Forests* **2023**, *14*, 937. [CrossRef]
43. Ferreira, M.J.; da Rocha, H.R. Green roof infrastructure outperforms grey technology in flood mitigation in São Paulo's urbanized region. *Front. Built Environ.* **2023**, *9*, 1254942. [CrossRef]
44. Pumo, D.; Alongi, F.; Cannarozzo, M.; Noto, L.V. Climate adaptive urban measures in Mediterranean areas: Thermal effectiveness of an advanced multilayer green roof installed in Palermo (Italy). *Build. Environ.* **2023**, *243*, 110731. [CrossRef]
45. Brom, P.; Engemann, K.; Breed, C.; Pasgaard, M.; Onaolapo, T.; Svenning, J.C. A Decision Support Tool for Green Infrastructure Planning in the Face of Rapid Urbanization. *Land* **2023**, *12*, 415. [CrossRef]
46. Cárdenas, M.L.; Wilde, V.; Hagen-Zanker, A.; Seifert-Dähnn, I.; Hutchins, M.G.; Loiselle, S. The circular benefits of participation in nature-based solutions. *Sustainability* **2021**, *13*, 4344. [CrossRef]

47. van Oorschot, J.; Sprecher, B.; van 't Zelfde, M.; van Bodegom, P.M.; van Oudenhoven, A.P.E. Assessing urban ecosystem services in support of spatial planning in the Hague, the Netherlands. *Landsc. Urban Plann.* **2021**, *214*, 104195. [CrossRef]
48. Lai, S.; Leone, F.; Zoppi, C. Assessment of municipal masterplans aimed at identifying and fostering green infrastructure: A study concerning three towns of the metropolitan area of Cagliari, Italy. *Sustainability* **2019**, *11*, 1470. [CrossRef]
49. López-Maciel, M.; Roebeling, P.; Llewellyn, R.; Figueiredo, E.; Matos, F.A.; Mendonça, R.; Bastos, M.I.; Mendes, R.; Postmes, L.; Van Dinter, M. Adoption and Diffusion of Nature-Based Solutions by Property Owners in Urban Areas: The Case of Green Roofs in Eindhoven, The Netherlands. *Resources* **2023**, *12*, 133. [CrossRef]
50. Savić, S.; Krstić, H.; Šećerov, I.; Dunjić, J. Decreasing the energy demand in public buildings using nature-based solutions: Case studies from Novi Sad (Republic of Serbia) and Osijek (Republic of Croatia). *Energy Sustain. Soc.* **2024**, *14*, 23. [CrossRef]
51. Reu Junqueira, J.; Serrao-Neumann, S.; White, I. Developing and testing a cost-effectiveness analysis to prioritize green infrastructure alternatives for climate change adaptation. *Water Environ. J.* **2023**, *37*, 242–255. [CrossRef]
52. Shrestha, S.; Cui, S.; Xu, L.; Wang, L.; Manandhar, B.; Ding, S. Impact of land use change due to urbanisation on surface runoff using GIS-based SCS–CN method: A case study of Xiamen city, China. *Land* **2021**, *10*, 839. [CrossRef]
53. Szopińska-Mularz, M.; Lehmann, S. Balancing Increased Urban Density with Green Spaces: The Marketing of New Housing Estates in Poland. *Buildings* **2023**, *13*, 777. [CrossRef]
54. Rice, L. Nature-based solutions for urban development and tourism. *Int. J. Tour. Cities* **2020**, *6*, 431–448. [CrossRef]
55. Zuniga-Teran, A.A.; Staddon, C.; de Vito, L.; Gerlak, A.K.; Ward, S.; Schoeman, Y.; Hart, A.; Booth, G. Challenges of mainstreaming green infrastructure in built environment professions. *J. Environ. Plann. Manag.* **2020**, *63*, 710–732. [CrossRef]
56. Neumann, V.A.; Hack, J. A methodology of policy assessment at the municipal level: Costa Rica's readiness for the implementation of nature-based-solutions for urban stormwater management. *Sustainability* **2020**, *12*, 230. [CrossRef]
57. Pancewicz, A.; Bednarz, D.; Drożdż, D.; Marszoł, M.; Suchy, N. The Use of Nature-Based Solutions in the Adaptation of Large Polish Cities to Climate Change and Energy Transformation: A Comparative Analysis. *Energies* **2023**, *16*, 5189. [CrossRef]
58. Dorst, H.; van der Jagt, A.; Toxopeus, H.; Tozer, L.; Raven, R.; Runhaar, H. What's behind the barriers? Uncovering structural conditions working against urban nature-based solutions. *Landsc. Urban Plan.* **2022**, *220*, 104335. [CrossRef]
59. Cui, M.; Ferreira, F.; Fung, T.K.; Matos, J.S. Tale of two cities: How nature-based solutions help create adaptive and resilient urban water management practices in singapore and lisbon. *Sustainability* **2021**, *13*, 10427. [CrossRef]
60. Mejía, C.V.; Shirotova, L.; De Almeida, I.F.M. Green infrastructure and German landscape planning: A comparison of approaches. *Urbani Izziv* **2015**, *26*, S25–S37. [CrossRef]
61. Ibric, A.; Bostenaru Dan, M.; Crăciun, C. Adapting Street Profile Design by Using Nature-Based Solutions in New Neighbourhoods and the Retrofit of Buildings. *Buildings* **2024**, *14*, 1920. [CrossRef]
62. Muwafu, S.P.; Rölfer, L.; Scheffran, J.; Costa, M.M. A framework for assessing social structure in community governance of sustainable urban drainage systems: Insights from a literature review. *Mitig. Adapt. Strateg. Glob. Change* **2024**, *29*, 42. [CrossRef]
63. Lemes de Oliveira, F.; Mare'e, S.; Khattab, R.; Alqatamin, H.; Younis, A.; Crinion, K.; Zhong, H.; Kaddour, I.Z.; Sharmin, T. Conceptualising nature-based solutions: Addressing environmental challenges in the city of Amman, Jordan. *Urban Res. Pract.* **2024**, *18*, 6–18. [CrossRef]
64. Perrotti, D.; Stremke, S. Can urban metabolism models advance green infrastructure planning? Insights from ecosystem services research. *Environ. Plann.* **2020**, *47*, 678–694. [CrossRef]
65. Adams, C.; Frantzeskaki, N.; Moglia, M. Mainstreaming nature-based solutions in cities: A systematic literature review and a proposal for facilitating urban transitions. *Land Use Policy* **2023**, *130*, 106661. [CrossRef]
66. Camerin, F.; Longato, D. Designing healthier cities to improve life quality: Unveiling challenges and outcomes in two Spanish cases. *J. Urban Des.* **2024**, *30*, 1–30. [CrossRef]
67. Mell, I.; Clement, S.; O'Sullivan, F. Mainstreaming Nature-Based Solutions in City Planning: Examining Scale, Focus, and Visibility as Drivers of Intervention Success in Liverpool, UK. *Land* **2023**, *12*, 1371. [CrossRef]
68. Buffam, I.; Hagemann, F.A.; Emilsson, T.; Gamstetter, D.; Pálsdóttir, A.M.; Randrup, T.B.; Yeshitela, K.; Ode Sang, Å. Priorities and barriers for urban ecosystem service provision: A comparison of stakeholder perspectives from three cities. *Front. Sustain. Cities* **2022**, *4*, 838971. [CrossRef]
69. Brokking, P.; Mörtberg, U.; Balfors, B. Municipal practices for integrated planning of nature-based solutions in urban development in the stockholm region. *Sustainability* **2021**, *13*, 10389. [CrossRef]
70. Vaňo, S.; Stahl Olafsson, A.; Mederly, P. Advancing urban green infrastructure through participatory integrated planning: A case from Slovakia. *Urban For. Urban Green.* **2021**, *58*, 126957. [CrossRef]
71. O'Sullivan, F.; Mell, I.; Clement, S. Novel Solutions or Rebranded Approaches: Evaluating the Use of Nature-Based Solutions (NBS) in Europe. *Front. Sustain. Cities* **2020**, *2*, 572527. [CrossRef]
72. Dushkova, D.; Haase, D. Not simply green: Nature-based solutions as a concept and practical approach for sustainability studies and planning agendas in cities. *Land* **2020**, *9*, 19. [CrossRef]

73. Song, J.; Hemingway, J.; Park, C.S. Perspective Swap from Central Europe to East Asia: How Relevant Is Urban Environmental Acupuncture in Small-Scale Green Space Development in the Context of the Republic of Korea? *Land* **2024**, *13*, 298. [CrossRef]
74. Moosavi, S.; Browne, G.R.; Bush, J. Perceptions of nature-based solutions for Urban Water challenges: Insights from Australian researchers and practitioners. *Urban For. Urban Green.* **2021**, *57*, 126937. [CrossRef]
75. Yazar, M.; Baykal Fide, E.; Daloglu Cetinkaya, I. The nested hierarchy of urban vulnerability within land use policies fails to address climate injustices in Turkey. *J. Environ. Policy Plann.* **2024**, *26*, 30–46. [CrossRef]
76. Matter, N.M.; Gado, N.G. Constructed wetlands nature-based solutions to enhance urban resilience in Egyptian cities. *HBRC J.* **2024**, *20*, 231–255. [CrossRef]
77. Hoover, F.A.; Meerow, S.; Coleman, E.; Grabowski, Z.; McPhearson, T. Why go green? Comparing rationales and planning criteria for green infrastructure in U.S. city plans. *Landsc. Urban Plann.* **2023**, *237*, 104781. [CrossRef]
78. Bona, S.; Silva-Afonso, A.; Gomes, R.; Matos, R.; Rodrigues, F. Nature-Based Solutions in Urban Areas: A European Analysis. *Appl. Sci.* **2023**, *13*, 168. [CrossRef]
79. Zain, A.F.M.; Pribadi, D.O.; Indraprahasta, G.S. Revisiting the Green City Concept in the Tropical and Global South Cities Context: The Case of Indonesia. *Front. Environ. Sci.* **2022**, *10*, 787204. [CrossRef]
80. Hamel, P.; Hamann, M.; Kuiper, J.J.; Andersson, E.; Arkema, K.K.; Silver, J.M.; Daily, G.C.; Guerry, A.D. Blending Ecosystem Service and Resilience Perspectives in Planning of Natural Infrastructure: Lessons from the San Francisco Bay Area. *Front. Environ. Sci.* **2021**, *9*, 601136. [CrossRef]
81. Shi, L. Beyond flood risk reduction: How can green infrastructure advance both social justice and regional impact? *Socio-Ecol. Pract. Res.* **2020**, *2*, 311–320. [CrossRef]
82. Zwierzchowska, I.; Fagiewicz, K.; Poniży, L.; Lupa, P.; Mizgajski, A. Introducing nature-based solutions into urban policy-facts and gaps. Case study of Poznań. *Land Use Policy* **2019**, *85*, 161–175. [CrossRef]
83. Raymond, C.M.; Frantzeskaki, N.; Kabisch, N.; Berry, P.; Breil, M.; Nita, M.R.; Geneletti, D.; Calfapietra, C. A framework for assessing and implementing the co-benefits of nature-based solutions in urban areas. *Environ. Sci. Policy* **2017**, *77*, 15–24. [CrossRef]
84. Wendling, L.A.; Huovila, A.; zu Castell-Rüdenhausen, M.; Hukkalainen, M.; Airaksinen, M. Benchmarking nature-based solution and smart city assessment schemes against the sustainable development goal indicator framework. *Front. Environ. Sci.* **2018**, *6*, 69. [CrossRef]
85. Reeve, A.C.; Desha, C.; Hargreaves, D.; Hargroves, K. Biophilic urbanism: Contributions to holistic urban greening for urban renewal. *Smart Sustain. Built Environ.* **2015**, *4*, 215–233. [CrossRef]
86. Latasa, I.; Laurenz, A.; Sádaba, J. Urban green infrastructure as a strategy to address urban energy efficiency and sustainability. A case study of milagrosa (pamplona). *Sustainability* **2022**, *14*, 28. [CrossRef]
87. Curran, W.; Hamilton, T. Nature-based solutions in hiding: Goslings and greening in the still-industrial city. *Socio-Ecol. Pract. Res* **2020**, *2*, 321–327. [CrossRef]
88. Blau, M.L.; Luz, F.; Panagopoulos, T. Urban river recovery inspired by nature-based solutions and biophilic design in Albufeira, Portugal. *Land* **2018**, *7*, 141. [CrossRef]
89. Capotorti, G.; Mollo, B.; Zavattero, L.; Anzellotti, I.; Celesti-Grapow, L. Setting priorities for urban forest planning. A comprehensive response to ecological and social needs for the metropolitan area of Rome (Italy). *Sustainability* **2015**, *7*, 3958–3976. [CrossRef]
90. Zulian, G.; Polce, C.; Maes, J. ESTIMAP: A GIS-based model to map ecosystem services in the European Union. *Ann. Bot.* **2014**, *4*, 1–7. [CrossRef]
91. Pérez, G.; Reyes, M.; Coma, J.; Alva, A.; Berigüete, F.E.; Lacasta, A.M. Methodological framework for impact evaluation of Building-Integrated Greenery (BIG-impact). *MethodsX* **2024**, *13*, 102961. [CrossRef]
92. Olgun, R.; Cheng, C.; Coseo, P. Nature-Based Solutions Scenario Planning for Climate Change Adaptation in Arid and Semi-Arid Regions. *Land* **2024**, *13*, 1464. [CrossRef]
93. Li, Y.B.; Wang, Q.C.; Chang, R. Nature-based solutions for fast-growing city regions: A new spatial equilibrium model for complementary urban green space planning. *J. Clean. Prod.* **2024**, *462*, 142671. [CrossRef]
94. Van der Borght, R.; Pallares-Barbera, M. Greening to shield: The impacts of extreme rainfall on economic activity in Latin American cities. *Glob. Environ. Change* **2024**, *87*, 102857. [CrossRef]
95. Hanna, E.; Felipe-Lucia, M.R.; Comín, F.A. Scenario Analysis of Green Infrastructure to Adapt Medium-Size Cities to Climate Change: The Case of Zaragoza, Spain. *Land* **2024**, *13*, 280. [CrossRef]
96. Staccione, A.; Essenfelder, A.H.; Bagli, S.; Mysiak, J. Connected urban green spaces for pluvial flood risk reduction in the Metropolitan area of Milan. *Sustainable Cities Soc.* **2024**, *104*, 105288. [CrossRef]
97. Meenar, M.; Rahman, M.S.; Russack, J.; Bauer, S.; Kapri, K. "The Urban Poor and Vulnerable Are Hit Hardest by the Heat": A Heat Equity Lens to Understand Community Perceptions of Climate Change, Urban Heat Islands, and Green Infrastructure. *Land* **2023**, *12*, 2174. [CrossRef]

98. Anderson, V.; Zgela, M.; Gough, W.A. Building Urban Resilience with Nature-Based Solutions: A Multi-Scale Case Study of the Atmospheric Cleansing Potential of Green Infrastructure in Southern Ontario, Canada. *Sustainability* **2023**, *15*, 14146. [CrossRef]
99. Goličnik Marušić, B.; Dremel, M.; Ravnikar, Ž. A frame of understanding to better link nature-based solutions and urban planning. *Environ. Sci. Policy* **2023**, *146*, 47–56. [CrossRef]
100. Biasin, A.; Masiero, M.; Amato, G.; Pettenella, D. Nature-Based Solutions Modeling and Cost-Benefit Analysis to Face Climate Change Risks in an Urban Area: The Case of Turin (Italy). *Land* **2023**, *12*, 280. [CrossRef]
101. Teotónio, I.; Oliveira Cruz, C.; Matos Silva, C.; Lopes, R.F.R. Bridging CBA and MCA for evaluating green infrastructure: Proposal of a new evaluation model (MAGICA). *Socio-Econ. Plann. Sci.* **2023**, *85*, 101446. [CrossRef]
102. Shih, W.Y. Socio-ecological inequality in heat: The role of green infrastructure in a subtropical city context. *Landsc. Urban Plann.* **2022**, *226*, 104506. [CrossRef]
103. Salata, S.; Erdoğan, B.; Ayruş, B. Designing Urban Green Infrastructures Using Open-Source Data—An Example in Çiğli, Izmir (Turkey). *Urban. Sci.* **2022**, *6*, 42. [CrossRef]
104. Lampinen, J.; García-Antúnez, O.; Olafsson, A.S.; Kavanagh, K.C.; Gulsrud, N.M.; Raymond, C.M. Envisioning carbon-smart and just urban green infrastructure. *Urban For. Urban Green.* **2022**, *75*, 127682. [CrossRef]
105. Banzhaf, E.; Anderson, S.; Grandin, G.; Hardiman, R.; Jensen, A.; Jones, L.; Knopp, J.; Levin, G.; Russel, D.; Wu, W.; et al. Urban-Rural Dependencies and Opportunities to Design Nature-Based Solutions for Resilience in Europe and China. *Land* **2022**, *11*, 480. [CrossRef]
106. Gantar, D.; Kozamernik, J.; Erjavec, I.Š.; Koblar, S. From Intention to Implementation of Vertical Green: The Case of Ljubljana. *Sustainability* **2022**, *14*, 3198. [CrossRef]
107. Voskamp, I.M.; de Luca, C.; Polo-Ballinas, M.B.; Hulsman, H.; Brolsma, R. Nature-based solutions tools for planning urban climate adaptation: State of the art. *Sustainability* **2021**, *13*, 6381. [CrossRef]
108. Rędzińska, K.; Piotrkowska, M. Urban Planning and Design for Building Neighborhood Resilience to Climate Change. *Land* **2020**, *9*, 387. [CrossRef]
109. Langemeyer, J.; Wedgwood, D.; McPhearson, T.; Baró, F.; Madsen, A.L.; Barton, D.N. Creating urban green infrastructure where it is needed—A spatial ecosystem service-based decision analysis of green roofs in Barcelona. *Sci. Total Environ.* **2020**, *707*, 135487. [CrossRef]
110. Shade, C.; Kremer, P. Article predicting land use changes in Philadelphia following green infrastructure policies. *Land* **2019**, *8*, 28. [CrossRef]
111. Stürck, J.; Schulp, C.J.E.; Verburg, P.H. Spatio-temporal dynamics of regulating ecosystem services in Europe- The role of past and future land use change. *Appl. Geogr.* **2015**, *63*, 121–135. [CrossRef]
112. Danilina, N.; Tsurenkova, K.; Berkovich, V. Evaluating urban green public spaces: The case study of Krasnodar region cities, Russia. *Sustainability* **2021**, *13*, 14059. [CrossRef]
113. Essuman-Quainoo, B.; Jim, C.Y. Understanding the drivers of green roofs and green walls adoption in Global South cities: Analysis of Accra, Ghana. *Urban For. Urban Green.* **2023**, *89*, 128106. [CrossRef]
114. Kwartnik-Pruc, A.; Droj, G. The Role of Allotments and Community Gardens and the Challenges Facing Their Development in Urban Environments—A Literature Review. *Land* **2023**, *12*, 325. [CrossRef]
115. IUCN. *The IUCN Programme 2013–2016*; International Union for the Conservation of Nature: Gland, Switzerland, 2013.
116. Moorish, W.R.; Brown, C.R. *Infrastructure for the New Social Compact*; Routledge: London, UK, 2008.
117. Benedict, M.A.; McMahon, E.T. *Green Infrastructure: Linking Landscapes and Communities*; Island Press: Washington, DC, USA, 2012.
118. Fainstein, S.S. The just city. *Int. J. Urban Sci.* **2014**, *18*, 1–18. [CrossRef]
119. Ahmadpoor, N.; Shahab, S. Urban form: Realising the value of green space: A planners' perspective on the COVID-19 pandemic. *Town Plan. Rev.* **2021**, *92*, 49–55. [CrossRef]
120. Aram, F.; Shahab, S.; Solgi, E. Editorial: Impacts of urban green spaces on environmental perceptions and social life. *Front. Environ. Sci.* **2022**, *10*, 1050597. [CrossRef]
121. *Green Infrastructure*; European Union Commission: Brussels, Belgium, 2017.
122. Bruno, E.; Falco, E.; Shahab, S.; Geneletti, D. Integrating ecosystem services in transfer of development rights: A literature review. *Land Use Policy* **2023**, *131*, 106694. [CrossRef]
123. Jerome, G.; Sinnett, D.; Burgess, S.; Calvert, T.; Mortlock, R. A framework for assessing the quality of green infrastructure in the built environment in the UK. *Urban For. Urban Green.* **2019**, *40*, 171–182. [CrossRef]
124. Byahut, S.; Ghosh, S.; Masilela, C. Urban Transformation for Sustainable Growth and Smart Living: The Case of the Atlanta Beltline. In *Smart Living for Smart Cities: Community Study, Ways and Means*; Vinod Kumar, T.M., Ed.; Springer: Singapore, 2020; pp. 73–100.
125. Ambrose-Oji, B.; Lawrence, A.; Stewart, A. Community based forest enterprises in Britain: Two organising typologies. *For. Policy Econ.* **2015**, *58*, 65–74. [CrossRef]
126. Nolan, P. Twenty five years of creating the Mersey Forest. *Environ. Law Rev.* **2019**, *21*, 265–268. [CrossRef]

127. Carter, J.; Labib, S.M.; Mell, I. Understanding and Assessing Climate Change Risk to Green Infrastructure: Experiences from Greater Manchester (UK). *Land* **2024**, *13*, 697. [CrossRef]
128. Al-Kodmany, K. Chicago as a Placemaking Model. In *Tall Buildings and the City: Improving the Understanding of Placemaking, Imageability, and Tourism*; Al-Kodmany, K., Ed.; Springer: Singapore, 2020; pp. 407–441.
129. Gabbianelli, A. Formal and formless in Madrid Río Park. *J. Landsc. Archit.* **2020**, *15*, 60–73. [CrossRef]
130. Crupi, F. Urban Regeneration and Green and Blue Infrastructure: The Case of the "Acilia–Madonnetta" Urban and Metropolitan Centrality in the Municipality of Rome. *Urban. Sci.* **2022**, *6*, 56. [CrossRef]
131. Cassarino, M.; Shahab, S.; Biscaya, S. Envisioning Happy Places for All: A Systematic Review of the Impact of Transformations in the Urban Environment on the Wellbeing of Vulnerable Groups. *Sustainability* **2021**, *13*, 8086. [CrossRef]
132. Dellinger, F.; Romier, A.-C.; Saint-Onge, G. Biodiversity in the Day-to-Day Practice of the Landscape Architect. In *Urban Biodiversity and Ecological Design for Sustainable Cities*; Ito, K., Ed.; Springer: Tokyo, Japan, 2021; pp. 35–76.
133. Back, P.; Collins, A.M. Negotiating the green obstacle course: Ranking priorities and problems for municipal green infrastructure implementation. *Urban For. Urban Green.* **2022**, *67*, 127436. [CrossRef]
134. Antony, J.; Ghadge, A.; Ashby, S.A.; Cudney, E.A. Lean Six Sigma journey in a UK higher education institute: A case study. *Int. J. Qual. Reliab. Manag.* **2018**, *35*, 510–526. [CrossRef]
135. Dickinson, J.; Bennett, E.; Marson, J. Challenges facing green space: Is statute the answer? *J. Place Manag. Dev.* **2019**, *12*, 121–138. [CrossRef]
136. Ryu, C.; Kwon, Y. How Do Mega Projects Alter the City to Be More Sustainable? Spatial Changes Following the Seoul Cheonggyecheon Restoration Project in South Korea. *Sustainability* **2016**, *8*, 1178. [CrossRef]
137. Newman, P. Green Urbanism and its Application to Singapore. *Environ. Urban. ASIA* **2010**, *1*, 149–170. [CrossRef]
138. Jang, N.; Doyon, A. Equitable Implementation of Green Infrastructure: An Analysis of the City of Vancouver. *Can. Plan. Policy* **2023**, *2023*, 110–132. [CrossRef]
139. O'Brien, P.; Pike, A. City Deals, Decentralisation and the Governance of Local Infrastructure Funding and Financing in the UK. *Natl. Inst. Econ. Rev.* **2015**, *233*, R14–R26. [CrossRef]
140. Cureton, P. The history of the future of a new town: Milton Keynes, the 'Forest City'. *J. Landsc. Archit.* **2023**, *18*, 84–99. [CrossRef]
141. Moore, L.; Mell, I. Examining the role of business improvement districts (BIDs) as contemporary alternatives for Green Infrastructure funding in London, UK. *Urban For. Urban Green.* **2023**, *83*, 127904. [CrossRef]
142. Doustaly, C. Participatory Budgeting and Progressive Cities: Are London and Paris Listening to Their Own Voices? In *The Rise of Progressive Cities East and West*; Douglass, M., Garbaye, R., Ho, K.C., Eds.; Springer: Singapore, 2019; pp. 117–136.

land

MDPI

Article

Enhancing Urban Resilience: Stormwater Retention and Evapotranspiration Performance of Green Roofs Under Extreme Rainfall Events

Marc Breulmann *, Amelie Merbach, Katy Bernhard and Lucie Moeller *

Department Systemic Environmental Biotechnology, Helmholtz Centre for Environmental Research—UFZ, Permoserstraße 15, 04318 Leipzig, Germany

* Correspondence: marc.breulmann@ufz.de (M.B.); lucie.moeller@ufz.de (L.M.)

Abstract: Rapid urbanisation and climate change have intensified extreme rainfall events, exacerbating stormwater runoff and overwhelming urban drainage systems. Nature-based solutions, such as green roofs with integrated retention capacity, offer promising strategies to mitigate these challenges. This study investigates the influence of substrate thickness and retention volume on the stormwater retention and evapotranspiration (ET) performance of three green roof variants under extreme rainfall scenarios (natural and 5-, 30- and 100-year events). Using lysimeter-based experimental setups, we show that the overall retention capacity is highly dependent on the filling status of the retention layer. Near full capacity, retention performance decreases significantly, resulting in runoff behaviour similar to that of conventional green roofs, while empty systems store up to 99% of rainfall. In addition, ET rates tend to decrease in systems with higher substrate layers and larger retention spaces due to reduced surface evaporation and greater thermal insulation. However, higher substrate layers store more water, allowing plants to maintain transpiration during dry periods, potentially increasing total cumulative ET over time. Overall, this study highlights the importance of designing intensive retention green roofs with dynamic water management to optimise both rainwater retention and ET, thereby increasing urban resilience to increasing rainfall extremes caused by climate change.

Keywords: blue–green infrastructure; green roofs; urban stormwater management; evapotranspiration; stormwater retention; weather extremes

1. Introduction

The rapid urbanisation of the world's population, exacerbated by the twin crises of climate change and water scarcity, poses critical challenges for sustainable urban water management [1]. Since the 19th century, global temperatures have risen by 1.1 °C, leading to an increase in extreme weather events such as heat waves, heavy rainfall and droughts [2]. Urban areas dominated by impervious surfaces are particularly vulnerable, intensifying surface runoff and increasing the risk of flooding during heavy rainfall [3]. These environmental and economic challenges, together with the impacts of climate change, have exposed the inadequacies of conventional wastewater systems and highlighted the need for innovative and multifunctional solutions in urban water management [4].

In response, nature-based solutions such as blue–green infrastructure (BGI)—including green roofs, permeable pavements and constructed wetlands—offer promising strategies to mitigate flooding and enhance urban resilience [5–7]. Integrated urban water management (IUWM) frameworks, along with smart water technologies, further contribute to managing the growing complexity of urban water systems [8]. These solutions require urban planners to integrate water management with broader urban development, pollution control and waste management strategies [9].

An IUWM strategy improves resource efficiency by treating the urban water cycle as an interconnected system. This approach addresses the water needs of the residential, industrial, agricultural and ecological sectors, while serving as a comprehensive framework for planning and managing urban water systems [10]. Successful implementation depends on collaboration across jurisdictions within the urban region, enabling cities and utilities to create resilient water systems that meet current and future demands driven by urban growth [11]. Ultimately, IUWM is reshaping the way cities interact with water, offering innovative perspectives on the design and management of water resources and infrastructure [12].

Urban densification and expansion, coupled with extreme rainfall events, have intensified stormwater runoff, putting significant strain on existing infrastructure, including networks and centralised treatment plants, and prompting cities to consider countermeasures such as the integration and implementation of urban BGI [13]. Urban BGI refers to managed and engineered infrastructures that use vegetation for decentralised urban water management and to provide sustainable solutions for climate resilience [14]. Examples include swales, infiltration trenches, green facades and green roofs that support water sensitive urban design [15].

Green roofs offer multiple benefits, including reducing urban heat island effects through evapotranspiration (ET) [16], enhancing biodiversity [17], and improving air quality [18]. They also play a critical role in stormwater management [19], making them a promising tool for mitigating the impacts of extreme weather events [20].

Numerous studies have investigated the performance of green roofs under conditions of extreme heat [21–23] or heavy rainfall [24–27]. However, there remains a notable gap in understanding the performance of these systems under extreme rainfall conditions, especially in European climates. In addition, many studies predominantly focus on extensive green roofs as these are more common in practice [28]. Extensive green roofs are a low-maintenance form of roof greening characterised by drought-tolerant plants, such as succulents, and a relatively low substrate layer, typically not exceeding 15 cm [29]. In contrast, intensive green roofs provide superior rainwater retention [30]. These systems are characterised by a substrate layer of at least 15 cm and vegetation with higher water requirements, ranging from grasses and perennials to shrubs and small trees [31,32]. Despite their potential benefits, intensive green roofs have been comparatively underexplored in previous research.

Recent studies have investigated retention green roofs, which provide enhanced water retention through raised substrate layers and integrated water storage layers [33]. In addition, water can be transported into the substrate layer by capillary forces [34], acting as an indirect irrigation mechanism. This stored water can also be used to irrigate other green spaces in a process known as BGI cascading [13,35]. Few studies have investigated the effect of substrate thickness [30,36] and retention volume [34,35] on the runoff and ET performance of green roofs. However, experimental studies and the consideration of intensive green roofs often receive insufficient attention in the context of retention green roofs.

A key component of the post-rainfall water cycle is ET, which includes both evaporation from plants and the land surface and transpiration from plants [37,38]. It significantly influences the effectiveness of vegetated BGI systems. As ET plays a major role in the overall water balance, it must be considered in the design of BGI such as green roofs, which aim to reduce the overall volume of runoff [39]. Therefore, it is important to integrate ET into the design of volume reduction strategies and account for vegetated BGI systems.

In most cities, green roofs are designed for smaller, more frequent rainfall events (e.g., 1-year events) [40]. Due to climate change and the occurrence of more frequent extreme rainfall events, BGI will need to be designed for much larger precipitation intensities, such as 30-year and 100-year events.

Therefore, this study investigates the influence of substrate thickness and retention volume on the performance of intensive green roof structures during extreme weather events. Specifically, we analyse three intensive green roof variants and assess their ET and stormwater retention capabilities under varying rainfall return periods (natural heavy rainfall event and 5-, 30- and 100-year events).

2. Materials and Methods

Three experimental green roofs were constructed to investigate rainwater retention and ET performance under extreme rainfall events. The experimental setups were based on lysimeter systems (Umwelt-Geräte-Technik GmbH, Müncheberg, Germany), which accurately measure weight changes and quantify ET by recording water inputs (precipitation and irrigation, if applicable) and outputs (runoff).

2.1. Experimental Setup

The three lysimeter-based green roofs were designed as variations of intensive retention systems with surface areas of about 0.6 m^2, with a sensitivity of the weight systems of about 0.1 kg, and a maximum weight load of 120 kg for Green Roof 1, 400 kg for Green Roof 2 and 800 kg for Green Roof 3 (Figure 1). The main difference between the green roofs was the thickness of the substrate and the retention volume.

Figure 1. The three lysimeters with intensive green roof designs and the weather station in the background. From left to right: Green Roof 1 with a drainage layer and substrate height of 15 cm, Green Roof 2 with an 8.5 cm retention layer and substrate height of 15 cm, and Green Roof 3 with a 17 cm retention layer and substrate height of 30 cm.

Figure 2 provides a detailed overview of the green roof structures. All three roof segments were vegetated with turf grass and the same type of substrate was used. Green Roof 1, designed as a conventional system, included a 15 cm substrate layer and a drainage layer with a water retention capacity of 8.7 L m^{-2} (Figure 2). Green Roof 2 had also a 15 cm substrate layer and an 8.5 cm retention layer instead of a drainage layer, allowing it to store up to 80 L m^{-2}. Green Roof 3 had a 30 cm substrate layer and a 17 cm retention layer, with a maximum water storage capacity of 161 L m^{-2} (Figure 2).

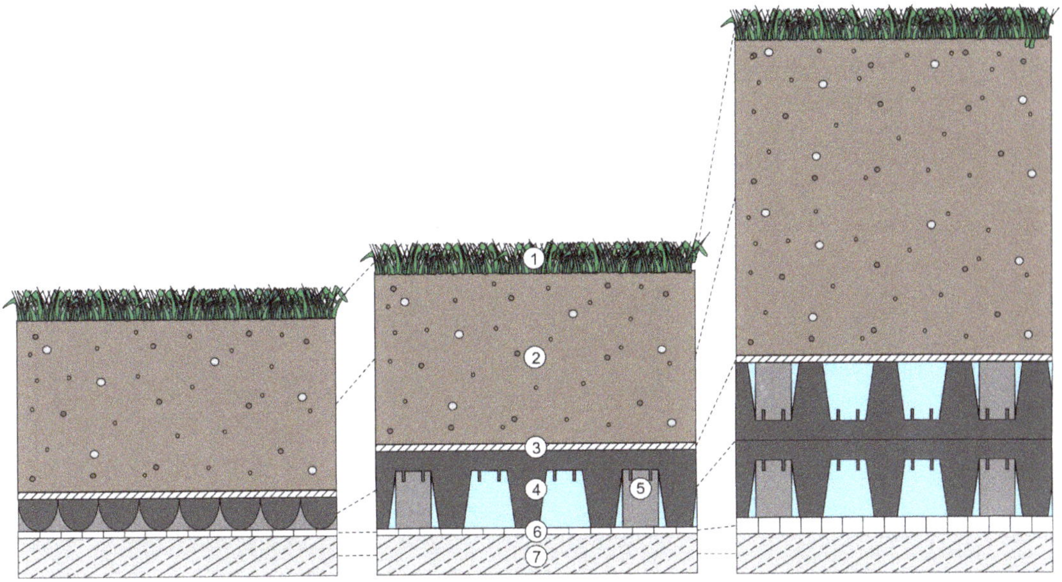

No.	Layer	Green Roof 1	Green Roof 2	Green Roof 3
1	Vegetation	Turf grass	Turf grass	Turf grass
2	Substrate	15 cm lawn substrate (90.7 dm³)	15 cm lawn substrate (84.9 dm³)	30 cm lawn substrate (181.4 dm³)
3	Separation fleece	Model: FIL 105 (Optigrün International AG)	Model: RMS 500K (Optigrün International AG)	Model: RMS 500K (Optigrün International AG)
4	Retention/ drainage layer	Storage/drainage Board (FKD 40) 8.7 L m^{-2} storage volume	Water Retention Box (WRB 85) 80 L m^{-2} storage volume	Water Retention Box (WRB 170) 161 L m^{-2} storage volume
5	Capillary fleece	–	3	3
6	Separation fleece	Model: RMS 300	Model: RMS 300	Model: RMS 900
7	Substructure	High-grade steel of lysimeter	High-grade steel of lysimeter	High-grade steel of lysimeter

Figure 2. Cross sections and functional layers of the three experimental green roof structures.

All functional layers were supplied by Optigrün International AG (Krauchenwies-Göggingen, Germany). Images of the drainage layers and retention boxes used in the study are in the Supplementary Materials (Figure S1).

A soil moisture sensor (model GS1 by METER Group GmbH, München, Germany), based on the measurement of the volumetric water content, was placed 10 cm below the turf layer of each experimental plot.

A tilt counter (Umwelt-Geräte-Technik GmbH, Müncheberg, Germany) was connected to each outlet to determine the volume of the overflow. The tilt counters used registered the outflow in 100 mL increments at a volume flow rate of up to 5 L min^{-1}.

All sensors and the weather station were connected to a measurement and control datalogger (CR1000X, Campbell Scientific, INC., Logan, UT, USA).

2.2. Experimental Site Conditions

The three experimental green roofs are located on the campus of the Helmholtz Centre for Environmental Research—UFZ in Leipzig, Germany (51°21′12.8″ N, 12°25′58.4″ E) and were installed at ground level, approximately 30 cm apart and fully exposed to direct sunlight (see Figure 1). A compact digital weather station (ClimaVUE 50, Campbell Scientific, INC., Logan, UT, USA) continuously monitored temperature and precipitation data with a resolution of one minute. The experiments were carried out between 8 July and 18 October 2024.

The weather data and timing of the three simulated rain events are shown in Figure 3. The average daily temperature for the study period was 18.8 °C. The highest daily temperature was recorded on 4 September with 26.4 °C, while the lowest daily temperature was recorded on 15 October with 7.5 °C.

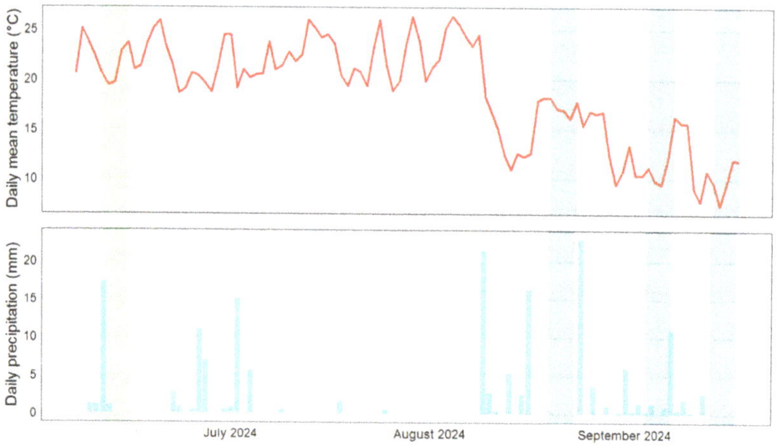

Figure 3. Daily precipitation (mm) and daily mean temperature (°C) during the experimental period. The natural heavy rainfall event is marked in green, while the simulated heavy rainfall events are indicated with dark turquoise stripes.

2.3. *Experimental Set-Up*

2.3.1. Natural Heavy Rainfall Event

On 12 July 2024, a heavy rainfall of 17.7 mm occurred between 12:57 and 13:43 p.m. According to KOSTRA-DWD-2010R (German Weather Service, Offenbach am Main, Germany), this event can be classified as a 2-year heavy rainfall event and was used to investigate the stormwater retention and ET performance of green roofs. Four days before the natural heavy rainfall event, the substrate of three experimental green roofs was saturated and all retention layers were filled to their maximum capacity.

2.3.2. Simulated Heavy Rainfall Events

The regional heavy rainfall data in Germany are available as open data and are provided by KOSTRA-DWD; they have been used to simulate heavy rainfall events in this study. Once a decade, the German Weather Service publishes a breakdown of heavy rainfall intensities for different return periods (1 year–100 years) and duration levels (5 min–72 or 168 h), divided into grid squares. The KOSTRA-DWD-2010R evaluation published in 2017 for the grid square of the south-eastern part of the city of Leipzig (grid square 51055) was used for the simulations.

Heavy rainfall events with different return periods (5, 30 and 100 years) were simulated from 19 September to 18 October 2024. All heavy rainfall simulations were carried out in the morning between 10:00 and 12:00 a.m. To prepare the experimental installations for the 30-year and 100-year rainfall events, a temporary roof was installed prior above the lysimeters to the start of the experiments and during the predicted rainfall events to prevent substrate saturation.

The Euler type II rainfall distribution, which is considered as the standard in Germany [41], was used. Here, the rainfall intensity, typically at 5 min intervals, reaches a maximum at one third of the time of the test duration [42]. The intensities are distributed in descending order, first to the left and then to the right of the maximum (see Supplementary Figure S2). Return periods of 5, 30 and 100 years were chosen for a duration level of 2 h each. The simulated rainfall for a duration of 2 h was 33 mm for the 5-year return period, 50 mm for the 30-year return period and 61 mm for the 100-year return period.

Each simulation was run once, with the same initial conditions for all cases: (i) a completely emptied retention layer, and (ii) a saturation deficit in the substrate layer. To create the saturation deficit, the retention chamber was emptied a few days before the experiment and, if rain was expected, a temporary roof was installed as described above. The test periods were chosen to ensure the highest possible outdoor temperatures, and no precipitation was expected to interfere with the experiment.

2.4. Data Analysis

Determining Water Balance Parameters

A water mass balance equation of the experimental green roof systems is expressed as Equation (1):

$$\Delta S + \Delta R = P + P_{Sim} - ET - F \tag{1}$$

where the water inflow variables of the three experimental green roofs are precipitation P and/or simulated precipitation P_{Sim}. Water outflow leaves the experimental system either by runoff F or by ET. Changes in the water content of the substrate ΔS and in the retention/drainage layer ΔR are also considered (Figure 4). All values are expressed in L m^{-2}. For simplicity, ΔR also includes the water that can be stored in the geotextile (fleece).

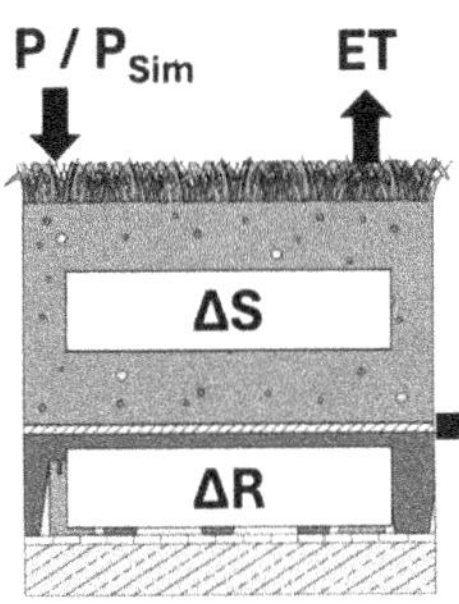

Figure 4. Overview of water balance of the three experimental green roofs. P: precipitation, *ET*: evapotranspiration, ΔS: changes in the water content of the substrate, ΔR: changes in the water content of the retention layer and *F*: runoff.

Furthermore, weight changes ΔW (of the lysimeters) were expressed as stated in Equation (2):

$$\Delta W = P + P_{Sim} - ET - F \tag{2}$$

Thus, evapotranspiration was determined by a rearrangement of Equation (2).

The performance parameters shown in Table 1 were calculated to compare the three experimental green roof designs.

A *t*-test with a significance level of $\alpha = 0.05$ was used to determine the statistical significance of the differences in the performance parameters having at least three values per green roof design.

Table 1. Calculated performance parameters for three experimental green roof designs.

Performance Parameter	Unit	Definition	Evaluation
Peak evapotranspiration ET_p	$\frac{mm}{h}$	Highest hourly ET performance in a certain time frame (absolute value)	$ET_p = max(ET_i)$
Daily evapotranspiration ET_d	$\frac{mm}{d}$	Sum of the hourly ET values over 24 h	$ET_d = \sum_{i=0}^{t=24h} ET_i$
Mean runoff coefficient C_m according to DWA [43]	-	Ratio between the total runoff and total precipitation of a rainfall event	$C_m = \frac{F}{P+P_{Sim}}$
Peak runoff coefficient C_S according to DWA [44]	-	Ratio between maximum runoff and precipitation intensity	$C_S = \frac{F_{max}}{P_{max}}$ or $C_S = \frac{F_{max}}{P_{Sim,\ max}}$
Stormwater retention rate RR	%	Percentage of stormwater retained in the system	$RR = \frac{\Delta S+\Delta R}{P+P_{Sim}} \times 100$
Runoff delay t_{lag}	min	Time difference between the start of the rain event and the first runoff event	$t_{lag} = t_{F,0} - t_{P,0}$ or $t_{lag} = t_{F,0} - t_{PSim,0}$

3. Results and Discussion

3.1. Natural Heavy Rainfall Event

On 12 July 2024, a heavy rainfall event of 17.7 mm occurred, corresponding to a two-year return period event. The substrates of the three experimental green roof designs were almost saturated and all retention layers were almost filled to their maximum capacity just before the rainfall event started.

Figure 5 presents the ET performance and meteorological data following the event. Among the three green roof designs, Green Roof 1 exhibited the highest peak ET rate (0.8 mm h^{-1}), followed by Green Roof 2 (0.7 mm h^{-1}) and Green Roof 3 (0.6 mm h^{-1}). Daily ET rates showed similar trends across all experimental green roof designs. The average daily ET over the measurement period was 4.2 mm d^{-1} for Green Roof 1, 4.1 mm d^{-1} for Green Roof 2 and 4.3 mm d^{-1} for Green Roof 3, respectively, whereas there were no statistically significant differences among them. On the day of the heavy rainfall event, ET rates were notably lower, with Green Roof 1 recording 2.4 mm d^{-1}, Green Roof 2 showing 2.3 mm d^{-1} and Green Roof 3 reaching the highest value of 2.9 mm d^{-1}. On 14 July 2024, the rates slightly increased to 3.7 mm d^{-1}, 3.9 mm d^{-1} and 4.1 mm d^{-1}, respectively. By 15 July 2024, the ET further increased to 5.2 mm d^{-1} for Green Roof 1, 5.3 mm d^{-1} for Green Roof 2 and 5.1 mm d^{-1} for Green Roof 3, respectively.

The ET of green roofs is influenced by several factors, including weather conditions, vegetation type and soil moisture [45]. During heavy rainfall events, the substrate of green roofs becomes saturated, resulting in reduced ET rates. The reason for this is the reduced capacity for further water uptake by plants when the substrate is fully saturated, thereby limiting the ET process. Xu, Chen [45] investigated the ET characteristics of different vegetation types on an intensive green roof under varying weather conditions. The authors found that ET rates and evaporative cooling effects differed significantly among vegetation types, with shrubs exhibiting the highest ET rates, followed by arbors, and grasslands exhibiting relatively lower ET rates. Importantly, the study highlighted that solar radiation and air temperature are the most important meteorological parameters inducing ET on green roofs. Evaporative cooling was observed to be highest in summer and on sunny days, while it was reduced on rainy days, indicating a decrease in ET during these periods. In addition, a study of the rainwater retention of various green roofs found that no ET effects were observed during periods of no rainfall due to the lack of water availability and the limited water retention capacity of the green roof [46]. This suggests that soil moisture availability is a key factor influencing ET rates on green roofs. In conclusion, saturation of the green roof substrate during heavy rainfall events leads to reduced ET rates as the water uptake capacity of the plants is reduced under such conditions.

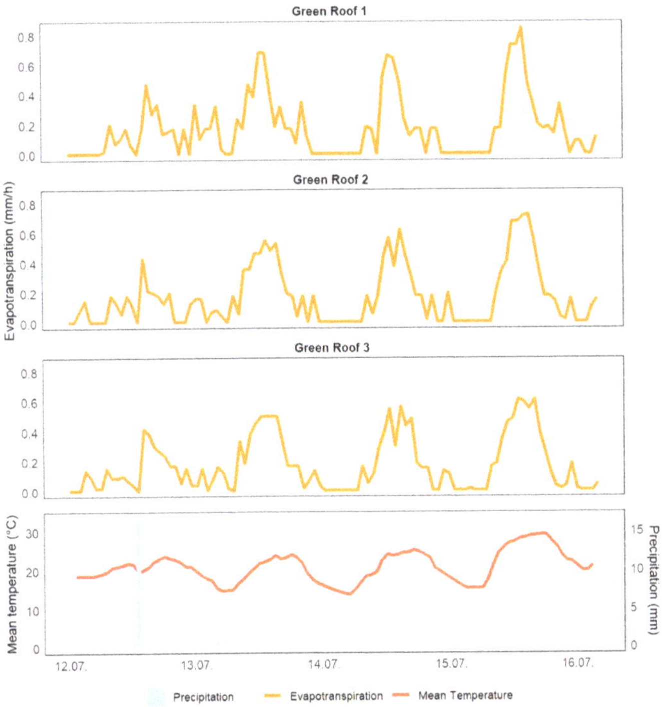

Figure 5. ET, mean temperature and precipitation during and after the natural heavy rainfall event for the three experimental green roof designs.

Due to a malfunction, the tipping bucket of Green Roof 1 failed to record runoff, resulting in a lack of data for evaluation and visualization. In contrast, Green Roof 2 and Green Roof 3 showed significant delays in runoff (Figure 6). The first runoff from Green Roof 3 was recorded 19 min after the onset of rainfall, while for Green Roof 2, it was recorded after 22 min. The total runoff volume was 8.7 mm for Green Roof 2 and 9.6 mm for Green Roof 3, corresponding to mean runoff coefficients C_m of 0.49 and 0.54, respectively. Peak runoff coefficients C_s were 0.49 for Green Roof 2 and 0.31 for Green Roof 3.

The performances of the three experimental green roof designs differed in their runoff characteristics. Although the total runoff delay was similar, Green Roof 2 showed a lower total runoff over a shorter period (64 min) when compared to Green Roof 3 (498 min). Green Roof 3 provided a greater runoff delay, while Green Roof 2 demonstrated better retention, as reflected in the runoff parameters. The lower peak runoff coefficient of Green Roof 3 suggested more effective peak flow attenuation. However, the difference in mean runoff coefficients between the two roof designs remains unclear. Possible reasons can be the variability in substrate composition or measurement uncertainties.

A direct comparison with previous studies is challenging due to differences in experimental conditions. Richter and Dickhaut [33] reported a mean runoff coefficient of 0.2 for a green roof similar to Green Roof 2. However, the authors analysed 453 rainfall events, 75% of which had less than 8 mm of rainfall, making direct comparisons difficult. Palla, Gnecco [47] observed mean runoff coefficients between 0.3 and 0.5 for high-intensity rainfall events (108–194 mm h^{-1} over 15–21 min) on an extensive green roof with a 12 cm substrate layer. The mean runoff coefficients of Green Roofs 2 and 3 were slightly higher, differing by approximately 20–50%, which may be attributed to variations in rainfall intensity, roof design or substrate properties.

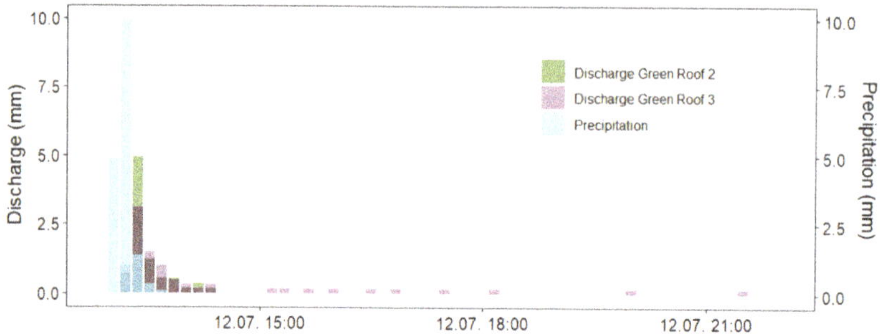

Figure 6. Delay in runoff after the natural rain event for Green Roofs 2 and 3. The rainfall intensity is shown in light blue, while the runoff of Green Roof 2 is shown in green and that of Green Roof 3 in purple.

Figure 7 shows the water balance of two experimental green roof designs, due to a malfunction, the tipping bucket. The greatest runoff was recorded for Green Roof 3 (53%). Only 2.9 mm was retained in the retention layer. Considering a free volume of 13.2 mm, the green roof design could theoretically have retained more than four times as much water in the retention layer. Even considering the maximum error of the soil moisture sensors of ±2 Vol%, the free volume in the retention layer was at least 7.2 mm. However, we suppose that inhomogeneous moisture distribution in the substrate may be the decisive factor. Green Roof 2 also retained less water in the retention layer than expected. With a free storage volume of 14.1 mm, only about one half, i.e., 7.4 mm, was retained. Even considering the error of the soil moisture sensors, the free storage volume of 12.6 mm was significantly higher than the actual volume retained. The high intensity of the rain event, during which the soil was unable to absorb the water, may explain why significantly more water was drained. For Green Roof 3, the higher volume retained in the substrate layer could also play a role (15%). However, considering the water stored in both retention and substrate layers, the rainwater retention was 54% for Green Roof 1, 41% for Green Roof 2 and 31% for Green Roof 3. This result seems to be reasonable, because the highest water deficit in the system ($\Delta S + \Delta R$) was recorded for Green Roof 1 during the drying period, meaning that the system was able to absorb more water. However, the rainwater retention in Green Roof is only an estimate, in contrast to that in Green Roofs 2 and 3. In the case of the two retention green roof designs, Green Roof 2 achieved a higher stormwater retention rate (RR).

Several publications have investigated the limitations of green roofs during heavy rainfall events, focusing on the water retention capacity of the substrate [46,48]. These studies show that while green roofs are effective in managing stormwater, their retention capacity can be exceeded during intense storms, leading to increased runoff. For example, Wang, Garg [49] studied the effect of different green roof configurations on rainwater retention capacity. The results suggest that while green roofs can retain a significant amount of rainfall, their effectiveness is reduced during heavy rainfall events, resulting in increased runoff. Similar, a study of the performance of green roofs for stormwater control by Raimondi and Becciu [50] found that while green roofs can be effective in reducing and delaying runoff, their retention capacity is limited during heavy rainfall events, resulting in increased overflow.

To sum up, the results indicate that while green roofs are beneficial for stormwater management, their retention capacity is finite and can be exceeded during heavy rainfall events, resulting in increased runoff.

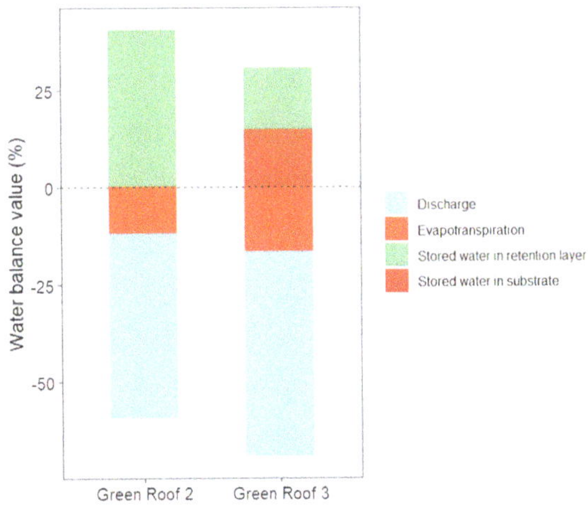

Figure 7. Overall water balance after the natural rain event for two experimental green roof designs.

3.2. Simulated Heavy Rainfall Events

3.2.1. Simulation of the 5-Year Event

Figure 8 shows the ET alongside the weather data, including simulated rainfall. During the two-hour simulation period, the ET at the start was set to zero due to significant weight fluctuations, which caused excessive measurement error. However, during heavy rainfall events, the ET is generally negligible [48,51], underlining the validation of this approach.

The highest ET peak was recorded for Green Roof 1 at 0.7 mm h^{-1}, followed by Green Roof 2 (0.4 mm h^{-1}) and Green Roof 3 (0.4 mm h^{-1}). The daily ET was similar for each test site with 2.8 mm d^{-1}, 2.6 mm d^{-1} and 2.4 mm d^{-1} for Green Roof 1, Green Roof 2 and Green Roof 3, respectively. No statistically significant differences were found among all the test sites. The overall lower daily ET rates compared to the natural rain event can be attributed to significantly lower daily temperatures during the time period of the experiments that were carried out in September.

Similar results were reported for studies conducted in September in previous studies. Gößner, Mohri [52] recorded an ET of approximately 1.5 mm d^{-1} for both a green roof with a design similar to Green Roof 1 and a retention green roof similar to Green Roof 2. Feng, Burian [23] reported an average daily ET of approximately 3 mm d^{-1} for an intensive green roof.

Following the 5-year heavy rainfall event, the ET increased on the second day (21 September) at all experimental sites. For Green Roofs 2 and 3, this increase was already apparent on the first day (20 September). This trend can be attributed to greater overall water availability, particularly in the substrate layer, which dried more slowly due to lower temperatures and retained a greater volume of water.

Runoff was only observed for Green Roof 1, while for Green Roofs 2 and 3, all precipitation was either retained or evaporated within the system. Figure S3 shows a correlation between runoff and precipitation, consistent with the findings of Wang, Garg [53]. The highest mean runoff rate recorded 40 min after the start of the rain simulation was 0.7 mm min^{-1}, which coincides with the peak precipitation rate of 1.1 mm min^{-1}. This gives a peak runoff coefficient C_s of 0.6. Zhang et al. (2021) reported a peak runoff coefficient of 0.6–0.7 for extreme rainfall events (>20 mm) on green roofs with drainage, which is close to the value observed for Green Roof 1 [48]. The total runoff from Green Roof 1 was 26.5 L m^{-2}, giving an average runoff coefficient C_m of 0.80. The first runoff from Green Roof 1 was recorded 18 min after starting the irrigation.

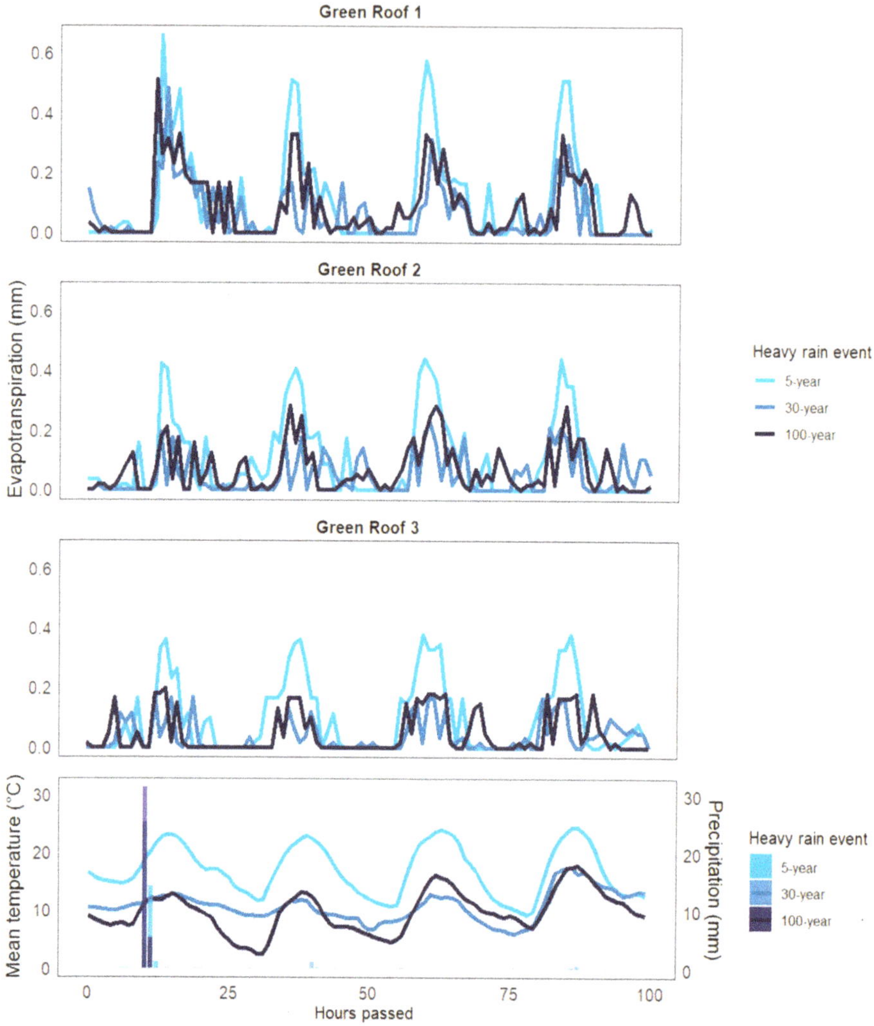

Figure 8. ET of the three experimental green roof designs for three simulated rain events and corresponding weather data (precipitation displayed in a form of a bar chart).

The water balance after 20 h illustrates the distribution of heavy rainfall within the systems (Figure 9). For Green Roof 1, 80% of the water input exited as runoff, while approximately 9% evaporated. The rainwater RR was 11% in this case, with most of the retained water (7%) stored in the drainage and fleece layers. When comparing Green Roofs 2 and 3, a similar pattern was observed, where about 91% of the retained water was held in the retention layer. A smaller proportion remained in the substrate layer—2% in Green Roof 2 and 4% in Green Roof 3, respectively. The RR, calculated as the sum of ΔR and ΔS, was highest in Green Roof 3 (95%), which was closely followed by Green Roof 2 (93%).

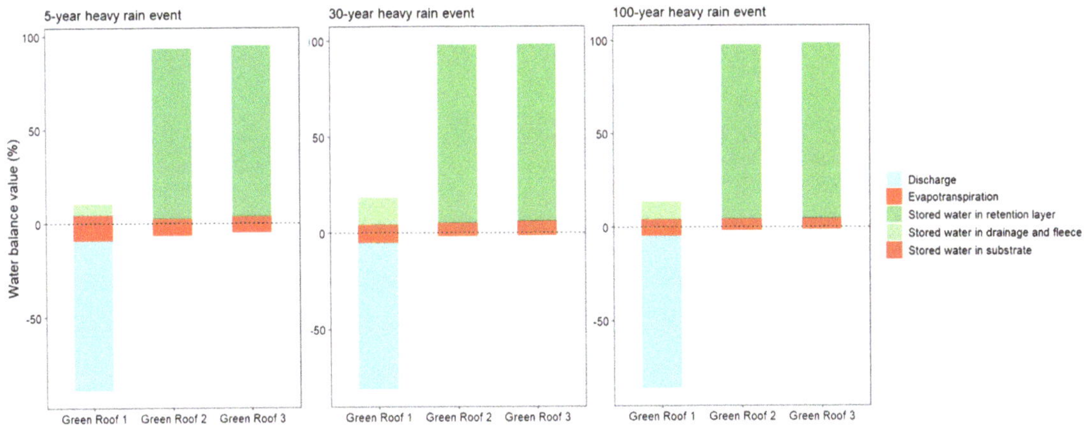

Figure 9. Overall water balance of the three simulated rain events for all three green roof designs.

These results are consistent with those of previous studies. Busker, de Moel [34] reported a RR of 12% during heavy rainfall events (>20 mm h^{-1}) for green roofs without a dedicated retention layer. Similarly, Gong, Yin [26] found that a 50% increase in substrate depth resulted in a 5% increase in retention. In contrast, our study showed that doubling the substrate depth resulted in only a 1.5% retention increase, suggesting that other factors such as substrate composition or drainage dynamics may play a more significant role. Wang, Garg [53] showed that substrate depth had a relatively small effect compared to substrate composition and the presence of a retention layer. However, other studies suggest that increasing the total retention volume is strongly correlated with a higher RR [54]. This pattern is evident in the present study, where Green Roofs 2 and 3, which had larger retention volumes, showed more than an eightfold increase in retention compared to Green Roof 1.

3.2.2. Simulation of the 30-Year Event

The ET in Figure 8 shows a significant decrease in case of a 30-year event when compared to a 5-year event, possibly due to the significantly lower temperatures during the 30-year experiment. Among the green roofs, Green Roof 1 had the highest ET peak of 0.5 mm h^{-1}, while Green Roofs 2 and 3 showed much lower peaks of 0.2 mm h^{-1} each. The pronounced peak in Green Roof 1 occurred immediately after the 30-year heavy rainfall event, suggesting a possible correlation with the saturated substrate. The daily ET was much lower than in the 5-year simulation, with the highest ET_d at 1.6 mm by Green Roof 1, followed by 1.2 mm and 1.0 mm for Green Roof 2 and Green Roof 3, respectively. No significant differences were observed in the daily ET_d of the green roofs. Similar results were reported by Jahanfar, Drake [55], who documented a mean daily ET of approximately 1.8 mm d^{-1} for October.

Currently, no studies have directly examined the behaviour of green roofs with saturated substrates, and such conditions are often excluded from models [25]. A commonly observed phenomenon of green roofs is surface runoff, typically between 5 and 15 mm [56]. However, if surface runoff occurred in Green Roof 1, it was not detectable but could have affected the ET data.

The first runoff of Green Roof 1 occurred 12 min after the irrigation started, i.e., 6 min earlier than in the case of the 5-year rainfall event. Wang, Garg [53] observed a similar reduction in runoff delay (5 min) with an increase in rainfall from 30 mm to 50 mm, but at higher flow rates. They also found that runoff always started at the same cumulative rainfall volume. In the present study, in the case of Green Roof 1, this volume was 2.9 L (5-year event) and 2.7 L (30-year event). The total runoff was 38.4 L m^{-2}, and the mean runoff coefficient C_m of 0.77 was close to the value of the 5-year simulation. The highest mean runoff was reached with 0.7 mm min^{-1} after 40 min (Figure S4). This value was also reported by Wang, Garg [53]. The peak runoff coefficient C_s was, at 0.46, somewhat lower when compared to the previous simulation.

Richter and Dickhaut [33] investigated the runoff behaviour of retention green roofs during a 30-year heavy rainfall event (35 mm in one hour, August 2019). They reported a mean runoff coefficient of 0.48 for a roof design comparable to Green Roof 1, and 0.15 for a roof design comparable to Green Roof 2. The highest runoff contribution of the drainage roof was 1.4 mm min^{-1}. Green Roof 1 had a higher mean runoff coefficient but a lower maximum runoff rate, indicating a greater lag. Green Roof 2 outperformed the comparable retention green roof in Richter and Dickhaut [33] by exhibiting no runoff. This could be due to its larger retention volume (80 mm vs. 30 mm) and its substrate layer being twice as high (15 cm vs. 7 cm). The lack of information on water saturation and pre-filling of the retention layer makes direct comparison difficult. Additionally, Richter and Dickhaut [33] investigated extensive green roofs, which generally store less water than intensive roofs. Nevertheless, both studies confirmed the superior runoff control of retention green roofs.

The water balance after 20 h of the 30-year rainfall experiment is shown in Figure 9. The distribution of runoff water was similar to that of the 5-year event. For Green Roof 1, the percentage of runoff was slightly lower (76%), while the percentage of water stored in the drainage system or in the fleece was slightly higher (14%). This could be due to the drainage system being largely empty following a two-week irrigation break prior to the simulation. However, as the storage volume of the drainage system could not be measured, no definitive conclusions can be drawn in this regard. The amount of evaporated water was slightly lower in Green Roof 1 (5%), while the amount of water retained in the substrate was identical to the first simulation (4%). For Green Roofs 2 and 3, the water distribution was the same as it was in case of the 5-year rainfall event. In both designs, slightly more water was stored in the retention layer than it was in the first experiment (93% for Green Roof 2 and 92% for Green Roof 3). A slight increase in water retention was also observed in the substrate layer (5% for Green Roof 2 and 6% for Green Roof 3). The ET was significantly lower in both green roof designs (2% for each). The RR increased for each test site, with the highest RR of 99% for Green Roof 3, followed by 98% for Green Roof 2 and 19% for Green Roof 1, respectively.

3.2.3. Simulation of the 100-Year Event

The results of the simulation of the 100-year event are shown in Figures 8 and 9. Regarding ET, the results were similar to those for the 30-year simulation (Figure 8). The highest peak ET was recorded for Green Roof 1 at 0.5 mm h^{-1} (vs. 0.3 mm h^{-1} for Green Roof 2 and 0.2 mm h^{-1} for Green Roof 3, respectively). Daily ET was slightly higher than in the previous experiment. No statistically significant differences were found among the experimental sites.

Similar to the 30-year simulation, the higher ET observed for Green Roof 1 may be attributed to greater substrate saturation, which likely reduced runoff and enhanced ET. A slight increase in daily ET was also observed for Green Roof 2 (1.7 mm d^{-1}, compared to 1.2 mm d^{-1} for the 30-year event), especially on the day of the simulation. No significant change was observed for Green Roof 3 (1.2 mm d^{-1}, compared to 1.0 mm d^{-1} for the 30-year event). These results suggest that temporary substrate saturation resulted in a transient increase in ET. An increase in daily ET was observed at all sites on the second day after the simulation, which was consistent with the findings from the 5-year heavy rainfall event.

As shown in Figure S5, which depicts runoff delay patterns for the 100-year event, runoff was recorded only for Green Roof 1, beginning 11 min after rainfall onset. At this time, the cumulative rainfall volume was 3.2 L, which was as high as other cumulative rainfall volumes at runoff initiation. Thus, the 100-year heavy rainfall event also demonstrated that runoff begins at approximately the same cumulative rainfall volume, regardless of the rainfall intensity. The total runoff of Green Roof 1 was 49 L, yielding an average runoff coefficient C_m of 0.82. As in previous simulations, the maximum runoff was recorded 40 min after rainfall initiation, at 0.9 mm min^{-1}. The peak runoff coefficient C_s of 0.42 was comparable to that of the 30-year heavy rainfall event (0.46). In addition to the peak runoff coefficient C_s, the runoff delay t_{lag} (12 min for the 30-year event and 11 min for the 100-year event), as well as the mean runoff coefficient C_m (0.77 for the 30-year event and 0.82 for the 100-year event) exhibited similarities. The mean runoff coefficient for the 5-year event was also in the same range (0.80). A possible explanation is that the percentage increase in precipitation from the 30-year to the 100-year event (23%) was significantly lower than the increase from the 5-year to the 30-year event (50%).

Silva, K. Najjar [57] simulated a heavy rainfall event of 116 mm h^{-1} over 30 min to investigate the runoff behaviour of extensive green roofs (14 cm substrate, 5 cm drainage). The total rainfall (57.9 mm) was comparable to that in the present study but occurred over a shorter duration. The mean and peak runoff coefficients reported were 0.3 and 0.4, respectively, with the latter showing the same value as in the case of Green Roof 1, while the mean runoff coefficient was higher. This suggests that the green roofs in Silva, K. Najjar [57] absorbed more water, possibly due to differing initial moisture or climatic conditions as the simulation was conducted in Brazil.

The water balance, illustrated in Figure 9, remained consistent with the first two simulations. No significant differences were observed for Green Roof 1 compared to previous simulations. Runoff was 82%, close to previous values (80% for the 5-year event and 76% for the 30-year simulation). ET accounted for 5%, similar to the 30-year simulation. The proportion of water in drainage decreased from 14% to 10% compared to the 30-year event, while the relative volume ΔS stored in the substrate remained nearly constant at 4%. No significant changes were detected in the water balance for Green Roofs 2 and 3. Rainwater retention was 14% for Green Roof 1, 98% for Green Roof 2 and 99% for Green Roof 3, respectively.

Silva, K. Najjar [57] reported an increase in RR from 0.68 to 0.82 with an increase in rainfall intensity from 115.8 mm h^{-1} to 145.4 mm h^{-1}. Green Roof 1 exhibited a similar trend, with an increased RR from the 5-year to the 100-year heavy rainfall event. However, a slight decrease was observed between the 30-year and 100-year events. Due to differences in initial conditions and climatic factors, as well as the lack of replication, no definitive correlation between RR and rainfall intensity could be established, consistent with the findings of Silva, K. Najjar [57]. Future studies should incorporate repeated trials and standardized initial conditions to better quantify these relationships.

3.2.4. Summary and Comparison of Heavy Rainfall Events

The heavy rainfall simulations for Green Roof 1 showed similar runoff behaviour and retention capacity. The mean runoff coefficient C_m was consistent across all simulations at approx. 0.80. According to DIN 1986-100 [58], this corresponds to that of an unvegetated flat roof (0.8–0.9, depending on the material). No reference value is given for intensive green roofs with less than 30 cm of substrate. Based solely on the mean runoff coefficient, the intensive green roof without additional retention capacity performed similarly to a conventional flat roof. However, as this study focused only on extreme rainfall events, the long-term mean runoff coefficient is expected to be lower. The peak runoff coefficient C_s of 0.50 on average is in line with the FLL [59] guideline, which specifies a value of 0.4 for substrates of 10–15 cm. Higher values observed in practice, as in this study, can be attributed to structural differences such as the installed drainage system. Compared to an unvegetated flat roof (0.8–1.0, DIN 1986-100 [58)], a significant reduction was achieved. The average RR across all simulations was 15%, excluding the natural extreme event, which was estimated to be higher. This is lower than the 24% reported by Wang, Garg [53] for a green roof with no retention capacity. Runoff retardation decreased with increasing rainfall intensity, which is consistent with the findings of Lee, Lee [27]. Runoff initiation occurred at a similar rainfall volume in all simulations, averaging 2.9 L (2.9, 2.7 and 3.2 L, respectively, for T = 5, 30 and 100 years), consistent with Wang, Garg [53].

For Green Roof 2, an average RR of 96% was achieved with an empty retention layer. As the storage capacity was never exceeded, this result was expected. The high RR during the 100-year heavy rainfall event highlights the efficiency of the green roof. Despite the temporary oversaturation of the substrate, no runoff was recorded. However, as the measurement setup did not capture surface runoff, potential bias cannot be ruled out. This result is consistent with the model-based study by Busker, de Moel [34] of retention green roofs with intelligent runoff control, where retention volumes were actively managed based on weather forecasts. The reported RR ranged from 68% to 99% depending on the accuracy of the forecasts, similar to the present results. In contrast, a natural heavy rainfall event with an 82% filled retention layer resulted in a significantly lower RR of 41%. Busker, de Moel [34] reported an average RR of 59% for green roofs without intelligent runoff control. Raimondi, Marchioni [60] highlighted the strong influence of infiltration on runoff performance and retention capacity. As many studies do not consider infiltration, direct comparisons remain challenging. The peak runoff coefficient C_s of 0.49 was similar to that of Green Roof 1, suggesting that Green Roof 2 had a runoff delay effect comparable to that of a conventional green roof under high prefilling conditions. However, its retention performance was superior, with a mean runoff coefficient of 0.49—significantly lower than the 0.80 for Green Roof 1.

Green Roof 3 had an average rainwater RR of 98%, which was only slightly higher than that of Green Roof 2. A high RR was expected due to the large retention volume. However, during the natural heavy rainfall event—when the retention layer was already 92% filled—only 31% of the water was retained, which was significantly less than in the other experimental setups. Nevertheless, Green Roof 3 showed the best peak runoff coefficient C_s (0.31). According to FLL [59], a peak runoff coefficient of about 0.2 is expected for a substrate height of 25–50 cm, which was almost achieved in this case. The mean runoff coefficient C_m was comparable across all green roof designs, indicating similar runoff behaviour at high initial saturation. In contrast, an empty retention space significantly improved water retention. No clear differences in delay effects were observed between the green roof variants, particularly as no runoff delay was recorded for Green Roof 1 during the natural heavy rainfall event. A t-test revealed significant differences in RR between Green Roof 1 and both Green Roofs 2 and 3, while no significant difference was found between Green Roofs 2 and 3.

Daily ET was statistically analysed with respect to return periods and showed significant differences between simulated and natural heavy rainfall events, as well as between 5-year and 100-year events. However, the correlation between ET and temperature suggests that this significance is primarily due to temperature variations rather than rainfall intensity or return period (Figure 10). Further repeated experiments are needed to refine this relationship.

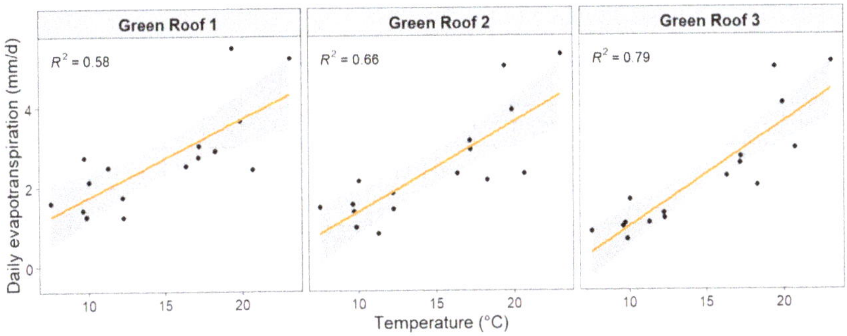

Figure 10. Correlation between ET_d and daily temperature during the heavy rainfall studies. The grey area corresponds to the standard error (confidence interval = 95%).

In addition, the trend shows that ET rates decrease in substrates with higher layers and in large retention layers (Table 2). Intuitively, higher substrates should hold more water, allowing plants to sustain transpiration longer, potentially increasing total ET over a dry period. In addition, deeper soils can support more extensive root systems. However, per unit time, ET rates may decrease in higher substrates due to lower evaporative fluxes from deeper, cooler and more insulated soil layers. Higher soil layers also hold moisture deeper and are less exposed to atmospheric demand (i.e., radiation, wind), thus reducing direct evaporation.

Table 2. Summary of the calculated performance parameters for the observed rainfall events. ET_P: peak evapotranspiration, ET_d: daily evapotranspiration, t_{lag}: runoff delay, C_m: mean runoff coefficient, C_S: peak runoff coefficient and RR: stormwater retention rate.

Green Roof	ET_P (mm h^{-1})	ET_d (mm d^{-1})	t_{lag} (min)	C_m	C_S	RR (%)
			Natural heavy rain event			
1	0.8	4.2	n.a.	n.a.	n.a.	54
2	0.7	4.1	22	0.49	0.49	41
3	0.6	4.3	19	0.54	0.31	31
			Simulated 5-year heavy rain event			
1	0.7	2.8	18	0.80	0.61	11
2	0.4	2.6	-	-	-	93
3	0.4	2.4	-	-	-	95
			Simulated 30-year heavy rain event			
1	0.5	1.6	12	0.77	0.46	19
2	0.2	1.2	-	-	-	98
3	0.2	1.0	-	-	-	99
			Simulated 100-year heavy rain event			
1	0.5	2.1	11	0.82	0.42	14
2	0.3	1.7	-	-	-	98
3	0.2	1.2	-	-	-	99

Several studies have shown that substrate depth is a critical factor influencing the dynamics of ET in green roofs. Voyde, Fassman [61] observed that shallow green roofs with substrate heights of less than 100 mm tended to dry out rapidly, resulting in high initial ET rates but limited overall water retention and reduced ET over time. In contrast, deeper substrates had lower peak ET rates but were able to maintain ET for longer periods. Similarly, Li and Babcock [62] reported that while green roofs with deeper substrates exhibited reduced daily ET rates, total cumulative ET over longer periods could be comparable to or even higher than that of shallower systems due to prolonged plant transpiration. Furthermore, Silva, Paço [63] emphasised that water availability, rather than substrate depth alone, is a key driver of ET. Their results suggest that although higher substrate layers may buffer drought stress more effectively, the magnitude of ET also depends on the adaptations of plant species to water-limited conditions, meaning that additional depth does not necessarily lead to proportionally higher ET rates.

Several mechanisms may explain the observed decrease in ET rates with increasing substrate thickness. Higher substrate layers tend to retain moisture deeper in the profile, thereby reducing exposure to surface evaporation. In addition, greater substrate depths provide better thermal insulation, reducing surface temperatures and, hence, evaporative losses. Furthermore, unless root systems can fully exploit deeper substrate layers, a significant proportion of stored water may remain unavailable for transpiration, limiting plant water use from these zones.

In conclusion, current evidence suggests that while daily ET rates may decrease with increasing substrate depth—particularly in the early stages due to reduced surface evaporation and thermal insulation effects—total cumulative ET may remain stable or even increase over longer periods. This is primarily due to the increased water storage capacity of deeper substrates, which can sustain plant transpiration during prolonged dry periods and, thus, contribute to long-term ET processes.

4. Conclusions

The results show that near-maximum filling of the retention layer leads to a significant reduction in retention efficiency, causing the runoff behaviour of the retention green roof designs (Green Roofs 2 and 3) to resemble that of a conventional green roof design (Green Roof 1) with respect to key runoff parameters. Both Green Roofs 2 and 3 retained significantly less water than theoretically expected, particularly within the substrate layer, in line with the simulation results. Notably, although Green Roof 3 achieved the most effective peak flow reduction and delay—reflected in lower peak runoff coefficient and longer time to last runoff event—it had the lowest overall retention rate of all the variants. When the retention chamber was almost empty, it successfully retained water during all simulated rainfall events (5-, 30- and 100-year return periods). In addition, spreading the inflow over a two-hour period resulted in a slower saturation of the substrate and a more even distribution of water.

These findings underscore a key challenge: Although Green Roofs 2 and 3 theoretically offer sufficient storage capacity in both the substrate and retention layers, the substrate cannot absorb water anymore if saturated. This causes excess water to bypass the substrate and flow directly into the retention layer. If this layer is already full, immediate runoff occurs. An active control mechanism is, therefore, essential to regulate and drain the retention layer, ensuring its capacity is available for subsequent rainfall events.

Naturally, some runoff exists in any green roof system. The goal is not to eliminate runoff entirely—as even the densest natural forests produce some runoff—but to reduce the volume and rate of flow into urban drainage networks during extreme rainfall events. Green roofs are, thus, intended to mitigate peak flows and delay runoff, rather than act as total retention systems.

ET rates tend to decrease in green roofs with higher substrates and larger retention layers. Higher substrates naturally hold more water, allowing plants to sustain transpiration for longer periods and potentially increasing total ET during dry periods. As deeper layers of soil are cooler and more insulated, they experience lower evaporation rates. In addition, moisture stored deeper in the substrate is less exposed to atmospheric influences such as sunlight and wind, which helps to reduce direct evaporation.

This study evaluated runoff behaviour using small-scale lysimeters over a short observation period. While the findings provide valuable insights, several limitations should be acknowledged: The use of small-scale setups may introduce oasis effects, the results may vary under different weather conditions due to the lack of repeated simulations and measurement accuracy poses potential constraints. Future research on larger scales and over longer durations is recommended to validate and expand on these findings.

Supplementary Materials: The following supporting information can be downloaded at: https://www.mdpi.com/article/10.3390/land14050977/s1. Figure S1: Images of the drainage layers and retention boxes. Figure S2: Example of a 5-year heavy rainfall simulation with Euler type II rainfall distribution. Figure S3: Delay in runoff after the 5-year event for Green Roof 2 and 3. Figure S4: Delay in runoff after the 30-year event for the three experimental green roofs. Figure S5: Delay in runoff after the 100-year event for the three experimental green roofs.

Author Contributions: Conceptualization, M.B., A.M. and L.M.; methodology, M.B., A.M. and L.M.; investigation, A.M. and K.B.; writing—original draft preparation, M.B. and A.M.; writing—review and editing, M.B., A.M. and L.M.; visualization, A.M.; supervision, M.B., L.M. and K.B.; project administration, L.M. All authors have read and agreed to the published version of the manuscript.

Funding: This work was funded by the Federal Ministry of Education and Research in frame of the project—LeipzigerBlauGrün—Blau-grüne Quartiersentwicklung in Leipzig "Leipziger BlauGrün II" (BMBF-FKZ: 033W110AN) as part of the funding initiative: BMBF-Fördermaßnahme "Ressourceneffiziente Stadtquartiere für die Zukunft—RES:Z".

Data Availability Statement: The original contributions presented in this study are included in the article. Further inquiries can be directed to the corresponding authors.

Conflicts of Interest: The authors declare no conflicts of interest.

References

1. He, C.; Liu, Z.; Wu, J.; Pan, X.; Fang, Z.; Li, J.; Bryan, B.A. Future global urban water scarcity and potential solutions. *Nat. Commun.* **2021**, *12*, 4667. [CrossRef] [PubMed]
2. IPCC. *Climate Change 2023: Synthesis Report. Contribution of Working Groups I, II and III to the Sixth Assessment Report of the Intergovernmental Panel on Climate Change*; IPCC: Geneva, Switzerland, 2023.
3. Skougaard Kaspersen, P.; Høegh Ravn, N.; Arnbjerg-Nielsen, K.; Madsen, H.; Drews, M. Comparison of the impacts of urban development and climate change on exposing European cities to pluvial flooding. *Hydrol. Earth Syst. Sci.* **2017**, *21*, 4131–4147. [CrossRef]
4. Larsen, T.A.; Hoffmann, S.; Luthi, C.; Truffer, B.; Maurer, M. Emerging solutions to the water challenges of an urbanizing world. *Science* **2016**, *352*, 928–933. [CrossRef] [PubMed]
5. Ariyarathna, I.S.; Abeyrathna, W.P.; Jamei, E.; Chau, H.-W. A Review of the Application of Blue–Green Infrastructure (BGI) as an Effective Urban Flood Mitigation Strategy for Livable and Healthy Cities in Australia. *Architecture* **2023**, *3*, 461–476. [CrossRef]
6. Kaur, R.; Gupta, K. Blue-Green Infrastructure (BGI) network in urban areas for sustainable storm water management: A geospatial approach. *City Environ. Interact.* **2022**, *16*, 100087. [CrossRef]
7. Almaaitah, T.; Appleby, M.; Rosenblat, H.; Drake, J.; Joksimovic, D. The potential of Blue-Green infrastructure as a climate change adaptation strategy: A systematic literature review. *Blue-Green Syst.* **2021**, *3*, 223–248. [CrossRef]
8. Choi, S.H.; Shin, E.; Kim, D.; Song, Y.; Zandaryaa, S.; Makarigakis, A.K.; Kim, J.-y.; Sohn, O.; Clench, C.; Trudeau, M. *Water Security and Cities: Integrated Urban Water Management*; United Nations Educational, Scientific and Cultural Organization (UNESCO), International Centre for Water Security and Sustainable Management: Paris, France, 2023; Volume 4.

9. Stoker, P.; Chang, H.; Wentz, E.; Crow-Miller, B.; Jehle, G.; Bonnette, M. Building Water-Efficient Cities. *J. Am. Plan. Assoc.* **2019**, *85*, 511–524. [CrossRef]
10. Nieuwenhuis, E.; Cuppen, E.; Langeveld, J.; de Bruijn, H. Towards the integrated management of urban water systems: Conceptualizing integration and its uncertainties. *J. Clean. Prod.* **2021**, *280*, 124977. [CrossRef]
11. Furlong, C.; Brotchie, R.; Considine, R.; Finlayson, G.; Guthrie, L. Key concepts for Integrated Urban Water Management infrastructure planning: Lessons from Melbourne. *Util. Policy* **2017**, *45*, 84–96. [CrossRef]
12. United Nations. General Assembly, Transforming Our World: The 2030 Agenda for Sustainable Development. 2015. Available online: https://sdgs.un.org/2030agenda (accessed on 17 February 2025).
13. Breulmann, M.; Müller, R.A.; van Afferden, M. Modelling urban stormwater and irrigation management with coupled blue-green infrastructure in the context of climate change. *Blue-Green Syst.* **2024**, *6*, 100–113. [CrossRef]
14. Almeida, A.P.; Liberalesso, T.; Silva, C.M.; Sousa, V. Dynamic modelling of rainwater harvesting with green roofs in university buildings. *J. Clean. Prod.* **2021**, *312*, 127655. [CrossRef]
15. Moeller, L.; Knapp, S.; Schmauck, S.; Otto, P.; Schlosser, D.; Wick, L.Y.; Georgi, A.; Friesen, J.; Ueberham, M.; Trabitzsch, R.; et al. Gründächer im urbanen Raum und ihre Ökosystemleistungen. In *Die Resiliente Stadt*; Kabisch, S., Rink, D., Banzhaf, E., Eds.; Springer: Berlin/Heidelberg, Germany, 2024; pp. 165–180.
16. Santamouris, M. Cooling the cities—A review of reflective and green roof mitigation technologies to fight heat island and improve comfort in urban environments. *Sol. Energy* **2014**, *103*, 682–703. [CrossRef]
17. Donati, G.F.A.; Bolliger, J.; Psomas, A.; Maurer, M.; Bach, P.M. Reconciling cities with nature: Identifying local Blue-Green Infrastructure interventions for regional biodiversity enhancement. *J. Environ. Manag.* **2022**, *316*, 115254. [CrossRef] [PubMed]
18. Rowe, D.B. Green roofs as a means of pollution abatement. *Environ. Pollut.* **2011**, *159*, 2100–2110. [CrossRef]
19. Akther, M.; He, J.; Chu, A.; Huang, J.; Van Duin, B. A Review of Green Roof Applications for Managing Urban Stormwater in Different Climatic Zones. *Sustainability* **2018**, *10*, 2864. [CrossRef]
20. Twohig, C.; Casali, Y.; Aydin, N.Y. Can green roofs help with stormwater floods? A geospatial planning approach. *Urban For. Urban Green.* **2022**, *76*, 127724. [CrossRef]
21. Zheng, X.; Yang, Z.; Yang, J.; Tang, M.; Feng, C. An experimental study on the thermal and energy performance of self-sustaining green roofs under severe drought conditions in summer. *Energy Build.* **2022**, *261*, 111953. [CrossRef]
22. Wollschlager, N.; Schlink, U.; Trabitzsch, R.; Moeller, L. Weather dynamics affect the long-term thermal and hydrological performance of different green roof designs. *Sci. Total Environ.* **2024**, *957*, 177376. [CrossRef]
23. Feng, Y.; Burian, S.; Pardyjak, E. Observation and Estimation of Evapotranspiration from an Irrigated Green Roof in a Rain-Scarce Environment. *Water* **2018**, *10*, 262. [CrossRef]
24. Czemiel Berndtsson, J. Green roof performance towards management of runoff water quantity and quality: A review. *Ecol. Eng.* **2010**, *36*, 351–360. [CrossRef]
25. Chen, P.-Y.; Hong, X.-F.; Lo, W.-H. Evaluating the stormwater reduction of a green roof under different rainfall events and antecedent water contents with a modified hydrological model. *Ecohydrol. Hydrobiol.* **2024**, *24*, 112–127. [CrossRef]
26. Gong, Y.; Yin, D.; Li, J.; Zhang, X.; Wang, W.; Fang, X.; Shi, H.; Wang, Q. Performance assessment of extensive green roof runoff flow and quality control capacity based on pilot experiments. *Sci. Total Environ.* **2019**, *687*, 505–515. [CrossRef] [PubMed]
27. Lee, J.Y.; Lee, M.J.; Han, M. A pilot study to evaluate runoff quantity from green roofs. *J. Environ. Manag.* **2015**, *152*, 171–176. [CrossRef] [PubMed]
28. Cook, L.M.; Larsen, T.A. Towards a performance-based approach for multifunctional green roofs: An interdisciplinary review. *Build. Environ.* **2021**, *188*, 107489. [CrossRef]
29. Eksi, M.; Rowe, D.B. Effect of Substrate Depth and Type on Plant Growth for Extensive Green Roofs in a Mediterranean Climate. *J. Green Build.* **2019**, *14*, 29–44. [CrossRef]
30. Mentens, J.; Raes, D.; Hermy, M. Green roofs as a tool for solving the rainwater runoff problem in the urbanized 21st century? *Landsc. Urban Plan.* **2006**, *77*, 217–226. [CrossRef]
31. Fernandez-Canero, R.; Emilsson, T.; Fernandez-Barba, C.; Herrera Machuca, M.A. Green roof systems: A study of public attitudes and preferences in southern Spain. *J. Environ. Manag.* **2013**, *128*, 106–115. [CrossRef]
32. Kaiser, D.; Köhler, M.; Schmidt, M.; Wolff, F. Increasing Evapotranspiration on Extensive Green Roofs by Changing Substrate Depths, Construction, and Additional Irrigation. *Buildings* **2019**, *9*, 173. [CrossRef]
33. Richter, M.; Dickhaut, W. Long Term Performance of Blue Green Roof Systems—Results of a Building Scale Monitoring Study in Hamburg, Germany. *Water* **2023**, *15*, 2806. [CrossRef]
34. Busker, T.; de Moel, H.; Haer, T.; Schmeits, M.; van den Hurk, B.; Myers, K.; Cirkel, D.G.; Aerts, J. Blue-green roofs with forecast-based operation to reduce the impact of weather extremes. *J. Environ. Manag.* **2022**, *301*, 113750. [CrossRef]
35. Knappe, J.; van Afferden, M.; Friesen, J. GR2L: A robust dual-layer green roof water balance model to assess multifunctionality aspects under climate variability. *Front. Clim.* **2023**, *5*, 1–13. [CrossRef]

36. Ouedraogo, A.A.; Berthier, E.; Ramier, D.; Tan, Y.; Gromaire, M.C. Quantifying evapotranspiration fluxes on green roofs: A comparative analysis of observational methods. *Sci. Total Environ.* **2023**, *902*, 166135. [CrossRef] [PubMed]
37. Muerdter, C.P.; Wong, C.K.; LeFevre, G.H. Emerging investigator series: The role of vegetation in bioretention for stormwater treatment in the built environment: Pollutant removal, hydrologic function, and ancillary benefits. *Environ. Sci. Water Res. Technol.* **2018**, *4*, 592–612. [CrossRef]
38. Ebrahimian, A.; Wadzuk, B.; Traver, R. Evapotranspiration in green stormwater infrastructure systems. *Sci. Total Environ.* **2019**, *688*, 797–810. [CrossRef] [PubMed]
39. Zhang, Z.; Szota, C.; Fletcher, T.D.; Williams, N.S.G.; Farrell, C. Green roof storage capacity can be more important than evapotranspiration for retention performance. *J. Environ. Manag.* **2019**, *232*, 404–412. [CrossRef]
40. McPhillips, L.E.; Matsler, M.; Rosenzweig, B.R.; Kim, Y. What is the role of green stormwater infrastructure in managing extreme precipitation events? *Sustain. Resilient Infrastruct.* **2020**, *6*, 133–142. [CrossRef]
41. Neumann, J.; Scheid, C.; Dittmer, U. Potential of Decentral Nature-Based Solutions for Mitigation of Pluvial Floods in Urban Areas—A Simulation Study Based on 1D/2D Coupled Modeling. *Water* **2024**, *16*, 811. [CrossRef]
42. DWA (DWA Deutsche Vereinigung für Wasserwirtschaft, Abwasser und Abfall e.V.). *Arbeitsblatt DWA-A 118: Hydraulische Bemessung und Nachweis von Entwässerungssystemen, März 2006, korrigierte Fassung vom September 2011*; DWA Deutsche Vereinigung für Wasserwirtschaft, Abwasser und Abfall e.V.: Hennef, Germany, 2006; p. 35.
43. *Arbeitsblatt DWA-A 117*; Bemessung von Regenrückhalteräumen. Deutsche Vereinigung für Wasserwirtschaft, Abwasser und Abfall e.V.: Hennef, Germany, 2013.
44. *ATV-DVWK-A 198*; Vereinheitlichung und Herleitung von Bemessungswerten für Abwasseranlagen. Deutsche Vereinigung für Wasserwirtschaft, Abwasser und Abfall e.V.: Hennef, Germany, 2003.
45. Xu, H.; Chen, H.; Qian, C.; Li, J. The Evapotranspiration Characteristics and Evaporative Cooling Effects of Different Vegetation Types on an Intensive Green Roof: Dynamic Performance Under Different Weather Conditions. *Sustainability* **2024**, *16*, 10812. [CrossRef]
46. Juras, P. Rainwater Retention Test of Various Green Roofs: Influence on Membrane Temperatures and Evapotranspiration. *Buildings* **2023**, *13*, 2058. [CrossRef]
47. Palla, A.; Gnecco, I.; Lanza, L. Hydrologic Restoration in the Urban Environment Using Green Roofs. *Water* **2010**, *2*, 140–154. [CrossRef]
48. Zhang, S.; Lin, Z.; Zhang, S.; Ge, D. Stormwater retention and detention performance of green roofs with different substrates: Observational data and hydrological simulations. *J. Environ. Manag.* **2021**, *291*, 112682. [CrossRef] [PubMed]
49. Wang, J.; Garg, A.; Huang, S.; Mei, G.; Liu, J.; Zhang, K.; Gan, L. The rainwater retention mechanisms in extensive green roofs with ten different structural configurations. *Water Sci. Technol.* **2021**, *84*, 1839–1857. [CrossRef] [PubMed]
50. Raimondi, A.; Becciu, G. Performance of Green Roofs for Rainwater Control. *Water Resour. Manag.* **2020**, *35*, 99–111. [CrossRef]
51. Soulis, K.X.; Valiantzas, J.D.; Ntoulas, N.; Kargas, G.; Nektarios, P.A. Simulation of green roof runoff under different substrate depths and vegetation covers by coupling a simple conceptual and a physically based hydrological model. *J. Environ. Manag.* **2017**, *200*, 434–445. [CrossRef]
52. Gößner, D.; Mohri, M.; Krespach, J.J. Evapotranspiration Measurements and Assessment of Driving Factors: A Comparison of Different Green Roof Systems during Summer in Germany. *Land* **2021**, *10*, 1334. [CrossRef]
53. Wang, J.; Garg, A.; Liu, N.; Chen, D.; Mei, G. Experimental and numerical investigation on hydrological characteristics of extensive green roofs under the influence of rainstorms. *Environ. Sci. Pollut. Res. Int.* **2022**, *29*, 53121–53136. [CrossRef]
54. Li, S.-x.; Qin, H.-p.; Peng, Y.-n.; Khu, S.T. Modelling the combined effects of runoff reduction and increase in evapotranspiration for green roofs with a storage layer. *Ecol. Eng.* **2019**, *127*, 302–311. [CrossRef]
55. Jahanfar, A.; Drake, J.; Sleep, B.; Gharabaghi, B. A modified FAO evapotranspiration model for refined water budget analysis for Green Roof systems. *Ecol. Eng.* **2018**, *119*, 45–53. [CrossRef]
56. Gan, L.; Garg, A.; Wang, H.; Mei, G.; Liu, J. Influence of biochar amendment on stormwater management in green roofs: Experiment with numerical investigation. *Acta Geophys.* **2021**, *69*, 2417–2426. [CrossRef]
57. Silva, M.d.; Najjar, M.K.; WA Hammad, A.; Haddad, A.; Vazquez, E. Assessing the Retention Capacity of an Experimental Green Roof Prototype. *Water* **2019**, *12*, 90. [CrossRef]
58. *DIN 1986-100*; Technische Regeln für Trinkwasser-Installationen—Teil 100: Schutz des Trinkwassers, Erhaltung der Trinkwassergüte; Technische Regel des DVGW (Deutsche Verein des Gas- und Wasserfaches e. V.). DVGW: Bonn, Germany, 2016.
59. FLL. *Dachbegrünungsrichtlinien—Richtlinien für Planung, Bau und Instandhaltung von Dachbegrünungen*; Forschungsgesellschaft Landschaftsentwicklung Landschaftsbau e.V. (FLL): Bonn, Germany, 2018.
60. Raimondi, A.; Marchioni, M.; Sanfilippo, U.; Becciu, G. Vegetation Survival in Green Roofs without Irrigation. *Water* **2021**, *13*, 136. [CrossRef]

61. Voyde, E.; Fassman, E.; Simcock, R. Hydrology of an extensive living roof under sub-tropical climate conditions in Auckland, New Zealand. *J. Hydrol.* **2010**, *394*, 384–395. [CrossRef]
62. Li, Y.; Babcock, R.W., Jr. Green roof hydrologic performance and modeling: A review. *Water Sci. Technol.* **2014**, *69*, 727–738. [CrossRef] [PubMed]
63. Silva, J.; Paço, T.A.; Sousa, V.; Silva, C.M. Hydrological Performance of Green Roofs in Mediterranean Climates: A Review and Evaluation of Patterns. *Water* **2021**, *13*, 2600. [CrossRef]

land

Project Report

Tree Infiltration Trenches in the City of Leipzig—Experiences from Four Years of Operation

Lucie Moeller [1,*], Katy Bernhard [1], Sabine Kruckow [2], Sabine Wolf [3], Anett Georgi [4], Jan Friesen [1], Katrin Mackenzie [4] and Roland A. Müller [1]

1 Department Systemic Environmental Biotechnology, Helmholtz Centre for Environmental Research (UFZ), Permoserstrasse 15, 04318 Leipzig, Germany; katy.bernhard@ufz.de (K.B.); jan.friesen@ufz.de (J.F.); roland.mueller@ufz.de (R.A.M.)
2 Mobility and Civil Engineering Office, City of Leipzig, Prager Straße 118-136, 04317 Leipzig, Germany
3 Office of Green Space and Waters, City of Leipzig, Prager Straße 118-136, 04317 Leipzig, Germany
4 Department Technical Biogeochemistry, Helmholtz Centre for Environmental Research (UFZ), Permoserstrasse 15, 04318 Leipzig, Germany; anett.georgi@ufz.de (A.G.); katrin.mackenzie@ufz.de (K.M.)
* Correspondence: lucie.moeller@ufz.de; Tel.: +49-341-6025-1847

Abstract: Increasing climate change requires cities to adapt to changing weather conditions. New elements for decentralized stormwater management must be installed to protect the sewer system from overloading during heavy rainfall events and to keep water in the city for irrigation use. A pilot project was implemented in Leipzig in 2020, in which infiltration tree trench systems with three different designs were installed and equipped with measuring technology during a road renovation project. The catchment areas of these three tree trenches are between 215 and 300 m² each. In two of the systems, water retention was included to supply the tree with water during drought periods. The retention elements are sealed with clay in tree trench TT1 and bentonite in tree trench TT3. For tree trench TT2, no retention capacity was provided. This article presents the design, construction, and scientific monitoring of the three tree infiltration trenches. The conclusions after four years of operation from the perspective of two departments of the City of Leipzig are summarized. The tree trench TT1 with the clay pan for water storage shows the best performance in terms of water retention and tree fitness. For the next generation of such infiltration systems, improvements in the design of the street runoff inlets and the surface of the tree trench system's interior are discussed.

Keywords: tree trench; water-sensitive urban design; blue-green infrastructure; urban climate adaptation; stormwater management

1. Introduction

As climate change progresses, cities face the challenge of protecting their citizens and infrastructure from evolving climatic conditions. This primarily involves rethinking how rainwater is handled. Up to now, rainwater has been rapidly drained from urban areas, typically via the municipal combined sewer system and routed to centralized wastewater treatment plants. However, due to the shift in precipitation patterns in the future, heavy rain events will occur statistically more frequently [1]. Existing sewer networks and treatment facilities are not designed to handle such volumes, resulting in the discharge of untreated wastewater into receiving water bodies during storm events due to overflow. This effect is further amplified by the increasing surface sealing of urban areas [2]. In future, rainwater management in cities should therefore increasingly take place at the local level and on a small scale with elements of Water-Sensitive Urban Design (WSUD) [3].

So-called blue-green infrastructures are considered a suitable instrument for a water-sensitive city as stormwater control measures. These are characterized by their function as water management sites and habitats for plants. In the street space, for example, this means linking a water retention system to a tree site. In this way, the incoming rainwater is retained on site and cleaned by passing through the soil and is available to the tree in dry periods. Such a system is called a tree-integrated infiltration trench [4], or simply tree trench. Infiltration trenches consist of a water storage that may be located in small spaces, such as a grassed verge, receiving runoff from a nearby urban surface (e.g., a road or roof) [4]. Several publications show experiences acquired with such constructions. Burge et al. [3] described the installation and operation of bioretention tree pits in Melbourne docklands, Australia. The authors demonstrated diverse types of installed tree trenches with various inlet structures that are used for the treatment of street runoff before it is collected in the lower areas and lead into a river, specifically addressing the challenges that are connected with the construction and operation of the infiltration systems. Another study from Melbourne, Australia, described the functionality of infiltration trenches to provide irrigation water for street trees, although the tree location was not directly part of the infiltration system [5]. The authors showed that the inlet design has a significant effect on successful runoff retention. Caplan et al. [6] presented the monitoring of a tree trench in Philadelphia, USA, containing several trees in one trench. The street runoff inlet occurred at one end of the trench and the water was distributed through it via a perforated pipe. The goal was to provide the trees with consistent irrigation during dry periods. The tree trenches described by Caplan et al. [6] were also the subject of a study conducted by Tu et al. [7], who studied the hydrological dynamics of these systems. The authors pointed out the possible overdesigning of the tree trench systems, which may lead to physiological stress on the trees in the trenches.

The potential benefits of such infiltration systems connected with a tree location are better irrigation status of the trees during dry periods [6], connected with higher evapotranspiration during hot days compared to conventional tree installations, as well as stormwater retention during heavy rain [5]. However, some risks are feared, such as waterlogging, which harms the trees in the trenches [8] or, conversely, too low substrate moisture due to enhanced infiltration for young, still not established trees [6] or due to overdesign with respect to the capacity to store water [7]. In this case, the installation of a capillary column between the water retention and the tree could be a solution. Moreover, appropriate tree species should be selected that cope well with varying amounts of water [8], and especially with prolonged periods of increased substrate moisture.

In the city district Gohlis in Leipzig, the municipality installed three tree trench systems of varying designs as part of a road renovation in 2020. These three locations were equipped with sensors by the Helmholtz Centre for Environmental Research and monitored for a period of four years. This was the first location of such systems in a streetscape in Germany. The aim of this project report is to provide a description of the design of the tree trenches in Leipzig and to outline the comprehensive monitoring concept including the estimation of water-related parameters, tree vitality, and concentrations of pollutants. Finally, the experience gained from four years of operation is described to highlight the opportunities and the challenges involved in installing such innovative systems directly in street spaces.

2. Structure of Tree Trenches

The tree trenches are situated along Kasseler Straße (51°21′46.4″ N 12°21′33.2″ E) in Leipzig-Gohlis and were constructed in the course of a regular road renovation project in 2020 with several new tree planting locations (Figure 1). They were planned by Ingenieurgesellschaft Prof. Dr. Sieker mbH and put into operation in 2020. The trees were planted in late November 2020.

Figure 1. Location plan of tree trenches (TT1–TT3) and tree pits (TP1–TP5) in the street Kasseler Straße, Leipzig, Germany (based on [9]).

The tree trenches are constructions with a total surface area of 16.1 m^2 (7.00 m × 2.30 m) and an inner infiltration area of 12.9 m^2 (6.50 m × 1.99 m) that collect rainwater from the adjacent sidewalk through direct influence (the schemes of the tree trenches are displayed in Figure 2; site photographs are provided in Figure 3). On the road-facing side, there are seven perforated curbstones (each with a length of 0.99 m and a width of 0.25 m) on the long side, with a total of 28 holes (4 holes per curbstone) with a diameter of 50 mm each, enabling road runoff to enter the tree trench. The catchment area was 235 m^2 for TT1, 300 m^2 for tree trench TT2, and 215 m^2 for TT3, including the road surface, parking spaces, and adjacent sidewalks. There is no curb between the sidewalk and the tree trench, allowing unhindered lateral runoff from the pavement directly into the tree trenches. In the center of each tree trench, a ten-year-old small-leaved linden (*Tilia cordata* MILL.), cultivar 'Greenspire', was planted.

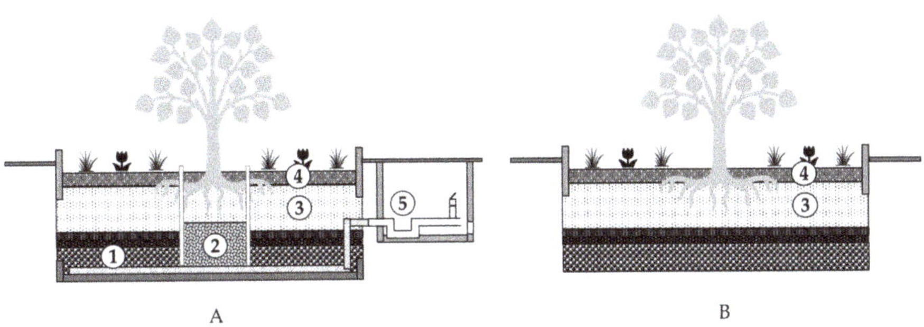

Figure 2. Schemes of the tree trenches TT1 and TT3 (**A**) and TT2 (**B**) with a water retention (1) (sealed with clay (TT1) and bentonite (TT3)), capillary column (2), tree substrate (3), topsoil (4), and overflow pipe (5).

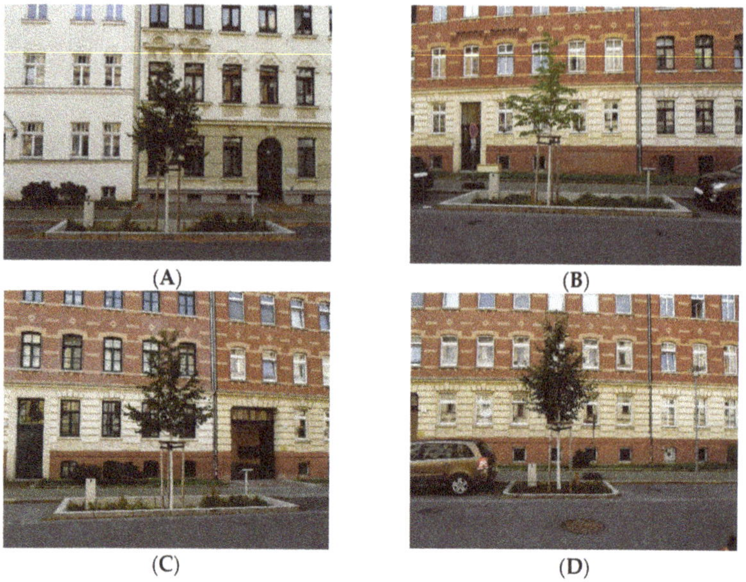

Figure 3. Site photographs of the tree infiltration trenches TT1 (**A**), TT2 (**B**), and TT3 (**C**), as well as the conventional tree pit TP2 (**D**) (taken on 18 October 2023 by author Katy Bernhard).

The substructure of the three tree trenches was designed differently. TT1 was optimized for water retention using a 0.3 m high clay pan with a wall thickness of 0.15 m at a depth of 2.4 m (measured from the top of the curb) that enables water accumulation (Figures 2A and 3A). Between the clay pan at the bottom and curb at the surface, a 0.98 m vertical space is available for infiltration within the trench. On the clay pan, a drainage pipe (DN200) was installed, allowing overflow to connect to the sewage system at a height of approximately 1.0 m above the trench bottom, thus preventing road flooding during heavy rainfall events. The overflow outlet is protected against sewer backflow by a non-return flap. Furthermore, a clayey (SI2) capillary column (1.5 m × 1.3 m × 0.9 m wdh) was placed on the clay pan directly under the future position of the tree to enable the tree to be supplied with water from the clay pan during dry periods. The area surrounding the capillary column was filled with a granite gravel (8/16 mm) layer with a thickness of 0.7 m, forming the primary water retention body. On top of this layer, a separating layer with granite gravel (3/8 mm) with a thickness of 0.2 m was placed. The subsequent layer comprised 0.68 m of tree substrate FLL Type 2 ("substrates of the planting pit construction method 2—covered planting pit", according to FLL [10]; selected specifications are summarized in Table 1), covered with 0.3 m of humus-rich topsoil. In the first year, lawn grass was seeded in the tree trenches around the linden tree; in the second year, it was replaced by a ground cover mat. The mats contained the following plants with 75% coverage by delivery: *Buglossoides purpurcaeruleum* (L.) HOLUB, *Convallaria majalis* L., *Lysimachia punctata* L., *Epimedium* × *versicolor* C. MORREN, *Hemerocallis* sp. L. (three varieties of early, mid-, and late flowering in different shades), *Euphorbia cyparissiens* L., *Gypsophila repens* L., *Bergenia cordifolia* (HAW.) STERND., *Geranium renardii* TRAUTV., *Allium hollandicum* R.M.FRITSCH, and *Crocus flavus* WESTON.

TT2 was designed to optimize the infiltration of the road runoff (Figures 2B and 3B). The construction of this tree trench was simpler than TT1 and TT3: in a 1.7 m deep pit, 0.7 m granite gravel 8/16 mm, 0.2 m granite gravel 3/8 mm, 0.7 m tree substrate FLL Type 2 [10], and 0.3 m humus-rich topsoil were placed. Compared to TT1 and TT3, no water retention layer, no capillary column, and no drainage pipe connected to the sewage system were installed in TT2.

Table 1. Selected specifications for the substrates of tree pits (Type 1) and tree infiltration trenches (Type 2).

Parameter	Type 1	Type 2
Grain size [mm]	0/16	0/32
pH value [-]	5.0–8.5	6.0—7.0
Organic matter [mass %]	2.1–4.2	1.5–2.0
Total pore volume [vol %]	n.s. [1]	45–50
Maximum water capacity [vol %]	Min. 25	Min. 25
Salt content [mg/100 g DM [2]]	n.s. [1]	10–50

[1] n.s.: not specified; [2] DM: dry matter.

TT3 was constructed according to TT1 with a single difference: the bottom of the water retention pan was made from bentonite layer with a wall thickness of less than 0.01 m (Figures 2A and 3C). All tree trenches were designed to perform on a five-year rain event without overflow into the tree substrate. Assuming that the porosity of 8/16 mm gravel is 35% (equivalent to 350 L of water per cubic meter of gravel), each of TT1 and TT3 can be expected to have a storage capacity of 1.16 m^3 of water in their retention pans. On top of that, due to the deepness of the tree trenches (approximately 190 mm vertical distance between the inlet hole and the substrate layer), extended detention on the soil surface is possible to maximize the volume of stormwater that passes through the soil substrate.

All newly planted trees in the location were maintained by a contracted service provider until autumn 2023. Maintenance included irrigation with approximately 100 L per tree, applied in two intervals (2 × 50 L), performed seven times in 2021, four times in 2022, and eleven times in 2023. Moreover, in June 2022 the trees were fertilized. After three years, maintenance was transferred to the City of Leipzig. Watering was carried out four times in 2024. This watering regime is planned to be continued until 2030.

3. Monitoring of Diverse Functions of the Tree Trenches

The installation and operation of the tree trenches was organized by the City of Leipzig, Mobility and Civil Engineering Office. The Helmholtz Centre for Environmental Research used this opportunity to monitor the functions of the TTs. A common tree pit (3.3 m × 2 m × 1.20 m wdh) with the same tree species and tree substrate layer FLL Type 1 ("Substrate of the planting pit construction method 1—open, non-built-over planting pit", according to FLL [10]; selected specifications are summarized in Table 1) was used as a reference (Figure 3D).

3.1. Installed Sensors

Two vertical pipes (0.05 m in diameter) were installed in each tree trench at a distance of approximately 0.85 m to the left and right sides of the tree as seen from the street. The pipes serve to ventilate the lower layers of the trenches. A level sensor was placed in each right pipe to estimate the water level in the retention space (for TT1 and TT3) or in the lower space (in the case of TT2) of the trenches (Figure 4). Each tree trench and the reference were equipped with five soil moisture (SM) measuring sensors (SMT 100, Truebner GmbH, Neustadt, Germany). Three sensors were placed directly under the tree at depths of −1.4 m (SM1), −1.1 m (SM2), and −0.8 m (SM3) (if the surface of the topsoil is taken as the zero point), and two more sensors were placed at the same depth as SM3 at distances of 0.8 m (SM4), and 1.4 m (SM5) from the tree (Figure 4). In the cases of TT1 and TT3, therefore,

SM1 and SM2 were positioned directly in the capillary column. A rain sensor was installed in a private garden on Kasseler Straße to estimate the amount and intensity of rainfall.

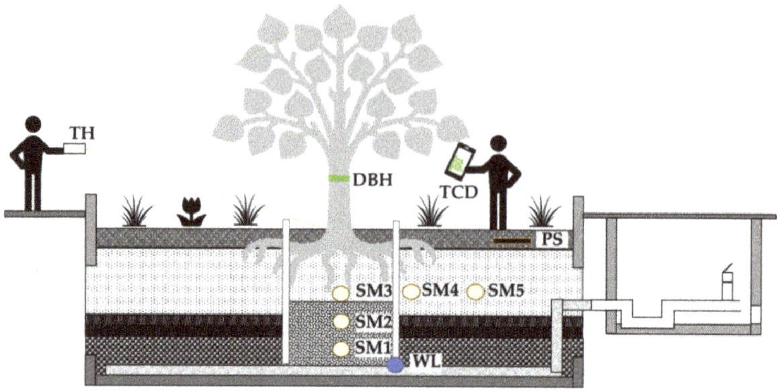

Figure 4. Locations of the passive sampler (PS) and the installed sensors for estimating soil moisture (SM, yellow circles) and water level (WL, blue circle) in the water retention pan. Positions of estimations of stem diameter at breast height (DBH), tree height (TH), and tree crown density (TCD).

3.2. Tree Vitality

All small-leaved linden trees planted as part of the road renovation in 2020—eight in total—were assessed for their vitality. Stem diameter at breast height, mean crown radius, and tree height were measured twice a year (except for 2023) (Figure 4). In addition, during the vegetation periods since their planting, the tree crowns were regularly photographed with a smartphone using a method developed by Sippel et al. [11] for this purpose.

As an example, we show the stem diameter (diameter at breast height, DBH) in this study (Figure 5). The stem diameter of trees in tree trenches that contain water retention is the same or slightly smaller than that of most linden trees in tree pits. However, the stem diameter of the tree in TT2 shows only a small increase throughout the estimated period. The reason for this bad growth is discussed in Section 4. It should be mentioned that tree pit TP4, which had the second-best increase in stem diameter, died in 2024 and had to be replaced.

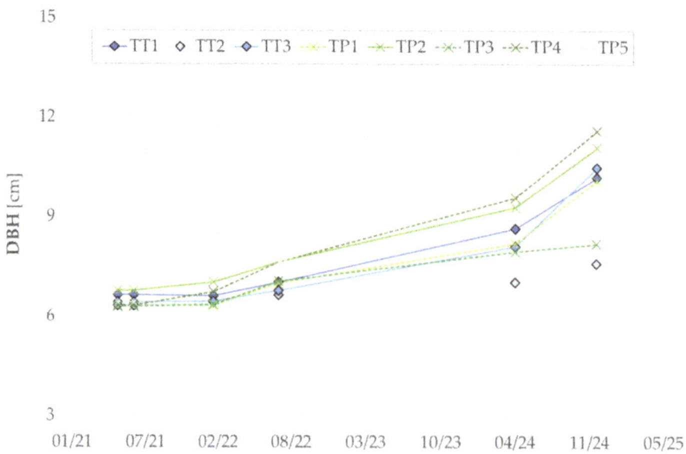

Figure 5. Stem diameter at breast height (DBH) for small-leaved linden trees in tree trenches (TT; curves with diamond symbols) and regular tree pits (TP; curves with cross symbols); TP2 is the reference tree equipped with soil moisture sensors.

3.3. Soil Pollution

The infiltration of surface runoff from an extended catchment area into tree trenches is also associated with an increased input of urban pollutants. As the tree trenches receive runoff from both roadways and pavements, including runoff from building surfaces, a variety of typical urban pollutants can be expected, such as traffic-related chemicals from car bodies, exhaust of combustion engines, tire wear particles, and leachate chemicals, as well as biocides from the paint and plaster of building surfaces [12]. Similarly to natural soils, the plant–soil system in the tree trenches acts as a natural barrier against the infiltration of pollutants into urban groundwater (TT2 type) or their release into the sewer system (TT1 and TT3 types). This barrier function is related to the filtration and sorption functions of the soil, as well as microbial transformation and degradation processes, depending on the biodegradability of the chemical and the actual microbial ecology and habitat function of the system [13]. Long-term experience regarding the transport and fate of urban runoff pollutants in tree trench systems and other blue-green infrastructures is still scarce [12]. The tree trenches in Kasseler Straße offer a valuable opportunity to investigate the retention and potential accumulation of contaminants, such as heavy metals or persistent organic pollutants, in the soils of tree trench systems over extended periods of time and in comparison to conventional street tree pits.

So far, two soil sampling events (12 August 2022 and 26 June 2023) have been carried out in the three tree trenches and one tree pit in the road, in which soil samples were taken from different horizontal (pavement and roadside at a distance of approximately 30 cm from the inflow side for upper-layer samples and 50 cm for lower-layer samples, respectively) and vertical positions (surface layer: soil directly under the plant mat for tree trenches or the gravel layer for tree pits; lower layer: 15 cm below the soil surface). These samples were analyzed for the presence of polycyclic aromatic hydrocarbons (PAHs), which are pollutants resulting from incomplete combustion processes, mainly from the exhaust gases of internal combustion engines and typical tire wear chemicals. Fiber mats, consisting of a commercially available activated carbon fiber felt (dimensions: 10 cm × 10 cm; felt thickness: 2 mm; fiber diameter: 10 μm) in a stainless-steel grid frame for stabilization, were placed at similar positions in the tree trenches (Figure 4) and the reference tree pit TP2. They serve as passive samplers to compare chemical concentrations in the different horizons of the tree trenches and tree pit, providing more integrated information than soil samples taken at different times.

Figure 6 presents selected results from the soil sampling events in the tree trenches and the tree pit in 2022 and 2023. A large number of pollutants typical of road runoff were included in the study. Here, only the detected concentrations of benzo[*a*]pyrene in the soil samples are presented as an indicator compound for PAH. At all sampling points, benzo[*a*]pyrene concentrations were found to be below the limit for residential areas (1 mg/kg) of the German Federal Soil Protection Ordinance [14], while the soil sampled from the surface layer of TT1 on the roadside is close to the limit for children's playgrounds (0.5 mg/L). The example of benzo[*a*]pyrene demonstrates that there can be enrichment effects in soil material over time that are especially relevant to infrastructural units collecting roadside runoff from a larger catchment area. This emphasizes the need to adjust the design of tree infiltration trenches and prevent the leaching of accumulated pollutants into receiving water bodies through structural adaptations. Investigations into this are underway, and the results will be presented in a separate paper.

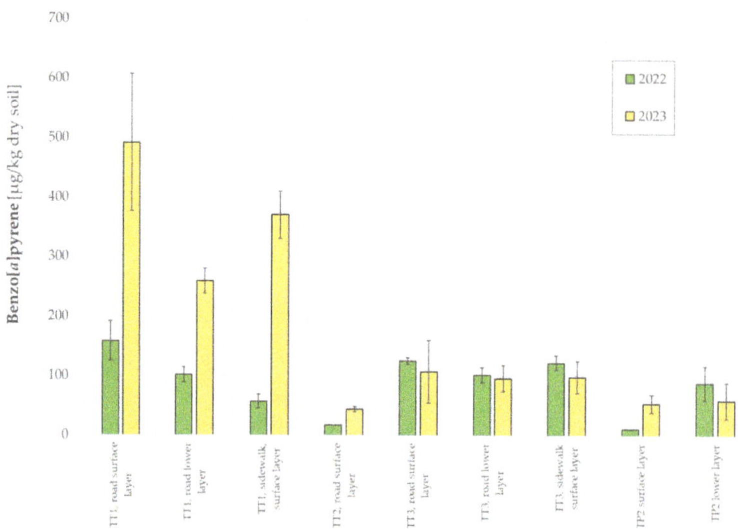

Figure 6. Concentrations of benzo[*a*]pyrene in soil determined at various sampling points of three tree trenches (TT1, TT2, and TT3) and a tree pit (TP2) in Kasseler Straße. The locations of sampling are explained in the text.

4. Lessons Learned

The tree trenches in Kasseler Straße were the first of their kind to be installed in a streetscape in Germany. For this reason, their construction and operational monitoring revealed a range of challenges that are important for the further spread of tree infiltration systems of this type and are summarized in this chapter.

1. Water inflow: The time series of the water levels in the bottom area of all three tree trenches during the entire operation time, from the beginning of the measurements until May 2025, are displayed in Figure 7. In TT1, the water level in the retention zone remained stable throughout the entire investigation period, showing peaks associated with the rainfall events. In contrast, the water level in TT3 over time indicates that the joints in the retention room were not properly sealed during construction, resulting in a reduced water-holding capacity. The water level in TT2 only shows small peaks during rainfall events, which disappear very quickly. In the case of both TT1 and TT3, water level reached heights of more than one meter during stormwater events, three times in the case of TT1 and twice for TT3 (Figure 7). This means that there was most likely overflow into the sewage system (no flow meter is installed in the overflow pipe). A rapid increase in water level was found in the period when the street side inlet holes were manually cleaned by UFZ staff using a bottle brush. The holes tend to become clogged by leaf litter and street debris so that no rainwater enters the tree trenches during smaller rainfall events. Therefore, in a future construction of such trench systems for road runoff, the problem of the water inflow into the tree trench needs to be addressed. For example, curb edges with wider interruptions (approx. 17 cm wide) are conceivable, which would allow for the free inflow of water into the infiltration trench. Problems with the inflow were also reported by Szota et al. [5]. The authors compared two types of inlet systems (lintel and pit) to divert stormwater from the curb and channel into the infiltration trenches. Initially, however, stormwater did not enter some of the trenches, particularly those with pit inlets, indicating that the inlets were restricted. To solve the problem, changes were made to the inlet design by replacing the permeable paver filters installed in the pit inlets with stainless steel

mesh, and to the regular maintenance by cleaning the inlets through emptying the basket filters in all inlets every six weeks.

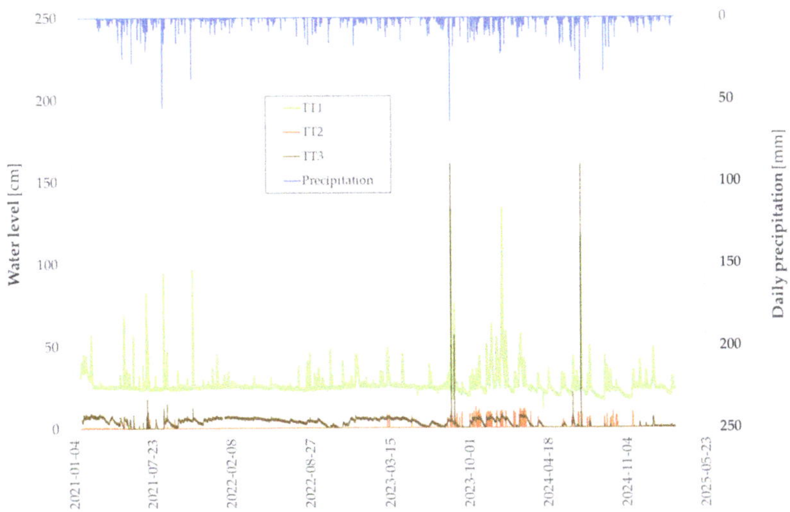

Figure 7. Water levels in the tree trenches and precipitation for the time period from 21 January 2021 to 7 March 2025. Target values of 30 cm (height of water storage layer) and levels of overflow of 100 cm (position of drainage pipe for the connection to the sewage system) are shown for TT1 and TT3 containing retention pans.

2. Construction defects: In addition to the suboptimal performance of the roadside inlets to the tree trenches, design flaws were also identified during operation. In the case of TT3, the bentonite layer was apparently not tested for impermeability, resulting in a significantly reduced water retention capacity within the storage space of the tree trench. For TT2, a very slow water infiltration was observed in infiltration tests carried out in 2021, during which the tree trenches were artificially filled to the top with water and the infiltration capacity was assessed (0.58 cm/h in comparison to 9.41 cm/h in the case of TT1 and 6.14 cm/h for TT3). This situation is caused by excessive compaction during the construction of TT2. A commissioned expert opinion determined very low water permeability coefficients of 8.2×10^{-8} and 9.1×10^{-8} m/s in the lower layers of −95 cm and −110 cm, respectively. In contrast, at the position of −70 cm, a water permeability coefficient of 3.7×10^{-6} was measured, indicating easier infiltration conditions in this layer. As a result of the poor infiltration in the lower part of TT2, the water was retained in the upper layer. This is also reflected in the average values measured by the soil water content sensors: 12.8 ± 0.5% in the case of SM1 vs. 25.9 ± 0.5% for SM3 (average values were calculated for the entire estimated period). Accordingly, the tree in TT2 visibly suffered from waterlogging, as already described by Sippel et al. [11]. The stem diameter of this tree is the smallest of all the trees assessed in Kasseler Straße (Figure 5). However, this tree was not the only one with vitality problems. A tree in a common tree pit in the same street (TD4 in Figure 5), which was planted at the same time as the trees in the infiltration trenches, dried out in its fourth year and had to be replaced. Tu and al. [7] reported that the implications of limited water availability in soil pits to the trees are strongly species-dependent; in their study, *Acer x freemanii* was more sensitive against water stress than *Plantanus x acerifolia*. Vitality problems in trees are also frequently reported as a consequence of road treatment in winter (e.g., in Ordóñez-Barona et al. [15]). Kasseler Straße is a

relatively quiet side street with low traffic volumes and parking spaces for cars on both sides of the road. As a result, there is no winter maintenance and therefore no use of de-icing salt, which could be detrimental to the vitality of the trees. Problems with design errors were also reported by Hanley et al. [4]: only one of three infiltration trenches receiving stormwater from a car park was functioning as designed. The authors showed that trees planted adjacent to the infiltration trenches grew larger than trees planted close to trenches without infiltration. Several issues could probably be solved through more targeted construction supervision, as classic road works (i.e., high soil compaction) are often not compatible with blue-green infrastructures.

3. Character of the pit: The lack of curbstones between the sidewalk and the tree trenches is criticized by residents who see the tree trenches as a potential accident site. To counter this, it would be important to adapt the groundcover vegetation planting accordingly, e.g., by using shrubs. This would also prevent the trees from being misused as bicycle racks and the trenches as rubbish dumps. The current planting using vegetation mats turned out to be suboptimal because the plants on the mats do not properly root into the depth. Another way to prevent accidents and the contamination of the infiltration trenches would be to cover them with a grate, panels, or lids, as described in Burge et al. [3].
4. Costs: As prototypes, the tree trenches in Kasseler Straße cost about 40k Euros each (without monitoring), much more than their successor models would be.
5. Communication: The first communication problems apparently occurred already during the construction. It is common practice to strongly compact all layers in the street space. In the case of TT2, however, this had detrimental consequences. Burge et al. [3] also emphasized: "Clear communication with construction contractors is important to ensure that the design intent of WSUD systems is captured in the built form." The second challenge was to clarify the responsibility for the maintenance of the tree trenches, which are in fact a combination of a tree site and an urban water management facility. A compromise had to be found that will also be a base for future WSUD constructions. Similar experiences have been reported by others. Burge et al. [3] described bioretention tree pits, similar to TT2, where the tree is planted into bioretention filter media and the treated stormwater is then collected via perforated pipes at the base of the cell, before being discharged into stormwater pipes that also act as an overflow. The authors concluded that strong communication between constructors and designers is important to ensure that the functional intent of the WSUD system is transferred from the design to the built form. They also found that in Melbourne, the maintenance responsibilities for the tree pits lie with a number of different departments, and thus the exact responsibility for a particular maintenance task is not always clear.

5. Conclusions

On the way to water-sensitive urban development, local pilot projects are very important for learning how to deal with stormwater through blue-green infrastructure. There are many aspects that can only be assessed using real objects and therefore such projects are of enormous importance. The present article shows the opportunities and challenges involved in implementing new structures in existing road areas. The inlet systems have to be designed in such a way that the runoff can fully enter the infiltration areas. During the construction process, the impermeability of the retention pans should be checked before filling, and the tree substrate should not be excessively compacted. Particular attention should be paid to the underground vegetation, which should cover the entire infiltration area and fill the pit optically. This would support the residents' acceptance of such new features in their street. New pilot systems are planned for the near future, which will be used to retain road runoff in combination with urban tree irrigation in Leipzig. The results of the monitoring outlined here will be used for the improvement of the design of future installations.

Author Contributions: Conceptualization, L.M., K.M., and R.A.M.; methodology, K.B., S.K., and S.W.; investigation, K.B., L.M., J.F., and A.G.; data curation, K.B., L.M., A.G., and J.F.; writing—original draft preparation, L.M., S.K., S.W., and A.G.; writing—review and editing, K.M., R.A.M., and J.F.; visualization, L.M. and A.G.; supervision, L.M., R.A.M., J.F., A.G., and K.M.; project administration, L.M.; funding acquisition, R.A.M. All authors have read and agreed to the published version of the manuscript.

Funding: This work was funded by the Federal Ministry of Education and Research under the frameword of the project Leipziger BlauGrün—Blau-grüne Quartiersentwicklung in Leipzig Leipziger BlauGrün II (BMBF-FKZ: 033W110A), as part of the funding initiative BMBF-Fördermaßnahme Ressource-neffiziente Stadtquartiere für die Zukunft—RES:Z. J.F. acknowledges support by the MULTISOURCE project, which received funding from the European Union's Horizon 2020 program under grant agreement 101003527.

Data Availability Statement: The raw data supporting the conclusions of this article will be made available by the authors on request.

Acknowledgments: The authors would like to acknowledge the additional support and outstanding assistance given by Grit Weichert, Jan Knappe, Ivo Sippel, Amelie Merbach, Emilia Münch, and Emilia Engelhardt (UFZ), as well as the representatives of the Office of Environmental Protection (Christiane Kawe), and Office of Mobility and Public Works (Ralf Czora) of the City of Leipzig, and Matthias Pallasch (Ingenieurgesellschaft Prof. Dr. Sieker mbH). This study was carried out as a part of the Integrated Platform Project "Technologies for Water and Heat Management in Urban Space (CityTech)" under the framework of the Helmholtz Center for Environmental Research (UFZ), Leipzig. We dedicate this article to the memory of Andreas Zehnsdorf, who was one of the initiators for the construction of the tree trenches.

Conflicts of Interest: The authors declare no conflicts of interest. The funders had no role in the design of the study; in the collection, analyses, or interpretation of data; in the writing of the manuscript; or in the decision to publish the results.

Abbreviations

The following abbreviations are used in this manuscript:

DBH	Diameter at breast height
DM	Dry matter
n.s.	Not specified
PAHs	Polycyclic aromatic hydrocarbons
PS	Passive sampler
SM	Soil moisture
TCD	Tree crown density
TH	Tree height
TP	Tree pit
TT	Tree trench
UFZ	Helmholtz Center for Environmental Research
WL	Water level
WSUD	Water-Sensitive Urban Design

References

1. Myhre, G.; Alterskjær, K.; Stjern, C.W.; Hodnebrog, Ø.; Marelle, L.; Samset, B.H.; Sillmann, J.; Schaller, N.; Fischer, E.; Schulz, M.; et al. Frequency of extreme precipitation increases extensively with event rareness under global warming. *Sci. Rep.* **2019**, *9*, 16063. [CrossRef] [PubMed]
2. Zangh, Y.; Carmin, J. Achieving urban climate adaptation in Europe and Central Asia. World Bank Policy Research Working Paper No. 5088. 1 October 2009. Available online: https://ssrn.com/abstract=1494826 (accessed on 20 November 2024).
3. Burge, K.; Allison, R.; Wong, T.; Breen, P. Water sensitive urban design in the Melbourne Docklands—Raingardens and bioretention tree pits. Environment Design Guide. *CAS* **2008**, *47*, 1–12.
4. Hanley, P.A.; Livesley, S.J.; Fletcher, T.D.; Grey, V.; Szota, C. Stormwater retention performance of tree integrated infiltration trenches designed for suburban streetscapes. *Sci. Total Environ.* **2024**, *954*, 176634. [CrossRef] [PubMed]
5. Szota, C.; Coutts, A.M.; Thom, J.K.; Virahsawmy, H.K.; Fletcher, T.D.; Livesley, S.J. Street tree stormwater control measures can reduce runoff but may not benefit established trees. *Landsc. Urban Plan.* **2019**, *182*, 144–155. [CrossRef]
6. Caplan, J.S.; Galanti, R.C.; Oshevski, S.; Eisenmann, S.W. Water relations of street trees in green infrastructure tree trench systems. *Urban For. Urban Green.* **2019**, *41*, 170–178. [CrossRef]
7. Tu, M.-c.; Caplan, J.S.; Eisenmann, S.W.; Wadzuk, B.M. When Green Infrastructure Turns Grey: Plant Water Stress as a Consequence of Overdesign in a Tree Trench System. *Water* **2020**, *12*, 573. [CrossRef]
8. Richter, M.; Heinemann, K.; Meiser, N.; Dickhaut, W. Trees in Sponge Cities—A Systematic Review of Trees as a Component of Blue-Green Infrastructure, Vegetation Engineering Principles, and Stormwater Management. *Water* **2024**, *16*, 655. [CrossRef]
9. Leipzig Giesst. Available online: https://giessdeinviertel.codeforleipzig.de/ (accessed on 5 June 2025).
10. FLL (Forschungsgesellschaft Landschaftsentwicklung Landschaftsbau e.V.). Empfehlungen für Baumpflanzungen—Teil 2: Standortvorbereitungen für Neupflanzungen. In *Pflanzgruben und Wurzelraumerweiterung, Bauweisen und Substrate (Recommendations for Tree Planting—Part 2: Site Preparation for New Plantings; Planting Pits and Root Space Expansion, Construction Methods and Substrates)*, 2nd ed.; FLL: Bonn, Germany, 2010; ISBN 978-3-940122-22-3.
11. Sippel, I.; Moeller, L.; Friesen, J. Cost-effective method for estimation of tree crown density in urban settings using a smartphone. *Blue-Green Syst.* **2023**, *5*, 121–134. [CrossRef]
12. Mutzner, L.; Zhang, K.; Luthy, R.P.; Arp, H.P.H.; Spahr, S. Urban stormwater capture for water supply: Look out for persistent, mobile and toxic substances. *Environ. Sci. Water Res. Technol.* **2023**, *9*, 3094–3102. [CrossRef]
13. Tedoldi, D.; Chebbo, G.; Pierlot, D.; Kovacs, Y.; Gromaire, M.-C. Impact of runoff infiltration on contaminant accumulation and transport in the soil/filter media of Sustainable Urban Drainage Systems: A literature review. *Sci. Total Environ.* **2016**, *569–570*, 904–926. [CrossRef] [PubMed]
14. BBodSchV. Bundes-Bodenschutz- und Altlastenverordnung. 2021. Available online: https://www.bmuv.de/WS6809 (accessed on 30 April 2025).
15. Ordóñez-Barona, C.; Sabetski, V.; Millward, A.A.; Steenberg, J. De-icing salt contamination reduces urban tree performance in structural soil cells. *Environ. Pollut.* **2018**, *234*, 562–571. [CrossRef] [PubMed]

land

MDPI

Article

Nature-Based Solutions and Public Participation: Unpacking Tensions in Sustainable City Development in Northern Europe

Bianka Plüschke-Altof [1,2], Bradley Loewen [3,4], Camilo Calderon [5,*], Mariia Chebotareva [1,4], Reeda Tuula-Fjodorov [1] and Julia Gäckle [6]

1 School of Natural Sciences and Health, Tallinn University, 10120 Tallinn, Estonia
2 Institute of Ecology and Earth Sciences, University of Tartu, 51003 Tartu, Estonia
3 Department of Interdisciplinary Studies of Culture, Norwegian University of Science and Technology, 7049 Trondheim, Norway
4 School of Economics and Business Administration, University of Tartu, 51009 Tartu, Estonia
5 Department of Urban and Rural Development, Swedish University of Agricultural Sciences, 75007 Uppsala, Sweden
6 Institute of Landscape Architecture, RWTH Aachen University, 52056 Aachen, Germany
* Correspondence: camilo.calderon@slu.se

Abstract: Securing the acceptance of nature-based solutions (NbS) in urban greening is central for addressing current social, environmental, and climate-related challenges. To achieve this, participatory planning processes are increasingly encouraged in NbS projects. However, challenges often arise, as NbS and public participation are not always seamlessly compatible. Based on a cross-case, qualitative content analysis of project organizers' experiences in seven urban NbS planning projects in Northern Europe, we unpack three social tensions that can limit the role of participatory NbS: (1) tensions between the openness or "closedness" of environmental and participatory objectives; (2) tensions between the increasing time demands for participatory NbS processes and limitations for largely project-based NbS delivery; and (3) tensions between the roles of expert and lay knowledge. We illustrate these tensions through our cases, showing the challenges that they create for project organizers, both in terms of implementing NbS projects, as well as the risks that they have for broader NbS and participatory goals. We use the experiences and learnings from our cases to suggest paths that can help planners balance these tensions and potentially lead to more inclusive and transformative NbS planning models.

Keywords: nature-based solutions (NbS); participatory NbS; NbS mainstreaming; NbS projectification; urban sustainability transition

1. Introduction

Nature-based solutions (NbS) are defined by the European Commission as actions "inspired by, supported by and copied from nature," which provide environmental, social, and economic benefits, while fostering broader societal transformations for sustainability through "locally adapted, resource-efficient and systemic interventions" [1] (p. 4). In urban green spaces, these interventions include green roofs, green walls, connected parks, and other ways to enhance the quality of natural habitats and increase human–nature interactions [2]. While NbS have made progress in becoming systematized, e.g., through cataloguing and standardization [2–4], scholars have called for a critical reframing centered on their ability to deliver their societal pledges [5].

Such calls advocate a shift from predominantly technical and normative approaches to a more critical, transdisciplinary view that better acknowledges the complexity of sus-

 https://doi.org/10.3390/land14081649

tainability discourses [6] and NbS' potential to support socio-ecological transformation [7]. This implies broadening the focus from technical expert knowledge to include other forms of knowledge (e.g., knowledge from different disciplinary perspectives, lay knowledge, tacit knowledge, etc.), and expanding the involvement of a wide range of actors in the lifecycle of NbS. In practice, this is achieved through participatory approaches to enhance reflexivity in NbS planning and foster long-term sustainability through lasting partnerships, management, and governance [8–13].

However, reframing NbS through participation is neither easy nor straightforward. Echoing broader debates in participatory planning [14–17], several NbS studies have highlighted challenges in relation to structural barriers [18,19], the depth and quality of participation [9,10,13], and the achievement of social outcomes [8]. These studies point to the need for critical research that moves beyond simply promoting the value and ideals of participation for reframing NbS. These demands look further than the implementation of participatory ideals in demonstration NbS projects—often highlighting short-term "successes"—towards understanding the conditions and strategies that support their long-term mainstreaming and impact.

This paper aims to advance the mainstreaming of participatory NbS by moving beyond idealized notions of participation and critically examining tensions that arise in the implementation of participatory ideals in NbS projects. For this purpose, we adopt a critical–pragmatic approach [20] and conduct cross-case analyses of participatory experiences from seven urban NbS projects in Northern Europe. These experiences are discussed in relation to the literature on planning sustainability transitions and participatory practices in NbS.

In the next section, we present the core elements of our conceptual framework for interpreting tensions in participatory NbS. We then introduce our cases followed by a brief description of the methodology and detailed discussion of the results according to three key tensions. Our results highlight the challenges associated with the project-based nature of NbS and their reliance on external funding. We argue that, while such "projectification" has helped to proliferate urban NbS interventions, it has also exacerbated challenges related to participation. We conclude with a discussion that advances ideas and practices of participation in NbS by questioning the projectification of NbS and emphasizing the need for participatory NbS approaches that reconcile long-term ecological sustainability with democratic approaches in planning.

2. Theoretical Overview

We base this paper on theoretical underpinnings and practical implications of NbS in the Northern European context of democratic planning approaches and principles, following the theoretical traditions of participatory, communicative, and collaborative planning [21–23], hereafter referred to as "participatory planning". For the application of NbS, we interpret participatory planning in the context of European Union (EU) initiatives including the "human-centered city," European Green Deal, and New European Bauhaus [24–26], which aim to support social and ecological sustainability. To operationalize participatory planning and EU principles of sustainability, co-creation is promoted as the preferred method for developing NbS in the EU, becoming a core concept for NbS and positioning NbS as a tool of participatory environmental governance and socio-ecological transformation [27–30]. According to the state of the art of social NbS research, two key areas identified for future research relate to mainstreaming NbS (e.g., [31–33]) and ensuring that NbS discourses remain focused on sustainability (e.g., [5,34–36]). Altogether, these set the theoretical orientation of this research.

The following sections expose the abovementioned theoretical strands, policy contexts, and key concepts that underpin the social analysis of NbS projects. We begin by introducing

NbS projects as technical and expert-based (i.e., technocratic) processes. This is followed by critiques prompting calls for more participatory models. We then describe a new wave of participatory NbS and the challenges encountered in their implementation.

2.1. NbS Projects as Technical and Expert-Driven Processes

Definitions of NbS have been developed by expert technical communities, academics, and policymakers, indicating a technocratic trajectory of development, despite often alluding to societal benefits. The European Commission defines NbS as "actions which are inspired by, supported or copied from nature" with potential to provide "co-benefits for health, the economy, society and the environment" as well as "innovation opportunities" [1] (p. 4). Similarly, the International Union for Conservation of Nature (IUCN) refers to NbS as "actions to protect, sustainably manage, and restore natural or modified ecosystems, that address societal challenges effectively and adaptively, simultaneously providing human well-being and biodiversity benefits" [37] (p. 1); see also [38]. Accordingly, NbS have reached the status of an umbrella concept relevant to a range of sustainability frameworks, as it relates to technical and expert-driven initiatives [5].

The documented cases of NbS range from the technical cases—related to farming and food production, green surfaces and building components, and extensive green and blue urban spaces and corridors for rainwater management, watercourse restoration, and renaturalization—to those that are more socially and economically oriented, including artistic installations, theatrical productions, and cultural and behavioral mapping [3,39]. To manage this variety, the IUCN has developed a global standard for NbS that, despite being a technocratic approach to structure the field, gives attention to contextual social and institutional factors for implementation and mainstreaming [4].

The IUCN positions NbS as "a way to operationalize the Ecosystem Approach" [37] (p. 1) (see also [40–43]), bringing a variety of benefits[1]. However, the benefits depend on the types of NbS applied and their interacting technical and environmental systems and communities. Applied to urban projects, NbS integrate the ecosystem approach into urban ecosystems and communities, taking ecological functions as a basis for urban development. Accordingly, the benefits of individual NbS projects are partial and incremental, although they are often designed to be replicable and scalable [1], leading to cumulative environmental and societal impacts.

Despite the claimed attention to social aspects, NbS are still dominated by normative goal setting and a technocratic approach to shift infrastructure from "gray" to "green". This often privileges technical expert knowledge and solutions [7,8,10,12,18], potentially reducing the model of NbS to one of technical feasibility and management rather than one of shared governance and participation. The "democratic and transformative potential" [6] (p. 1) of NbS may thereby go unharnessed. As a result, "key questions on the capacity of NbS to deliver on larger societal pledges remain wide open" [5] (p. 1). While the technical expertise of NbS continues to develop, the social aspects of implementation remain a hindrance to mainstreaming NbS in practice.

2.2. Reframing NbS Towards More Inclusive and Participatory Approaches

Drawing on experiences from the first NbS projects to be funded under the EU's participatory agenda [48], there is a need to focus on planning processes and the long-term sustainability of NbS rather than solely on technical solutions. This has prompted recent calls for a "critical reassessment of participation" [7] (p. 9) as a part of a wider reframing of NbS [34–36,49]. For this purpose, a critical stance must be adopted towards NbS, which marks a shift from the overtly positive framing as an umbrella solution [5]. Faced with the increasing knowledge of challenges and barriers of implementation—as with other areas of sustainability transitions research [50,51]—the direction of progress shifts from technical solutions to the human factors supporting the democratic and transformative potential of NbS. Calls to reframe NbS imply a shift towards more inclusive and participatory approaches, moving away from the established focus on technical ecological centered knowledge and processes. The co-creative context of NbS is based on theories of stakeholder engagement in environmental management [52,53] and participatory planning [22,23,54]. Drawing from these theories, genuine participation is commonly characterized by three core principles: (1) inclusiveness, ensuring the involvement of all relevant stakeholders; (2) power balance, providing participants with equal voice and influence; and (3) consensus building, fostering agreement through deliberation and facilitation. The ideals of participation are centrally positioned in the co-creation approach adopted by European NbS projects [48,55]. By incorporating participatory approaches, NbS projects can build trust and goodwill for collaboration and develop shared narratives and learning opportunities for replication [8,56].

While these represent positive social aspects, critical views also highlight the downsides of participation, where the need to achieve a balance with environmental aspects is a challenge [13]. This has much to do with the tendency of NbS projects towards privilege engineering (i.e., material, technical) knowledge [12,18] and pragmatism in planning [20], on the one hand, in contrast with the empowering aspects of participation that can raise uncertainty around the NbS outcomes [13,57], on the other. To achieve the long-term sustainability aims of NbS, however, "uncertainties should not always be eliminated," and emphasis should be placed on "'culturing' rather than 'controlling' transformations" [7] (p. 3). Such culturing through participation requires resources and expertise for coordination as well as a commitment to lead participants towards a sustainable outcome [13]. By opening up the participatory processes beyond technical actors, it is thought that participatory NbS could achieve a stronger position in urban sustainability actions and become embedded in planning practices [13]. Hence, reframing NbS to be more participatory and open-ended can be a strategy for mainstreaming. After the first phase of technical experimentation [56], mainstreaming is considered to be a long-term social challenge of NbS [11,31,32].

2.3. *Social and Participatory NbS: Co-Creation and Transformation*

The EU's Horizon 2020 and Horizon Europe programs have been instrumental in forwarding the ideas of participatory NbS by funding the first cohort of projects and requiring co-creation in their implementation [48]. Similar to the principles of genuine participation, co-creation refers to the inclusion of diverse actors in all phases of the NbS project cycle including design, production, and governance [55,58]. The evolving EU ideals of co-creation are best represented in the New European Bauhaus Compass guide, emphasizing the participatory process, multi-level engagement, and transdisciplinarity as its core working principles [26]. However, when extending beyond the earlier mentioned technical scope of NbS, meeting the expectations of co-creation remains a challenge. Project leaders must deal with known participation issues from environmental management [53], including the complexity of the ecosystem approach [41], and from communicative planning [22,23], including a lack of environmental awareness amongst participants and balancing different forms of knowledge in decision making [59–62].

Additionally, NbS projects' reliance on external EU funding mechanisms imposes constraints on their capacity for co-creation, whereby the extensive work required is limited by the resources available in the fixed timelines of project cycle stages (e.g., the agenda-setting project call phase, participatory phase, implementation, and project delivery). The constraints of such "projectification" of NbS exacerbate participation challenges, particularly the involvement of different groups including the hard-to-reach and keeping them engaged throughout the process. These difficulties stem from unbalanced trust and power relations, a lack of time to invest in relationships, and the complex facilitation of group dynamics and communication flows [63–66]. This, in turn, contributes to challenges of consensus building, including negotiating conflicting rationalities of stakeholders and difficulties in attaining consensus on controversial decisions [14–16,67].

These challenges point to additional tensions of participation in relation to their transformative potential and the limits of projectification. For the first, participation in NbS can be transformative in terms of actors' agency [68] and institutional capacities [69,70]. This influences the power balances amongst participants and the potential of NbS to stimulate cultural change towards sustainable practices, e.g., through increased human–nature interactions [30]. NbS can thus be viewed as a tool to address complex environmental and climate issues through specialized and co-produced knowledge [61,62,71]. As such, their implementation requires tangible time and knowledge resources that intensify already existing tensions between swift and slow planning processes [72].

This leads to the second point, i.e., the limits of projectification. The fact that NbS are often project-based and temporally limited to the project cycle has an impact on participatory ideals [18,31]: the incentives to participate are temporary and limited by external funding; the principles of inclusion risk not being sustained within existing power structures; and, without a guarantee of citizen empowerment, diverse sources of local knowledge risk not being cultivated and institutionalized over the long term [31,73].

In the next section, we draw on the abovementioned ideas to critically investigate how participatory NbS approaches have been implemented in the European context. Adopting a critical–pragmatic approach, inspired by [20], we emphasize what is realistically achievable in everyday NbS planning practice, rather than relying on normative or idealized visions on how participatory NbS ought to be. Accordingly, our analysis focuses on key tensions that emerged during the project-led implementation of participation in seven NbS projects across the Nordic–Baltic region and the strategies project organizers employed to address them.

3. Methods and Materials

3.1. Case Study Overview

This research follows a multi-sited, qualitative case study approach, drawing on original, primary empirical material—and, in some cases, secondary material for triangulation—to analyze and compare seven cases of participatory NbS project implementations in the Nordic–Baltic region. Mirroring the projectification of NbS, the analyzed cases were part of two EU-funded projects: the Central–Baltic Interreg project *Augmented Urbans* (2017–2021) and the H2020 Research and Innovation project *GoGreenRoutes* (2020–2024). Both EU projects utilized NbS as a central tool for socio-ecological transformation with diverse aims.

Augmented Urbans focused on urban renewal and ecological resilience, using biodiversity enhancement and green infrastructure development in five cases. *GoGreenRoutes* is represented by two cases with a focus on wellbeing enhancement via nature connectedness and active mobility. The NbS included foremost biodiversity—enhancing measures such as sponge gardens, health forests, urban meadows, pocket parks, green corridors, or the curation of biodiverse landscapes, as well as measures such as the retrofitting of existing buildings or light traffic enhancement according to ecological standards to foster sustainable city planning more widely[2]. The cases were situated in the countries of Sweden, Finland, Estonia, and Latvia. While facing similar environmental challenges and conditions for NbS, they differ in political, institutional, social, and economic contexts, e.g., in size and development of urban settlement or legacies affecting planning and participation. Despite these differences, common tensions were experienced.

All cases were implemented with ambitious participatory agendas set at the EU-funded project level, blending digital with non-digital means and co-creation methods. The projects referred to the criteria of the International Association for Public Participation (IAP2) [75], based on Arnstein's famous ladder of participation [21], to select methods to enhance the flow of information, consultations, and collaboration with stakeholders, including art-based methods. Some also aimed towards empowerment by co-creating biodiversity, for example, in the form of window gardens or insect hotels.

As project organizers became aware of the transformatory aspect of NbS as social interventions [68,70,76], many engaged in actions to foster a transformative change. These included transformative knowledge building with biodiversity apps, site walks or nature marathons, and advocacy for gaining legitimacy for their projects, e.g., with image campaigns or active expectation management (for a more in-depth analysis; see [13]).

Table 1 gives an overview of the cases, including their NbS objectives, participatory goals, and the data used in the analysis.

Table 1. Overview of NbS project cases and used data.

Environmental Theme	NbS Synonym	NbS Area	NbS Project Objective	Participatory Goals	Quoted Respondents and Form of Data Collection
Augmented Urbans: ecological resilience and urban renewal	Case 1: *Courtyard Meadows*		Transform courtyards of existing housing estate by engaging people in creating more biodiversity and pollinator-friendly outdoor areas with focus on meadows as an alternative to lawns	Information and consultation–collaboration- and empowerment-enhancing transformative change	Project organizer Helle, in-depth interview (online)
	Case 2: *Street Axis*		Develop underutilized street axis located in the city center into a work and cultural center by deepening the commitment of existing stakeholders to a sustainable planning and retrofitting process	Information and consultation–collaboration-enhancing transformative change	Project organizer Lily, in-depth interview (in-person)
	Case 3: *Bee Corridor*		Transform gray into green infrastructure, following the concept of a bee corridor; this included the construction of a light traffic road, an active linear park as well as the curation of biodiverse landscapes	Information and consultation; collaboration and empowerment; enhancing transformative change	Project organizers Nele and Stiina, in-depth group interview (online) Local resident Marten, supported by input from student works (field visit)
	Case 4: *Main Street*		Develop human-scale and resilient planning solutions for the main street that would decrease commuting to the city center and enhance sustainable mobility options	Information and consultations; enhancing transformative change	Project organizer Robin, in-depth interview (online)
	Case 5: *Riverfront*		Develop a green corridor connecting the urban core with the riverfront by combining tourism and recreation with biodiversity-enhancement and landscapes preservation	Information and consultations; collaboration; enhancing transformative change	Project organizer Zaiga, in-depth interview (online)

Table 1. *Cont.*

Environmental Theme	NbS Synonym	NbS Area	NbS Project Objective	Participatory Goals	Quoted Respondents and Form of Data Collection
GoGreenRoutes: nature connectedness and nearby greenery	Case 6: *Active Mobility*		Develop a street corridor to encourage active, non-motorized mobility and foster the development of safe and inclusive urban spaces, including small pocket parks and green interventions	Information and consultation; collaboration; enhancing transformative change	Project organizer Bella, focus group discussions and in-depth interviews (online and in-person)
	Case 7: *Nearby Greenery*		Maintain biodiversity at the site while diversifying a residential green area through a renewal in accordance with site-specific needs and the intention to make it more accessible and pleasant	Information and consultation; collaboration and empowerment; enhancing transformative change	Project organizers Epp, Maie, and Madle, in-depth group interview (in person) Local residents Annela, Anu, Merike, Tiit, Tarmo, Peeter, Irma, Jaan, and Karmen from open comment survey section and I360, supported by input from student works

Source: Authors' analysis, categorizations based on [10,13,21,53,75]; case and respondents' names have been anonymized.

3.2. Data and Empirics

This study builds on a qualitative research design utilizing in-depth interviews with project organizers of the local cases. From *Augmented Urbans* (15 interviews with six project organizers from five cases), the experiences of project organizers were studied with recurring in-depth interviews conducted in three rounds to allow room for reflection during NbS project implementation. From *GoGreenRoutes* (two interviews with four project organizers in two cases), in-depth interviews with project organizers were held towards the end of the project in 2024. All interviews lasted between 45 min and 1 h 45 min, were conducted in English, recorded with informed consent, and transcribed. To allow for more depth in the analysis, secondary data were available and used to complement the two *GoGreenRoutes* cases, giving insight into local perceptions of the projects. This additional data were collected by the researchers and researcher-supervised student groups (following the principle of research-based teaching [77]) and included fieldwork, in situ interviews with residents (analyzed as the so-called I360 interviews: [78]), and qualitative data from resident surveys (analyzed in more detail: [74]).

As the analysis focuses on the experiences of local project organizers, additional measures had to be taken to ensure the anonymity of interview partners. Pseudonyms are therefore assigned to both individual respondents and specific case locations. Table 1 gives, however, an overview of the central characteristics of the cases studied here.

3.3. Analytical Approach

While the studied projects have been analyzed in greater detail elsewhere [13,74], the qualitative interview data were re-examined with the aim to understand tensions evolving during the implementation of participatory NbS projects and how these projects were handled by the project organizers.

For the purpose of this paper, our use of tensions corresponds with the approach of qualitative content analysis, namely, representing emergent themes from the analysis. For us, tensions describe thematically where social conflicts arise that threaten the long-term sustainability outcomes of NbS. This aligns with prior uses investigating social sustainability aspects in planning, management, and transition fields, relating to similar terms such as ambiguities, contradictions, oxymorons, and paradoxes [79–82]. Of particular relevance to us, tensions convey a "strained dialectical relationship" [83] (p. 844) that can be found in planning politics, discourses, and rhetorics. Tensions are therefore highly contextual, emerging at multiple scales and levels, between various social groups and in connection with specific issues. Therefore, they are suitably applied within an inductive approach that allows them to emerge from and to be compared across cases.

Qualitative content analysis [84] was conducted for the seven cases following the conventional method of (1) manually coding interview transcripts and, if applicable, secondary material used for triangulation and (2) combining the results into analytical categories summarizing these codes. On the highest level of analysis, (3) analytical categories are abstracted into themes. The themes are then formulated as major tensions surrounding the implementation of participatory NbS:

- Tension 1: between open-ended participatory objectives and closed-ended NbS objectives.
- Tension 2: between increasing time demands for participatory processes and limited time for NbS delivery.
- Tension 3: between expert and lay knowledge and their role in participatory NbS.

The findings for each tension are structured around three main content–analytical categories. The first two categories are used to describe each of the three tensions in detail, while the third category discusses the tensions in relation to the dominant projectification of NbS. Altogether, they convey the subjective experiences of project organizers in these seven study cases. Table 2 gives an overview of the analysis, summing up categories and codes for the three tensions.

Table 2. Overview of analytical results.

Theme	Category	Codes
Tension 1: Between open-ended participatory and closed-ended NbS objectives	NbS objectives as too green	• Counteracting nature esthetics and human-centered planning norms. • Increasing risk of green gentrification. • Green objective vs. other pressing local issues and priorities.
	NbS objectives as not green enough	• Questioning the sincerity of green objectives and local government. • Embedded in general skepticism of government-led green transition. • Based on differing definitions of "green".
	Influence of NbS projectification on agenda setting	• Tension between expectations raised by participatory process vs. early-fixed and normative green focus of NbS project, leading to the following: ○ Delimiting participatory openness. ○ Balancing participatory openness with closed-endedness of NbS objectives. ○ Transformative planning measures to nudge acceptance of NbS objectives.
Tension 2: Between increasing time demands for participatory processes and limited time for NbS delivery	Visible time-consuming challenges of participatory NbS process	• Ambitious and exhausting participatory agendas. • Common participatory challenges: motivation, inclusion, and processing output.
	Invisible time-consuming challenges of implementing NbS project results	• Lengthy procurement processes. • Ensuring support for the transformative nature of NbS, including the following: ○ Securing green NbS land use against stakeholder interests with and beyond the government. ○ Securing co-funding.
	Influence of NbS projectification on time resource management	• Tension between time resources due to transformative nature of NbS vs. fixed timelines for NbS projects, leading to the following: ○ Lowering NbS ambitions in terms of implementation. ○ Lowering NbS ambitions in terms of alignment with participatory input.
Tension 3: Between expert and lay knowledge and their role in participatory NbS	Theoretical rationale to support lay knowledge	• Missing experience with new forms of knowledge production. • Need for negotiation of missing or conflicting knowledge.
	Practical drivers to support expert knowledge	• Complexity of NbS. • Need for transformative knowledge production. • Need for evidence-based advocacy.
	Influence of NbS projectification on knowledge production and use	• Tension between theoretical rationale to support lay knowledge and practical drivers to support expert knowledge, leading to the following: ○ Reliance on expert knowledge in decisive phases of the project cycle. ○ Reliance on expert knowledge to negotiate complexities of NbS as a tool and as a transformative project in need of local support.

Source: authors' illustration, based on inductive content analysis.

4. Results

4.1. Tension 1. Between Open-Ended Participatory Objectives and Closed-Ended NbS Objectives

Tension 1 is concerned with the mismatch between the open-ended ideals of participatory processes, in which participants are meant to be able to strongly influence the project, with the foundational nature-orientation of NbS [11,39–42], prescribing ecological outcomes that may not fully align with the priorities or expectations of participants. All of the analyzed NbS projects have a common focus on "green" solutions to foster urban sustainability [38]. While participatory NbS are used to enhance adoption and social acceptance [8,10,34], the project organizers experienced tensions evolving around the question of how green NbS ought to be, as well as critiques of whether the NbS objectives really are as green as claimed, and whether the objectives can be openly set within the limits of the project model. Core to this tension is the "negotiability" of green objectives of NbS.

4.1.1. NbS Objectives as "Too Green"

Particularly in the projects focused on biodiversity, residents critically challenged the NbS objectives as being "too green" and contrary to competing priorities. For many, such as in the *Courtyard Meadows* project, "nice and tidy" lawns still represent cultural norms of "strict and clean environments" (project organizer Helle). The idea of more natural lawns was perceived rather as elitist environmentalism or "laziness" and non-care of the local government [85]. In the *Bee corridor* case, a new management plan for the NbS area that foresaw reduced mowing elicited a series of complaints claiming the site looked unattended. These included "a lively discussion" in social media about "what is going on" (project organizer Nele). As NbS projects widen the implicit human-centered focus of city planning [43,86], they were critically assessed by residents as caring more "about the wellbeing of beetles" than of people who might have other priorities than biodiversity enhancement (resident Marten). Changes in mowing regimes were seen as "something scary, almost" with people "afraid of snakes, ticks and the like" (project organizer Helle). Residents also expressed concerns about socio-economic impacts that such greening could have in the areas surrounding the NbS. This dynamic is discussed more widely, urging for critical NbS impact assessments [11,87]. The highly marketed *Bee Corridor* NbS project was, for instance, seen as a green gentrification threat involving a change in land use that would limit the access of a particular local subculture to nature.

In cases where the green NbS focus was not resisted—or was even supported to "plan urban space with more connection to nature" or "keep as much wilderness as possible" (residents Tiit and Anu)—residents were concerned whether being "green" was really needed the most. This question has been raised by [88] who analyzed whether the green focus of NbS correctly interprets the needs of and benefits to participants. In the *Nearby Greenery* case, these were expressed as feelings of whether important local issues, such as heritage preservation or safety, were being sufficiently addressed in relation to the green objectives: "Unfortunately, the area is still very trashy, I have even seen syringes on the ground. The only good time to go there is in the morning when [...] there are no old drunkards drinking between the trees" (resident Tarmo). The feeling of safety concerns being overlooked was also relevant elsewhere (project organizer Bella), as well as "green" in relation to other priorities such as tourism development, infrastructural planning, or real-estate development (project organizers Zaiga, Stiina and Nele, Robin). All of this illustrates how participatory processes can challenge and potentially undermine NbS' green objectives, forcing project organizers to restrict participants' influence in decision making. This contradicts the open-ended ideals of participatory NbS.

4.1.2. NbS Objectives as "Not Green Enough"

Additionally, the sincerity of green NbS objectives was questioned in relation to greenwashing, as shown by residents' doubts, skepticism, and lack of trust in project owners. In the *Nearby Greenery* case, participating residents asked whether the NbS focus was driven by a desire to gain more votes in the upcoming local elections, with one resident remarking, "This is why they are trying so hard" (Karmen). They were also critical of EU project funding schemes shifting the focus away from "old-fashioned" large-scale parks and green space solutions to small-scale NbS such as green walls or roofs (resident Peeter). Former research [56] has pointed out that "NbS experiments require [. . .] trust between the city and its citizens both for the aim of the experiment and for the experimenting process itself" (p. 106). In this light, the doubts in the sincerity of green objectives seem to be embedded in the general political skepticism and lack of trust towards the city government. Similarly, residents' doubts about the "actual greenness" of the NbS projects were also shaped by connections they made between the initiatives and broader green transition agendas—and their skepticism towards the latter: "I think [climate change] does not even affect [us] very much, hence we need to treat it more soberly" (resident Epp), or: "it's just there to spend money. . ." (resident Jaan).

A part of this skepticism is embedded in different views of what type of "green" solutions should be prioritized in urban projects. As critically discussed in the *Street Axis* case, "green" can point towards different solutions, such as housing energy efficiency, green mobility, or urban ecology—all of them necessitating different priority settings or contradictory measures (e.g., cutting down trees for more sustainable infrastructures). Altogether, these critical voices question how "green" in NbS projects is normatively framed in terms of ecological solutions and whether these are sufficient for reaching broader sustainability goals [34].

4.1.3. Influence of NbS Projectification on Agenda Setting

The projectification of NbS tends to exacerbate the tensions around green objectives due to the requirement for fixed, early agenda setting in the funding application [6]. Only few NbS projects have sought to circumvent this by formulating the project objectives iteratively [12]. Discussions on actual greenness become particularly relevant in a context where the green focus is set as a cornerstone early in the project plan, even before the participatory process is designed or initiated. This was reported in the *Nearby Greenery* case, where project organizer Epp confirmed that the ecological NbS focus was fixed already in the application because it was a condition of EU funding and seemed "tangible to implement". To follow the open-ended idea of participatory processes, project organizers nevertheless asked people, "what do you like?", which generated expectations of including local foci such as safety and historical heritage into the project plan (project organizers Epp and Madle).

The pre-given ecological focus of the NbS created a feeling of non-recognition of residential interests. Some told how they had advocated before to build a heritage memorial, which was declined because "the city has no resources" (resident Annela). Once the resource problem seemed to be solved with new funding, it remained unclear to them why heritage preservation could not have been included as a project objective from the outset. Such a feeling of non-recognition of local needs by an overly green focus could lead to disengagement and lower acceptance of final solutions. At the same time, fully opening the participatory process risks a focus shift away from the green objectives, as conveyed in the critical attitudes towards NbS being "too green" or being green but not necessarily what was needed in the area.

The final NbS plan for the *Nearby Greenery* and *Active Mobility* cases had a focus on local heritage, improved safety, and nature connectedness, conveying that it is possible to go beyond such "either-or" perceptions. However, beyond actively keeping the (discursive) room open for a combination of NbS with local objectives, project organizers also drew boundaries for the NbS projects and the participatory process. This included the clarification of NbS purposes, such as in *GoGreenRoutes*, where iterative loops and diverse communication channels were used to manage expectations during the co-creation process [74]. Moreover, a realignment of the participatory input with NbS objectives was necessary. Convinced that it is "important to make people understand what their input is used for, what can really be done and, more importantly, what cannot," the project organizers of the *Street Axis* case actively communicated that the participatory process is "neither a vote nor a wish list" (Lily). This was also relevant in other cases, where project organizers constantly negotiated local objectives with the NbS focus set by the project plan: "We had to balance. So, if people say that the historical orchard is very valuable, but to preserve it, we need to take down more trees, which is again not really a green solution. Which part to prioritize?" (Maie).

Finally, project organizers also engaged in transformative planning attempts by nudging residents towards NbS objectives. This included mini-window garden workshops and the initiation of a community garden in the *Courtyard Meadows* and *Bee Corridor* cases, with the idea to involve residents early on and create a visible biodiversity-enhancing ecosystem. Developing transformative agency has been recognized as central to NbS implementation [13,57], even though projects often face challenges due to insufficient public resources or a lack of capacity to engage in these transformative attempts [18].

4.2. Tension 2: Between Increasing Time Demands for Participatory Processes and Limited Time for NbS Delivery

Tension 2 lies in the mismatch between the extended timeframes required for meaningful participatory processes to achieve the potential social and ecological benefits of NbS, and the often rigid, short-term deadlines imposed by funding requirements. This disparity reflects a broader tension of "swift" and "slow" planning [72], wherein the demand for efficient decision making and fast implementation comes into conflict with inherently time-intensive participatory processes. The pressure for efficiency in planning has increased with neoliberal and market-oriented planning agendas [6,8]. While time-consuming participatory processes are seen as a major obstacle to increasing time efficiency [72], planners are urged to include different forms of knowledge to address the complexity of emerging challenges [61,66] and to deepen the democracy and social acceptance of planning processes [89].

The case analysis shows that the timeframe tension of NbS projects derives from a combination of highly visible demands for allocating sufficient time to achieve participatory ideals with often less visible governance processes slowing down their implementation. Within the scope of projectification, the pressure for time efficiency creates situations where NbS are only partially implemented or stay at a planning document level. For residents, attempts to reconcile diverging time horizons may result in unmet expectations impeding the same adoption and social acceptance that participatory NbS seek to generate.

4.2.1. Visible Time-Consuming Challenges of Participatory NbS Process

All projects were implemented with ambitious participatory agendas planned at the application stage and steered on a project level: "Actually, for the engagement we had a specific format from the project, so we did not decide when [...] and how we do [it]." (project organizer Maie). The time dedicated to participatory activities led to positive experiences, as participants expressed being "thankful that this green area is finally dealt with" and "approving of the diverse events" (residents Anu and Merike). Moreover, new EU standards such as co-creation or the "leave-no-one-behind" principle helped to empower and monitor inclusion. This was the case in *Nearby Greenery*, where a multi-stakeholder and bilingual approach was used: "I do not feel that we are lacking anyone" (project organizer Maie). However, prolonged participatory processes also resulted in fatigue amongst organizers and participants: "This engagement process was so long. It was too long...".

Such ambitious participatory processes are undoubtedly time-consuming, as they account for the design of participatory activities, for the creation of engagement opportunities for diverse stakeholder groups, and for translation in multilingual meetings and contexts (project organizers Maie and Epp). The need for time resources increased even more when those organizers that wished to engage showed limited interest in the project. In the *Riverfront* case, the tourism sector was central but took little interest in the NbS project that sought to balance biodiversity with touristic development. Mainly, tourism stakeholders did not see biodiversity as attractive enough to create "revenue opportunities sometime in the future" (project organizer Zaiga). Others also had to dedicate extra time motivating participants. In the *Courtyard Meadows* case, project organizers addressed this with initially unplanned activities, nudging residents towards NbS objectives with pollinator-friendly window gardens. On top of that, the analysis of participatory input is time-intensive (project organizer Helle). This is not only true for technicalities and feasibility, but also regarding the communication and justification of decisions to include or exclude inputs from the final NbS solution.

4.2.2. Invisible Time-Consuming Challenges of Implementing NbS Project Results

While participatory processes need visible time-consuming efforts, significant time resources are also required for governance processes underlying NbS implementation [89]. These often remain invisible to participants who expect swift decision making and implementation, which raises the risk of participation fatigue: "I understand them because you have been engaged like a hundred times, but nothing is ready" (project organizer Madle). Among the less visible challenges are time-intensive procurement procedures, depending on local capacities and resources [18], as well as the need to gain political goodwill by negotiating stakeholder interests with and beyond the government. Lengthy procurement processes not only consume time but are also often difficult to communicate and to fit into the project logic, which tends to lack frameworks for the actual implementation of NbS plans [8,10]: "We work for the city government, and [the EU project organizers who wrote the application] have no clue how the city works. It's not like we're engaging people and then going to construct something. No, there's a lot of [...] papers to fill and licenses to gain, making those public procurements" (project organizer Madle).

Next to lengthy procurement processes, the transformative nature of NbS projects entails the need to ensure support for their actual implementation and long-term management. These include securing the green NbS land use against other stakeholder interests and securing co-funding—activities often invisible to the residents and partially unexpected for project organizers [13]. NbS projects in urban areas take place in the context of competing land use interests within a limited space. In cases where NbS were used as means for urban renewal, this contestation was particularly pronounced. In the *Main Street* case, situated in a suburb area, the threat of real estate development to green spaces has been documented over the last decades. During the participatory process, project organizers highlighted that "there seemed to be a strong bias towards the developers" (Robin) who have more resources to make their voices heard in the local council. For the project organizers, this meant using tangible time negotiating stakeholder interests "behind the scenes": "This is a tough nut to crack: how to communicate with real estate developers, how to make them understand?"

Other projects' green land uses were contested by infrastructural developments, such as when the transport department wanted to use the *Bee Corridor* area for a motor highway: "When the land ownership was finally transferred from the state to the municipality, which freed space [for the NbS], the transport department put the plans for a motorway road back on the table" (project organizer Stiina). When funding only covers part of the NbS, there is the need to secure extra local funds including political negotiations that come with it: "But that is the point. None of the small parts would mean anything there. And really you cannot do anything with [that sum of money] that people were saying that they want, such as accessibility. You cannot do accessibility with [that sum of money]. Sorry, you just cannot!" (project organizer Maie). As the extra money comes from local funds, it raises questions of "priority-setting inside the city" in a context where the project site might "not be a priority place in the city."

4.2.3. Influence of NbS Projectification on Time Resource Management

Projectification tends to exacerbate this tension due to relatively fixed timeframes and deadlines of NbS project plans, even when flexible and tangible time resources are needed for the participatory process and to govern the NbS development itself. How visible and less-visible time-consuming tasks are accounted for also plays a role, as project plans tend to focus more on the visible time resources required for participatory activities, while neglecting less visible processes, with frameworks, capacities, or resources oftentimes missing [8,10,18]. Moreover, the fact that the transformative nature of NbS is not sufficiently acknowledged in project plans [13] leads to a lack of time resources for activities to gain legitimacy for the NbS project within the local governance system, such as to secure land use. In the *Bee Corridor* case, an image campaign to "create a pool of supporters in and outside the government" was developed to tackle the land use conflict with the transport department (project organizer Stiina). In the *Street Axis* case, private owners and developers were convinced to build in accordance with environmental governing principles using a "give and take" approach that offered shorter procurement periods for environmentally oriented retrofitting projects (project organizer Lily). In other cases, such as *Nearby Greenery*, it was not possible to prevent competing land uses, leading to a time-intensive relocation of the project site. These invisible stakeholder negotiations are particularly time-intensive amongst changing government coalitions, where goodwill needs to be repeatedly ensured: "People are changing in the city government. You have to get somebody on board–I do not know, twice or so–explaining why you are doing things" (project organizer Madle).

The resulting lack of time can affect participatory processes, becoming less inclusive and influential at final project stages. Furthermore, it can lead to situations where ambitions for implementing NbS plans are either lowered (e.g., replaced by planning documents/architectural designs instead of building the actual NbS), or less ambitious NbS are implemented that meet the funder's demands but might be less aligned with the participatory input. This can conflict with the expectations raised during participatory processes.

An illustrative example is *Nearby Greenery*, which ran out of time due to several constraints, including the relocation of the site, lengthy procurements, and the need to secure additional local funds. Even though a landscape architectural solution was developed in close alignment with the participatory inputs, the final project output will only achieve an interim stage of the planned development, with the full implementation now pending on local follow-up funding: "Because our NbS will not be ready by the end of the project, we are having a small NbS before [in form of] a pollinator friendly sponge garden" (project organizer Madle). "[It] is not something that people asked for, but it is ticking all boxes. It's

NbS, inviting more pollinators, and the orchard is very well connected... It's something new that could be done anywhere in the city, not site specific" (project organizer Maie). In other cases, such as *Active Mobility* and *Bee Corridor*, the final NbS construction processes were delayed, causing pressures among project organizers on how to meet the set deadlines, and leading to applications for deadline prolongation and critical scrutiny among citizens (project organizers Bella, Nele, and Stiina).

4.3. Tension 3: Between Expert and Lay Knowledge and Their Role in Participatory NbS Processes

Tension 3 concerns the balancing of lay and expert knowledge in participatory NbS projects. While the inclusion of diverse, place-based, and tacit knowledge is widely acknowledged as a core component of participatory NbS [6,8,10,34], our findings reveal a persistent reliance on expert-oriented knowledge during implementation.

In the literature and policy documents, the essential role of lay knowledge for addressing complex ecological and climate challenges targeted in NbS is increasingly acknowledged, indicating a shift in traditional roles, where planners act as mediators between stakeholders and their knowledge rather than as decision makers [13,90,91]. This recognition underscores the need for collaborative, transdisciplinary learning between actors [61,62], with particular emphasis on tacit, place-based, and "wordless knowledge" [92] (p. 327) (see also [71]). Nevertheless, expert knowledge is required for NbS due to its technical and ecological complexity [57,93] and, moreover, to counteract populist movements and ill-informed or intentioned segments of the public [94]. Being faster to recruit than participatory lay knowledge, the inclusion of expert knowledge might be incentivized due to the time pressures in NbS projects [41,57]. The same dynamic is visible in the analyzed NbS cases. However, they also show that reliance on experts increases due to the need for transformative knowledge dissemination, where expert knowledge is strategically used to encourage residents' alignment with NbS objectives or facilitate evidence-based advocacy with stakeholders.

4.3.1. Theoretical Rationale to Include and Support Lay Knowledge

While there is a strong theoretical rationale to support lay knowledge, existing research points to the challenges of gaining access to lay knowledge due to a lack of skills in process design and facilitation methods [95,96] and practicalities of participatory processes, e.g., to align time schedules of experts and lay people, or to gain access to different knowledge at the right time [18,34]. Challenges in integrating diverse forms of knowledge also stem from the capacities of residents engaged in the participatory process, who might lack experience with "new" co-creative ways of inquiry. This happened in *GoGreenRoutes*, where the canvas method was used to "discuss, visualize, and write [...] opinions and wishes for the green spaces," which for some felt very unfamiliar: "at the beginning, the canvas was just a white sheet of paper [...] no one wanted to start" (project organizer Bella, [74] (p. 10)). "Just asking about the needs" without a concrete proposal to discuss yet was rather uncommon: "When we had those engagement meetings, people were always asking why you engaged us at that point, when you have your plans not finalized" (project organizers Epp and Madle).

Project organizers perceived a need to guide missing or conflicting lay knowledge with regard to NbS objectives in ways that would support the effective implementation of project results. This challenge was acknowledged in previous studies in urban ecology, where participatory processes aim to incorporate and transform lay knowledge, seeking to address disconnection from nature or limited environmental awareness [59,60]. Dealing with this involved monitoring the communication around the NbS project to provide input and information where misunderstandings could arise. In the *Nearby Greenery* case, major discussions evolved in social media around maintenance tree cutting and the wetland in the area, which generated quick reactions via different channels, also involving local experts (project organizer Maie).

4.3.2. Practical Drivers to Include and Support Expert Knowledge

While there is a theoretical rationale to increasingly include lay knowledge in NbS processes, the cases conveyed the manifold practical drivers to draw on expert knowledge. Dorst et al. [18] observed the same tendency, pointing out that "the dominance of one type of expertise is likely not incidental but reflects power relations between actors" (p. 5). Our interviews convey that a part of the dominance of expert knowledge stems from the different functions for which knowledge is used in NbS projects. On the one hand, expert knowledge is drawn on as technical, academic, ecological input, in order to explain the complexity of the issues at stake (e.g., [6,18,59,60]). In the projects studied here, expert knowledge was used to increase the environmental awareness of residents, e.g., in forest management, dendrology, or local biodiversity and pollinator-friendliness (*Riverfront*, *Nearby Greenery*, and *Courtyard Meadows* cases).

On the other hand, expert knowledge fulfilled important functions for transformative knowledge production to induce behavioral change [13,57,70] and to develop a feeling of ownership of NbS [90]. Following a learning-by-doing approach to overcome the "knowing-doing gap" relevant for transforming socio-ecological systems [97], e.g., in the *Courtyard Meadows* case, an app was initiated for residents and green space managers who could "walk on site with the app open, point to a flower bed and get information on the type of plant they are looking at, as well as the reason behind planting them [there], their characteristics, when they flower, what benefits they give to the ecosystems and what type of conditions they need" (project organizer Helle).

Other project organizers used in situ workshops with local experts, like nature observation marathons (*Bee Corridor* and *Riverfront* cases). This was seen as particularly beneficial to lower objections and raise ownership of NbS objectives. Project organizer Zaiga in the *Riverfront* case remembers participants saying, "'Ooh, I cannot see the river from here,' or, 'Why is this forest so messy?' and then the biologist could explain why it's important to not clean it out." In situ workshops were also intended to mediate the complexity and conflictuality of NbS measures by using experts as knowledge brokers: "We were the most concerned about the fact that we, at the end, are going to take down about 100 trees, [. . .] and we did a walking tour together with the dendrologist who explained that on the spot, so that people could understand that this is not a light decision" (project organizer Maie).

Beyond that, expert knowledge was strategically relied on for evidence-based advocacy towards other stakeholders, such as among other city departments, to gain legitimacy for the NbS project. In all cases, scientific research on landscape and biodiversity or pollinator mobility, as well as design solutions, was prepared. These functioned as inputs to planning processes but were also referred to in land use interest conflicts: "We needed less salesmanship for our vision after it had been approved by experts" (project organizer Robin); the research findings "worked as a strong argument during institutional discussions" (project organizer Stiina).

4.3.3. Influence of NbS Projectification on Knowledge Production and Use

Regarding knowledge production and use in participatory NbS processes, the projectification trend tends to have a twofold effect. On the one hand, the inclusion of different forms of knowledge is fostered by ambitious participatory designs, which are part of funders' requirements. Among these are the "leave-no-one-behind" principle central to the 2030 Agenda for Sustainable Development and the European Green Deal [24,98]. Such enhanced efforts on inclusion have the potential to leave participants with novel but positive experiences, e.g., as in *GoGreenRoutes* [74]. On the other hand, the project cycle and complexity of NbS written into project applications influences how expert and lay knowledge are balanced against each other [6,18].

Although the prevalence of expert input is common in the project call-creation and application stage, experiences of project organizers show that this is also true for other decisive phases, such as project delivery, where final decisions on NbS are often made under deadline pressures (Tension 2). In *Nearby Greenery*, "clear and repeating statements from residents" on the use of the area and their preferences for development informed the NbS design principles. Nevertheless, the approval of the final solution strongly depended on input from dendrologists, landscape architects, designers, and city departments whose weighing role was explained by the complex needs of the NbS model: "Of course, if we talk about the wetland, I imagine that some specialists would say that the best way would be to channelize it somehow" (project organizers Madle and Maie).

Across projects, it became clear that there is a high incentive to include expert knowledge due to the technical and ecological complexities of NbS as a tool and due to the transformative nature of NbS projects. The latter indicates the need for gaining local support among residents and other stakeholders, which is difficult since NbS have often proved to be a contested concept (Tension 1). Expert knowledge was heavily drawn upon to build legitimacy with the help of expert-mediated transformative knowledge production and evidence-based advocacy. This indicates that the question of balancing lay and expert knowledge is not only one of inclusion and social acceptance, but also one of addressing the complexities of NbS as a tool and as a contested concept, where the latter sets strong incentives for the actual prevalence of expert knowledge.

Figure 1 sums up the dynamics between the tensions with regard to the green objectives, process, and knowledge inclusion of participatory NbS and the projectification of its implementation.

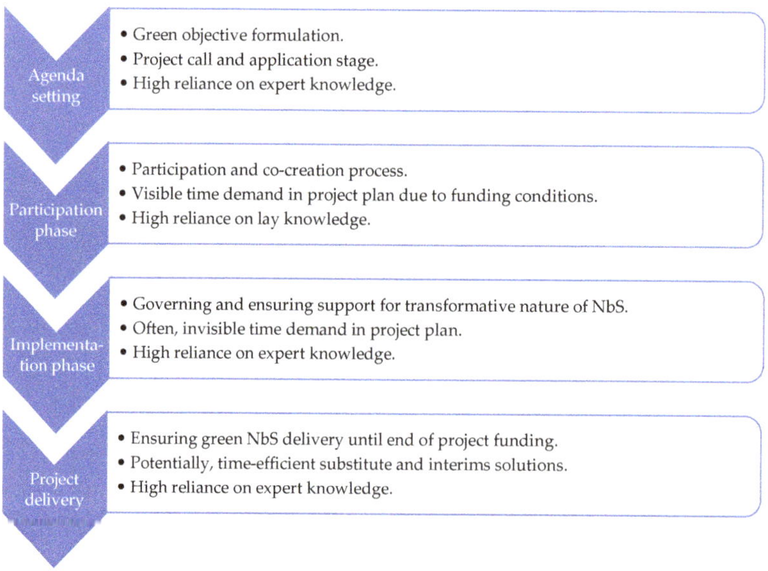

Figure 1. Experienced tensions of participatory NbS application under projectification conditions. Source: authors' illustration, based on stages of NbS project cycle and authors' analysis.

5. Discussion

The findings presented in this paper contribute to ongoing efforts to reframe NbS, going beyond the dominant technical and expert-based focus towards more participatory approaches [6,7,12,38,43]. Moreover, they offer a distinct and complementary critical–pragmatic perspective [20] by exploring tensions that can hinder the achievement of such reframing. In this sense, rather than promoting a normative or idealized vision of how participation in NbS ought to occur, our findings foreground what can be realistically achievable in everyday planning practice, laden with complexities, constraints, and conflicts.

In addition to distinct strategies used by project organizers to navigate tensions, the critical–pragmatic approach offers more general insights for advancing participatory NbS and enhancing its suitability for integration into mainstream planning practice.

First, there is an urgent need to move beyond the projectification of participatory NbS initiatives, which in the European context is characterized by a one-off, research-driven nature and dependence on international funding mechanisms. While our cases show that such projectification has been instrumental in proliferating the ecological orientation of NbS in planning and urban design contexts where such focus was previously absent, they also reveal significant limitations of this model for participatory NbS. Most notably, the impact of participatory NbS often remains confined within the temporal and institutional boundaries of individual projects, with limited continuity or institutionalization beyond their conclusion. Ideally, participatory NbS should be embedded into local planning frameworks and routines, rather than treated as standalone efforts. Admittedly, this remains a significant challenge, and NbS projectification may still be necessary in many contexts to continue advancing the development and mainstreaming of participatory NbS knowledge and practice. Hence, we suggest the need to set more open-ended objectives in project plans, which could be revised iteratively, as well as to provide more time and guidelines for "invisible" tasks associated with participatory NbS (see also [8,10,12]). We also suggest that funding mechanisms should better support the development of lasting partnerships with municipalities, invest in capacity building of planning practitioners, and foster the securing of local co-funding to ensure beyond project lifecycles.

Second, and more specifically in relation to the development of participatory ideas and practices within the NbS field, our findings point to two potential pathways. These alternatives concern trade-offs between the two foundational pillars of participatory NbS: (i) the democratic, social orientation of participation and (ii) the ecological focus at the core of NbS initiatives.

The first pathway involves fully embracing the reframing of NbS through democratic and participatory principles like inclusiveness, power balance, and consensus. While this alternative can help address many of the challenges associated with the technical dominance of NbS, particularly by improving social acceptance and enhancing long-term sustainability of project outcomes, it often requires compromises on ecological ambitions. These compromises reflect our findings, where participants in some projects challenged the projects' green solutions, often based on legitimate reasons regarding local values and more pressing issues. Pursuing this path would require NbS projects to invest greater effort in accommodating and integrating values, knowledge, and ultimately solutions that may not be fully aligned with ecological goals, potentially undermining the environmental focus central to NbS. Projects also need to recognize that, although not impossible, achieving the ideals of participation is very much conditioned by local contexts [16].

The second path involves rethinking participatory NbS in a way that more directly aligns with and actively supports the ecological goals and ecosystem approaches that set NbS projects apart from other urban interventions; goals that form their core contribution to today's pressing environmental challenges [7,34–36,49]. This path implies tempering ambitions around participatory ideals by setting clearer boundaries around the scope and purpose of participation. Rather than open-ended processes where participants have the possibility to directly influence decisions, this path emphasizes processes with more explicitly defined limits on what issues or solutions are open to input. It also requires process organizers and experts to develop and exercise their role in setting agendas and making final decisions (see [13]). While this implies granting more power to these actors, which participatory NbS approaches often seek to challenge, this does not need to be viewed as inherently problematic. Instead, it reflects evolving thinking in participatory planning, which acknowledges the inevitable and at times necessary "art of exclusion" [14] and the legitimate exercise of authority and power in guiding complex planning processes [17]. Our study suggests that this does not need to imply a return to fully expert-driven processes. Instead, the exercise of authority and power could be close to the knowledge breaker role identified earlier, in which project organizers and experts focus on guiding and nudging the participants towards transformative knowledge that supports ecological goals.

6. Conclusions

In this paper, we drew on the cross-case analyses of seven urban NbS projects in Northern Europe, to critically examine and advance ideas and practices of participation within the NbS field. Rather than promoting normative and idealistic views of participatory NbS, we adopted a critical approach to unpack three key social tensions. In addition to deepening the understanding of tensions, our findings also showed distinct strategies used by project organizers to navigate them. In our view, understanding these tensions is essential for advancing the mainstreaming of participatory NbS and strengthening their role in addressing current environmental, social, and climate-related challenges.

Despite our critical approach, the cases reinforce the view that participation can play a significant role in fostering urban sustainability, particularly by promoting social acceptance of NbS project outcomes [8,10,34]. However, this contribution was neither straightforward nor consistently aligned with the expectations of project organizers or the way that participatory activities had been initially conceived in the funding proposals. Most projects had to diverge from original participatory intentions, whether in terms of ambitions for broad inclusion or meaningful co-creation, planned timelines or the type of NbS that were ultimately implemented. In our view, such deviations were not necessarily a result of flaws in the ideals that participatory NbS seek to pursue, but rather reflect the broader challenges of adhering to normative and idealistic views of participation without critically engaging with their practical limitations and the complex, and at times conflict-laden, planning contexts in which they are embedded [13–17,54,72].

As a way forward, we suggested two pathways to address these inherent tensions. These two pathways do not need to be viewed as mutually exclusive or rigid alternatives. Rather, it is better to see them as two ends of a spectrum along which participatory NbS projects can position themselves, depending on their specific goals and local institutional constraints. NbS projects may also shift along this spectrum throughout different stages of their development, requiring greater openness and inclusion at certain points and more directive, goal-oriented approaches at others. Recognizing this fluidity allows for a more context-sensitive application of participatory NbS, one that can flexibly balance ecological ambitions with democratic ideals as projects and contexts evolve. Building on these, future research could explore how NbS projects navigate shifts along this spectrum over time, and what institutional, social, or environmental factors enable or constrain such movement. There is also a need for comparative studies that examine how different positioning along the spectrum affects both ecological outcomes and the quality of participation. Finally, further investigation into the evolving roles of project organizers and experts, as both facilitators and knowledge brokers, can deepen the understanding of how authority and inclusion can co-exist in participatory NbS.

Author Contributions: Conceptualization, B.L., C.C. and B.P.-A.; methodology, B.P.-A., J.G. and M.C.; validation, R.T.-F.; formal analysis, B.P.-A.; investigation, B.P.-A., M.C. and R.T.-F.; writing—original draft preparation, B.P.-A., B.L. and C.C.; writing—review and editing, B.L., B.P.-A., C.C., M.C. and R.T.-F.; visualization, B.P.-A.; project administration, M.C. All authors have read and agreed to the published version of the manuscript.

Funding: This research was funded by H2020 Societal Challenges [grant no. 869764]; Interreg Central Baltic Programme Augmented Urbans [grant no. 14.5.2018 1/26]; H2020 GoGreenRoutes project through the European Union's Horizon 2020 Research and Innovation Program under grant agreement No. 869764; and the Estonian Centre of Excellence for Well-Being Sciences (EstWell), through grant TK218 from the Estonian Ministry of Education and Research.

Data Availability Statement: The data presented in this study are available in a restricted form only upon individual request from the corresponding author due to ethical considerations (i.e., the anonymity of the respondents).

Acknowledgments: The authors would like to thank the interview partners, survey respondents, and workshop participants, as well as the students at Tallinn University for their valuable inputs.

Conflicts of Interest: The authors declare no conflicts of interest.

Abbreviations

The following abbreviations are used in this manuscript:

EU	European Union
NbS	Nature-based Solutions

Notes

1 The benefits of NbS have been well covered by the literature. For a variety of disciplinary aspects, see [27,38,44–47], as well as policy documents promoting the benefits of NbS [1,2].

2 Details of the specific NbS interventions have previously been presented in [13,74].

References

1. European Commission. *Towards an EU Research and Innovation Policy Agenda for Nature-Based Solutions & Re-Naturing Cities. Final Report of the Horizon 2020 Expert Group on "Nature-Based Solutions and Re-Naturing Cities"*; Publications Office of the European Union: Luxembourg, 2015.
2. IUCN France. *Nature-Based Solutions to Address Climate Change*; IUCN: Paris, France, 2016.
3. World Bank. *A Catalogue of Nature-Based Solutions for Urban Resilience*; World Bank Group: Washington, DC, USA, 2021.
4. IUCN. *Global Standard for Nature-Based Solutions. A User-Friendly Framework for the Verification, Design and Scaling up of NbS*, 1st ed.; IUCN: Gland, Switzerland, 2020; ISBN 978-2-8317-2058-6.
5. Sekulova, F.; Anguelovski, I.; Kiss, B.; Kotsila, P.; Baró, F.; Palgan, Y.V.; Connolly, J. The Governance of Nature-Based Solutions in the City at the Intersection of Justice and Equity. *Cities* **2021**, *112*, 103136. [CrossRef]
6. Remme, D.; Haarstad, H. From Instrumentalization to Commoning: A Critical Review of Participation in Urban Nature-Based Solutions. *Front. Sustain. Cities* **2022**, *4*, 917607. [CrossRef]
7. Hafferty, C.; Tomude, E.S.; Wagner, A.; McDermott, C.; Hirons, M. Unpacking the Politics of Nature-Based Solutions Governance: Making Space for Transformative Change. *Environ. Sci. Policy* **2025**, *163*, 103979. [CrossRef]
8. Kiss, B.; Sekulova, F.; Hörschelmann, K.; Salk, C.F.; Takahashi, W.; Wamsler, C. Citizen Participation in the Governance of Nature-Based Solutions. *Environ. Policy Gov.* **2022**, *32*, 247–272. [CrossRef]
9. Ferreira, V.; Barreira, A.P.; Loures, L.; Antunes, D.; Panagopoulos, T. Stakeholders' Engagement on Nature-Based Solutions: A Systematic Literature Review. *Sustainability* **2020**, *12*, 640. [CrossRef]
10. Puskás, N.; Abunnasr, Y.; Naalbandian, S. Assessing Deeper Levels of Participation in Nature-Based Solutions in Urban Landscapes—A Literature Review of Real-World Cases. *Landsc. Urban Plan.* **2021**, *210*, 104065. [CrossRef]

11. van der Jagt, A.P.N.; Buijs, A.; Dobbs, C.; van Lierop, M.; Pauleit, S.; Randrup, T.B.; Skiba, A.; Wild, T. With the Process Comes the Progress: A Systematic Review to Support Governance Assessment of Urban Nature-Based Solutions. *Urban For. Urban Green.* **2023**, *87*, 128067. [CrossRef]
12. van der Jagt, A.P.N.; Kiss, B.; Hirose, S.; Takahashi, W. Nature-Based Solutions or Debacles? The Politics of Reflexive Governance for Sustainable and Just Cities. *Front. Sustain. Cities* **2021**, *2*, 583833. [CrossRef]
13. Plüschke-Altof, B.; Loewen, B.; Müüripeal, A.; Sooväli-Sepping, H. Balancing Participatory Planning and Planning for Resilience in Nature-Based Solutions. A Case of Transformative Agency? *Geogr. Ann. B* **2024**, *106*, 422–440. [CrossRef]
14. Connelly, S.; Richardson, T. Exclusion: The Necessary Difference between Ideal and Practical Consensus. *J. Environ. Plann. Man.* **2004**, *47*, 3–17. [CrossRef]
15. Calderon, C.; Butler, A. Politicising the Landscape: A Theoretical Contribution towards the Development of Participation in Landscape Planning. *Landsc. Res.* **2020**, *45*, 152–163. [CrossRef]
16. Calderon, C.; Westin, M. Understanding Context and Its Influence on Collaborative Planning Processes: A Contribution to Communicative Planning Theory. *Int. Plan. Stud.* **2021**, *26*, 14–27. [CrossRef]
17. Westin, M. The Framing of Power in Communicative Planning Theory: Analysing the Work of John Forester, Patsy Healey and Judith Innes. *Plan. Theory* **2022**, *21*, 132–154. [CrossRef]
18. Dorst, H.; Van Der Jagt, A.; Toxopeus, H.; Tozer, L.; Raven, R.; Runhaar, H. What's behind the Barriers? Uncovering Structural Conditions Working against Urban Nature-Based Solutions. *Landsc. Urban Plan.* **2022**, *220*, 104335. [CrossRef]
19. Egusquiza, A.; Cortese, M.; Perfido, D. Mapping of Innovative Governance Models to Overcome Barriers for Nature Based Urban Regeneration. *IOP Conf. Ser. Earth Environ. Sci.* **2019**, *323*, 012081. [CrossRef]
20. Forester, J. On the Theory and Practice of Critical Pragmatism: Deliberative Practice and Creative Negotiations. *Plan. Theory* **2013**, *12*, 5–22. [CrossRef]
21. Arnstein, S.R. A Ladder of Citizen Participation. *J. Am. Inst. Plann.* **1969**, *35*, 216–224. [CrossRef]
22. Healey, P. The Communicative Turn in Planning Theory and Its Implications for Spatial Strategy Formation. *Environ. Plann. B Plann. Des.* **1996**, *23*, 217–234. [CrossRef]
23. Innes, J.E.; Booher, D.E. Collaborative Policymaking: Governance through Dialogue. In *Deliberative Policy Analysis: Understanding Governance in the Network Society*; Hajer, M.A., Wagenaar, H., Eds.; Cambridge University Press: Cambridge, UK, 2003; pp. 33–59. ISBN 978-0-521-53070-5.
24. European Commission. *The European Green Deal. (COM (2019) 640 Final)*; European Commission: Brussels, Belgium, 2019.
25. European Commission. *The Human-Centred City: Opportunities for Citizens Through Research and Innovation*; Publications Office of the European Union: Luxembourg, 2019.
26. European Commission. *New European Bauhaus Compass*; European Commission: Brussels, Belgium, 2023. Available online: https://new-european-bauhaus.europa.eu/tools-and-resources/use-compass_en (accessed on 26 June 2025).
27. Raymond, C.M.; Frantzeskaki, N.; Kabisch, N.; Berry, P.; Breil, M.; Nita, M.R.; Geneletti, D.; Calfapietra, C. A Framework for Assessing and Implementing the Co-Benefits of Nature-Based Solutions in Urban Areas. *Environ. Sci. Policy* **2017**, *77*, 15–24. [CrossRef]
28. IPBES. Summary for Policymakers of the Global Assessment Report on Biodiversity and Ecosystem Services. In *Proceedings of the IPBES Plenary at Its Seventh Session, Paris, France, 29 April–4 May 2019*; IPBES: Paris, France, 2019.
29. Pickering, J.; Hickmann, T.; Bäckstrand, K.; Kalfagianni, A.; Bloomfield, M.; Mert, A.; Ransan-Cooper, H.; Lo, A.Y. Democratising Sustainability Transformations: Assessing the Transformative Potential of Democratic Practices in Environmental Governance. *Earth Sys. Gov.* **2022**, *11*, 100131. [CrossRef]
30. Harms, P.; Hofer, M.; Artmann, M. Planning Cities with Nature for Sustainability Transformations—A Systematic Review. *Urban Transform.* **2024**, *6*, 9. [CrossRef]
31. Adams, C.; Frantzeskaki, N.; Moglia, M. Mainstreaming Nature-Based Solutions in Cities: A Systematic Literature Review and a Proposal for Facilitating Urban Transitions. *Land Use Policy* **2023**, *130*, 106661. [CrossRef]
32. Magni, F.; Musco, F.; Litt, G.; Carraretto, G. The Mainstreaming of NBS in the SECAP of San Donà Di Piave: The LIFE Master Adapt Methodology. *Sustainability* **2020**, *12*, 10080. [CrossRef]
33. Wamsler, C.; Pauleit, S.; Zölch, T.; Schetke, S.; Mascarenhas, A. Mainstreaming Nature-Based Solutions for Climate Change Adaptation in Urban Governance and Planning. In *Nature-Based Solutions to Climate Change Adaptation in Urban Areas: Linkages between Science, Policy and Practice*; Kabisch, N., Korn, H., Stadler, J., Bonn, A., Eds.; Theory and Practice of Urban Sustainability Transitions; Springer International Publishing: Cham, Switzerland, 2017; pp. 257–273. ISBN 978-3-319-56091-5.
34. Cooper, C.; Cunningham, N.; Bracken, L.J. Exploring Different Framings of Nature-Based Solutions with Respect to Governance, and Citizen Participation, Beneficiaries, and Quality of Life Outcomes. *Environ. Sci. Policy* **2023**, *150*, 103592. [CrossRef]
35. Chausson, A.; Bugre, A.; Spiegelenberg, F.; Welden, E.A.; Melanidis, M. *Nature-Based Solutions: Narratives, Frames, and Future Horizons*; Unearthodox: Versoix, Switzerland, 2024.

36. Seddon, N.; Smith, A.; Smith, P.; Key, I.; Chausson, A.; Girardin, C.; House, J.; Srivastava, S.; Turner, B. Getting the Message Right on Nature-Based Solutions to Climate Change. *Glob. Change Biol.* **2021**, *27*, 1518–1546. [CrossRef] [PubMed]
37. Meyer, K.; Hessenberger, D. *Prototype Database of International Nature-Based Solutions Case Studies: Supplementary Report to the CCICED Special Policy Study on Value Assessment of Nature-Based Solutions*; IUCN: Gland, Switzerland, 2022.
38. Kabisch, N.; Korn, H.; Stadler, J.; Bonn, A. Nature-Based Solutions to Climate Change Adaptation in Urban Areas—Linkages Between Science, Policy and Practice. In *Nature-Based Solutions to Climate Change Adaptation in Urban Areas: Linkages Between Science, Policy and Practice*; Kabisch, N., Korn, H., Stadler, J., Bonn, A., Eds.; Springer International Publishing: Cham, Switzerland, 2017; pp. 1–11.
39. URBiNAT. NBS Catalogue. Available online: https://urbinat.eu/nbs-catalogue/ (accessed on 30 June 2025).
40. Wamsler, C.; Niven, L.; Beery, T.H.; Bramryd, T.; Ekelund, N.; Jönsson, K.I.; Osmani, A.; Palo, T.; Stålhammar, S. Operationalizing Ecosystem-Based Adaptation: Harnessing Ecosystem Services to Buffer Communities against Climate Change. *Ecol. Soc.* **2016**, *21*, 31. [CrossRef]
41. Waylen, K.A.; Blackstock, K.L.; Holstead, K.L. How Does Legacy Create Sticking Points for Environmental Management? Insights from Challenges to Implementation of the Ecosystem Approach. *Ecol. Soc.* **2015**, *20*, 21. [CrossRef]
42. Díaz, S.; Demissew, S.; Carabias, J.; Joly, C.; Lonsdale, M.; Ash, N.; Larigauderie, A.; Adhikari, J.R.; Arico, S.; Báldi, A.; et al. The IPBES Conceptual Framework—Connecting Nature and People. *Curr. Opin. Environ. Sustain.* **2015**, *14*, 1–16. [CrossRef]
43. Raymond, C.M.; Rautio, P.; Fagerholm, N.; Aaltonen, V.A.; Andersson, E.; Celermajer, D.; Christie, M.; Hällfors, M.; Saari, M.H.; Mishra, H.S.; et al. Applying Multispecies Justice in Nature-Based Solutions and Urban Sustainability Planning: Tensions and Prospects. *npj Urban Sustain.* **2025**, *5*, 2. [CrossRef]
44. Aghaloo, K.; Sharifi, A.; Habibzadeh, N.; Ali, T.; Chiu, Y.-R. How Nature-Based Solutions Can Enhance Urban Resilience to Flooding and Climate Change and Provide Other Co-Benefits: A Systematic Review and Taxonomy. *Urban For. Urban Green.* **2024**, *95*, 128320. [CrossRef]
45. Dorst, H.; van der Jagt, A.; Raven, R.; Runhaar, H. Urban Greening through Nature-Based Solutions—Key Characteristics of an Emerging Concept. *Sustain. Cities Soc.* **2019**, *49*, 101620. [CrossRef]
46. Kabisch, N.; van den Bosch, M.; Lafortezza, R. The Health Benefits of Nature-Based Solutions to Urbanization Challenges for Children and the Elderly—A Systematic Review. *Environ. Res.* **2017**, *159*, 362–373. [CrossRef]
47. McQuaid, S.; Kooijman, E.; Rizzi, D.; Andersson, T.; Schanté, J. *The Vital Role of Nature-Based Solutions in a Nature Positive Economy*; Publications Office of the European Union: Luxembourg, 2022; p. 147.
48. European Research Executive Agency. *Nature-Based Solutions: EU-Funded NBS Research Projects Tackle the Climate and Biodiversity Crisis*; Publications Office of the European Union: Luxembourg, 2024.
49. Welden, E.A.; Chausson, A.; Melanidis, M.S. Leveraging Nature-Based Solutions for Transformation: Reconnecting People and Nature. *People Nat.* **2021**, *3*, 966–977. [CrossRef]
50. Loewen, B. From Social Barriers to Transformative Planning in Energy Transitions: Learning from Norwegian Planners' Perspectives. *Urban Plan.* **2024**, *9*, 8158. [CrossRef]
51. Baer, D.; Loewen, B.; Cheng, C.; Thomsen, J.; Wyckmans, A.; Temeljotov-Salaj, A.; Ahlers, D. Approaches to Social Innovation in Positive Energy Districts (PEDs)—A Comparison of Norwegian Projects. *Sustainability* **2021**, *13*, 7362. [CrossRef]
52. Reed, M.S. Stakeholder Participation for Environmental Management: A Literature Review. *Biol. Conserv.* **2008**, *141*, 2417–2431. [CrossRef]
53. Luyet, V.; Schlaepfer, R.; Parlange, M.B.; Buttler, A. A Framework to Implement Stakeholder Participation in Environmental Projects. *J Environ. Manag.* **2012**, *111*, 213–219. [CrossRef]
54. Calderon, C. Unearthing the Political: Differences, Conflicts and Power in Participatory Urban Design. *J. Urban Design* **2020**, *25*, 50–64. [CrossRef]
55. Sarkki, S.; Pihlajamäki, M.; Soini, K.; Ojala, A.; Kluvankova, T.; Spacek, M.; Mishra, H.; Hiedanpää, J. Integrative Literature Review on Co-Concepts in Connection with Nature-Based Solutions. *Environ. Sci. Policy* **2025**, *169*, 104073. [CrossRef]
56. Frantzeskaki, N. Seven Lessons for Planning Nature-Based Solutions in Cities. *Environ. Sci. Policy* **2019**, *93*, 101–111. [CrossRef]
57. Wamsler, C.; Alkan-Olsson, J.; Björn, H.; Falck, H.; Hanson, H.; Oskarsson, T.; Simonsson, E.; Zelmerlow, F. Beyond Participation: When Citizen Engagement Leads to Undesirable Outcomes for Nature-Based Solutions and Climate Change Adaptation. *Clim. Change* **2020**, *158*, 235–254. [CrossRef]
58. Frantzeskaki, N.; Collier, M.; Holscher, K.; Gaziulusoy, I.; Ossola, A.; Albulescu, P.; Bonneau, M.; Borgstrom, S.; Connop, S.; Dumitru, A.; et al. Premises, Practices and Politics of Co-Creation for Urban Sustainability Transitions. *Urban Transform.* **2025**, *7*, 7. [CrossRef]
59. Dunn, R.R.; Gavin, M.C.; Sanchez, M.C.; Solomon, J.N. The Pigeon Paradox: Dependence of Global Conservation on Urban Nature. *Conserv. Biol.* **2006**, *20*, 1814–1816. [CrossRef]

60. Remme, R.P.; Frumkin, H.; Guerry, A.D.; King, A.C.; Mandle, L.; Sarabu, C.; Bratman, G.N.; Giles-Corti, B.; Hamel, P.; Han, B.; et al. An Ecosystem Service Perspective on Urban Nature, Physical Activity, and Health. *Proc. Natl. Acad. Sci. USA* **2021**, *118*, e2018472118. [CrossRef]
61. Torfing, J.; Sørensen, E.; Røiseland, A. Transforming the Public Sector into an Arena for Co-Creation: Barriers, Drivers, Benefits, and Ways Forward. *Admin. Soc.* **2016**, *51*, 795–825. [CrossRef]
62. Schneider, F.; Giger, M.; Harari, N.; Moser, S.; Oberlack, C.; Providoli, I.; Schmid, L.; Tribaldos, T.; Zimmermann, A. Transdisciplinary Co-Production of Knowledge and Sustainability Transformations: Three Generic Mechanisms of Impact Generation. *Environ. Sci. Policy* **2019**, *102*, 26–35. [CrossRef]
63. Brand, R.; Gaffikin, F. Collaborative Planning in an Uncollaborative World. *Plan. Theory* **2007**, *6*, 282–313. [CrossRef]
64. Nightingale, A.J. Power and Politics in Climate Change Adaptation Efforts: Struggles over Authority and Recognition in the Context of Political Instability. *Geoforum* **2017**, *84*, 11–20. [CrossRef]
65. Noppenberger, V.; Chapman, E.; Gäckle, J.; Garcia, M.; Koff, T. Review of Existing Approaches to Collaboration in Research (GoGreenRoutes Deliverable 3.1). GoGreenRoutes Project; 2021. Available online: https://gogreenroutes.eu/publication?t=GoGreenRoutes%20Report%20on%20Co-creation (accessed on 26 June 2025).
66. Zingraff-Hamed, A.; Hüesker, F.; Lupp, G.; Begg, C.; Huang, J.; Oen, A.; Vojinovic, Z.; Kuhlicke, C.; Pauleit, S. Stakeholder Mapping to Co-Create Nature-Based Solutions: Who Is on Board? *Sustainability* **2020**, *12*, 8625. [CrossRef]
67. van den Hove, S. Between Consensus and Compromise: Acknowledging the Negotiation Dimension in Participatory Approaches. *Land Use Policy* **2006**, *23*, 10–17. [CrossRef]
68. Westley, F.; Tjornbo, O.; Schultz, L.; Olsson, P.; Folke, C.; Crona, B.; Bodin, Ö. A Theory of Transformative Agency in Linked Social-Ecological Systems. *Ecol. Soc.* **2013**, *18*, 27. [CrossRef]
69. Castán Broto, V.; Trencher, G.; Iwaszuk, E.; Westman, L. Transformative Capacity and Local Action for Urban Sustainability. *Ambio* **2019**, *48*, 449–462. [CrossRef]
70. Wolfram, M. Conceptualizing Urban Transformative Capacity: A Framework for Research and Policy. *Cities* **2016**, *51*, 121–130. [CrossRef]
71. Berglund, E. Science, Art and Other Ways of Knowing: A Proposal from a Struggle Over a Helsinki Green Space. In *Whose Green City? Contested Urban Green Spaces and Environmental Justice in Northern Europe*; Plüschke-Altof, B., Sooväli-Sepping, H., Eds.; Springer International Publishing: Cham, Switzerland, 2022; pp. 145–162.
72. Calderon, C.; Mutter, A.; Westin, M.; Butler, A. Navigating Swift and Slow Planning: Planners' Balancing Act in the Design of Participatory Processes. *Eur. Plan. Stud.* **2024**, *32*, 390–409. [CrossRef]
73. de Geus, T.; Wittmayer, J.M.; Silvestri, G. A Balancing Act: Radicality and Capture in Institutionalising Reflexive Governance for Urban Sustainability Transitions. *Urban Transform.* **2024**, *6*, 2. [CrossRef]
74. Gäckle, J.; Chebotareva, M.; Plüschke-Altof, B.; Meul, J.; Väänänen, I.; Utkarsh, S.; Timpe, A.; Lohrberg, F.; Suutari, T.; Lovheim, E.M.; et al. Co-Creating Change: Seedbed Interventions as Catalysts for Equitable Urban Planning—The Case of Umeå. *Urban Plan.* **2025**, *10*, 1–24. [CrossRef]
75. Bammer, G. Stakeholder Engagement: Learning from Arnstein's Ladder and the IAP2 Spectrum. Integration and Implementation Insights 2022. Available online: https://i2insights.org/2022/08/30/learning-from-arnsteins-ladder-and-iap2-spectrum/ (accessed on 5 August 2025).
76. Novalia, W.; Rogers, B.C.; Bos, J.J.; Brown, R.R.; Soedjono, E.S.; Copa, V. Transformative Agency in Co-Producing Sustainable Development in the Urban South. *Cities* **2020**, *102*, 102747. [CrossRef]
77. Aditomo, A.; Goodyear, P.; Bliuc, A.-M.; Ellis, R.A. Inquiry-Based Learning in Higher Education: Principal Forms, Educational Objectives, and Disciplinary Variations. *Stud. High. Educ.* **2013**, *38*, 1239–1258. [CrossRef]
78. Delbaere, B.; Pereira Barboza, E.; Plüschke-Altof, B.; Chebotareva, M.; Van Rafelghem, E.; Blockmans, L.; Keune, H. The Environmental Health Citizen Interview Tool: Towards an Inclusive Qualitative Environmental Wellbeing Approach in Support of Planetary Health. *Front. Public Health* **2024**, *12*, 1462561. [CrossRef]
79. Korsnes, M.; Loewen, B.; Dale, R.F.; Steen, M.; Skjølsvold, T.M. Paradoxes of Norway's Energy Transition: Controversies and Justice. *Clim. Policy* **2023**, *23*, 1132–1150. [CrossRef]
80. Martin, C.J.; Evans, J.; Karvonen, A. Smart and Sustainable? Five Tensions in the Visions and Practices of the Smart-Sustainable City in Europe and North America. *Technol. Forecast. Soc. Change* **2018**, *133*, 269–278. [CrossRef]
81. Manders, T.N.; Wieczorek, A.J.; Verbong, G.P.J. Complexity, Tensions, and Ambiguity of Intermediation in a Transition Context: The Case of Connecting Mobility. *Environ. Innov. Soc. Transit.* **2020**, *34*, 183–208. [CrossRef]
82. Skjølsvold, T.M.; Coenen, L. Are Rapid and Inclusive Energy and Climate Transitions Oxymorons? Towards Principles of Responsible Acceleration. *Energy Res. Soc. Sci.* **2021**, *79*, 102164. [CrossRef]
83. Granqvist, K.; Humer, A.; Mäntysalo, R. Tensions in City-Regional Spatial Planning: The Challenge of Interpreting Layered Institutional Rules. *Reg. Stud.* **2021**, *55*, 844–856. [CrossRef]
84. Braun, V.; Clarke, V. Using Thematic Analysis in Psychology. *Qual. Res. Psychol.* **2006**, *3*, 77–101. [CrossRef]

85. Lebowitz, A.; Trudeau, D. Digging in: Lawn Dissidents, Performing Sustainability, and Landscapes of Privilege. *Soc. Cult. Geogr.* **2017**, *18*, 706–731. [CrossRef]
86. Maller, C. *Healthy Urban Environments: More-than-Human Theories*; Routledge: London, UK, 2018.
87. Anguelovski, I. From Toxic Sites to Parks as (Green) LULUs? New Challenges of Inequity, Privilege, Gentrification, and Exclusion for Urban Environmental Justice. *J. Plan. Lit.* **2016**, *31*, 23–36. [CrossRef]
88. McHale, M.R.; Beck, S.M.; Pickett, S.T.A.; Childers, D.L.; Cadenasso, M.L.; Rivers III, L.; Swemmer, L.; Ebersohn, L.; Twine, W.; Bunn, D.N. Democratization of Ecosystem Services—A Radical Approach for Assessing Nature's Benefits in the Face of Urbanization. *Ecosyst. Health Sustain.* **2018**, *4*, 115–131. [CrossRef]
89. Anderson, C.C.; Renaud, F.G. A Review of Public Acceptance of Nature-Based Solutions: The 'Why', 'When', and 'How' of Success for Disaster Risk Reduction Measures. *Ambio* **2021**, *50*, 1552–1573. [CrossRef]
90. Coletta, V.R.; Pagano, A.; Pluchinotta, I.; Fratino, U.; Scrieciu, A.; Nanu, F.; Giordano, R. Causal Loop Diagrams for Supporting Nature Based Solutions Participatory Design and Performance Assessment. *J. Environ. Manag.* **2021**, *280*, 111668. [CrossRef]
91. Carr, J.; Dionisio, M.R. Flexible Spaces as a "Third Way" Forward for Planning Urban Shared Spaces. *Cities* **2017**, *70*, 73–82. [CrossRef]
92. Hoppe, E.E.; Holmegaard, H.T. Art-Based Research Methods in Science Education Research: A Systematic Review of Their Prevalence and an Analysis of Their Potentials in Addressing Complex Questions. *Nordina* **2022**, *18*, 323–336. [CrossRef]
93. Butzlaff, F. Consenting Participation? How Demands for Citizen Participation and Expert-Led Decision-Making Are Reconciled in Local Democracy. *Polit. Stud. Rev.* **2023**, *21*, 340–356. [CrossRef]
94. Nichols, T.M. *The Death of Expertise: The Campaign Against Established Knowledge and Why It Matters*; Oxford University Press: Oxford, UK, 2024; ISBN 978-0-19-776383-4.
95. Armstrong, M.; Derrien, M.M. Chapter 4: Language in the Recreation World. In *Igniting Research for Outdoor Recreation: Linking Science, Policy, and Action*; Selin, S., Cerveny, L.K., Blahna, D.J., Miller, A.B., Eds.; U.S. Department of Agriculture, Forest Service, Pacific Northwest Research Station: Portland, OR, USA, 2020; pp. 51–62.
96. Carpenter, J. Picture This: Exploring Photovoice as a Method to Understand Lived Experiences in Marginal Neighbourhoods. *Urban Plan.* **2022**, *7*, 351–362. [CrossRef]
97. Knight, A.T.; Cook, C.N.; Redford, K.H.; Biggs, D.; Romero, C.; Ortega-Argueta, A.; Norman, C.D.; Parsons, B.; Reynolds, M.; Eoyang, G.; et al. Improving Conservation Practice with Principles and Tools from Systems Thinking and Evaluation. *Sustain. Sci.* **2019**, *14*, 1531–1548. [CrossRef]
98. United Nations. *Transforming Our World: The 2030 Agenda for Sustainable Development. A/RES/70/1*; United Nations: New York, NY, USA, 2015.

land

Article

Clustering Comfort: A Cluster Analysis on Housing Conditions and Nature-Based Solutions in Polish Cities

Anita Walter, Alice Wanner * and Ulrike Pröbstl-Haider

Insistute of Landscape Development, Recreation and Conservation Planning, BOKU University, Peter-Jordan-Str. 82, 1190 Vienna, Austria; ulrike.proebstl@boku.ac.at (U.P.-H.)
* Correspondence: alice.wanner@boku.ac.at

Abstract: As cities struggle to balance affordable housing, sufficient public green space, climate change, and rising temperatures, the urgency for planners to integrate nature-based solutions into urban strategies is magnified. In Poland, social inequalities and neighborhoods with limited access to green areas often characterize cities. Urban strategies are needed to increase environmental justice and mitigate climate change impacts. In a survey conducted in Poland (n = 963), a cluster analysis identified four groups based on their living situations, rating of green spaces, and climate change. The results illustrate differences in perceived impacts of heat waves with one group demonstrating a high tolerance towards heat waves, although they live in unfavorable housing conditions. Green space use varies depending on accessibility and distribution in their neighborhoods. Context-sensitive planning is required to ensure more equitable urban planning in Polish cities, which allocates for both the environmental and social needs of the city.

Keywords: nature-based solutions; Poland; climate change; cluster analysis; housing conditions

1. Introduction

Many recent publications have emphasized the importance of urban green elements for reducing environmental hazards, air pollution, heat waves, and flooding. These elements, also called nature-based solutions (NbSs), provide further benefits for noise and stress reduction, health, and wellbeing benefits as well as carbon offsetting [1–6]. Furthermore, NbSs and biophilic designs [7–9] are often integrated to improve public spaces, while also providing measures to tackle climate change, urban heat islands, and biodiversity loss in cities [10–12]. The paper at hand focuses on functions of NbSs, which contribute to improving urban environments for residents. This includes improved air quality, temperature reduction, and increasing the number of green spaces in public places, to name only some benefits, but touches less on design principles.

The benefits of NbSs are known, but there is still uncertainty about the effectiveness [13] and an increasing demand for more stakeholder involvement to ensure greater social justice in their distribution across cities [14,15]. When implementing NbSs, trade-offs must be made to determine where they can fit without impeding other necessary infrastructure. Especially when it comes to new housing construction, green areas are faced with fierce competition [14]. NbSs can come in a variety of shapes and sizes with differing functions ranging from communal gardens to smaller rain gardens, large scale green corridors, or simple street trees. The design of these will greatly impact their functionality in a neighborhood.

Equal neighborhood conditions are not always equally experienced. Living conditions are shaped by both the objective, built environment as well as individual, subjective perceptions and experiences. The built environment is described by factors such as the location of the neighborhood, the density and type of buildings, the year of construction, and surrounding gray and green infrastructure [16]. The quality of the neighborhood is linked to both built environment but also functionality, accessibility, and quality of stay and use which are often subjective perceptions [17].

Subjective perceptions include preferences for green areas or types of NbSs or personal experiences with stressors such as noise, heat, or poor air quality [18,19]. Urban green areas are perceived differently and as Bai et al. and Grütter [17,20] state in their separate studies, the time spent here is individual and subjectively valued, adding to residents identification with their neighborhoods and feelings of belonging. NbSs are thus elements of neighborhoods that contribute to sense of place.

As climate change impacts in cities increase, impacts are unequally felt. Those living in densely built areas with little to no green areas and no access to private outdoor spaces such as balconies and terraces are more strongly impacted by rising temperatures [11,21]. Socio-demographics and especially age also play into how heat impacts health and wellbeing [22]. That being said, socio-economic factors such as education and income influence individuals' abilities to cope with and adapt to climate change impacts. NbSs, and, most predominantly, street greening and parks are considered some of the most important measures to tackle this climate change impacts in European cities [19,23,24] but these must also be justly implemented to benefit vulnerable and disadvantaged households and residents [25,26].

Aside from attractive and healthy living conditions, cities should also provide affordable housing, suitable infrastructure, and services for their residents. These requirements are often in conflict with the development of NbSs mostly due to increasing costs for real estate and space. Despite green spaces increasing quality of life and adding to social and ecological quality of urban spaces [17,23,27] the trade-offs made by decision-makers are often against greening and in favor of dense housing structures; even when planning entirely new developments. Therefore, we need to understand if and which consequences it has, if these green spaces and NbSs are not justly distributed and residents are lacking access to these spaces for their health and wellbeing [25].

The rapidly growing Eastern European cities are a good example for exploring this aspect. This study investigates the situation in Poland looking into cities with more than 20,000 residents. Poland has 36.6 million inhabitants of which nearly 60% live in cities [28,29]. Living conditions in urban areas are predominantly characterized by modern, often gated, communities growing in the outer districts, while block high-rises left over from the socialist era of the 20th century continue to remain prominent [30]. The older building stocks are less efficient concerning heating and cooling than newer constructions.

After the political shifts of 1989, housing became a consumable good, leading to rising prices. Currently, affordable housing is difficult to come by for many urban residents, which is leading to an intensification of social segregation across neighborhoods. Government programs designed to support younger people to rent or buy homes had limited success [30]. Social segregation is further strengthened through urban developments, which focus on building new constructions in the urban periphery for more wealthy households, while building stock in the central and urban districts are left to age and decay [30].

Past studies from Poland [31,32] have shown that the provision of green areas across Polish cities vary greatly: while some cities benefit from urban forests in the outskirts, which are easily accessible to residents, many other cities have very limited green areas of notable size for their residents to access. Wysmulek et al. [31] determined that on average, only 20.5% of residents have access to green areas in their neighborhoods and deem the situation to be insufficient and unevenly distributed. Suligowski et al. [32] highlight that the larger the city is, the worse provision of green areas per resident becomes. This is especially true for Polish cities, where several studies have already identified social and ecological injustice characterized by cramped living conditions with little access to green areas [30]. A comparison of the conditions in different European member states also demonstrates, that living conditions in Poland are more cramped and household sizes are slightly larger than in other European member states [33]. These factors, including the poor access to outdoor green areas, indicate that urban populations in Poland are more vulnerable to climate change impacts, especially heat, and are currently not sufficiently profiting from access to green areas or NbS.

The study area, Poland, provides an interesting study object analyzing the relevance of poor access to green areas for urban residents along with social segregation between neighborhoods characterized by building stock age and density. The paper at hand tries to understand the currently perceived benefits of NbSs by the urban population in the context of their living conditions. To achieve this, we put forward the following hypotheses to be tested by a representative national wide survey:

H1. *Residents of Polish cities experience heat stress independently from their objective living situation.*

H2. *Clusters with different living conditions differ in their desire for improving NbSs in their neighborhood.*

H3. *Differences in the subjective perception of heat can be explained by living conditions and socio-demographic characteristics.*

2. Materials and Methods

2.1. Survey

As part of the European Union funded project UPSURGE [18,34], a pan-European survey was conducted to investigate urban residents acceptance and expectations for NbSs in their neighborhoods. The study at hand focuses on the results obtained from the Polish respondents (n = 1021). The survey was distributed online via a panel in October 2022 among residents of cities with more than 20,000 inhabitants and over the age of 18 [18] and includes respondents from across Poland. Aside from trade-offs on NbS scenarios, the survey also included questions on living situation, building characteristics, surrounding green areas, experiences with climate change, and socio-demographics. After quality control, the final data set used for this analysis included 963 respondents.

For the analysis of the data, SPSS 29.0 was used.

2.2. Cluster Analysis

For a differentiated view of the housing situation a hierarchical cluster analysis was conducted using objective criteria of the built environment as well as subjective criteria concerning rating on green space provision and heat stress. For methodological reasons, all variables were dichotomized with the assignment of values 0 and 1 in order to take account of the different scale levels.

The following objective criteria relating to the living situation were used: area of city (dummy variables: 1 = city center, 0 = no; 1 = suburbs, 0 = no; omitted redundant reference variable = urban districts), type of housing (1 = high-rise, closed block, or cluster, 0 = house, duplex, or row house), building height (1 = 5 or more stories, 0 = less than 5 stories), year built (1= from 1990, 0 = before 1990), open spaces in the residential environment (gardens, public green spaces, traffic areas, paved areas: 1= many/predominantly, 0 = view/none), distance to shops (1 = 0–5 min, 0 = more), distance to green areas (parks, playground, city forest: 1 = 0–5 min to walk, 0 = more than 5 min to walk). Subjective criteria included the assessment of green space availability (1 = excellent/good, 0 = poor/nonexistent), experience with heat waves (1 = yes, 0 = no) and negative impact of heatwaves/heat stress (1 = yes, 0 = no).

The Ward algorithm was used as a clustering method, based on squared Euclidean distance as a measure of dissimilarities between cluster objects [35]. The dendrogram was applied in order to determine the appropriate number of clusters, suggesting a 3-cluster solution. Since the result was not sufficiently differentiated, further differentiation was carried out based on content considerations [36] resulting in a 4-cluster solution.

Kruskal–Wallis tests were applied to further test if the clusters differ in the desire for improving NbSs. Pairwise comparisons supported the exact differentiation between the clusters. To avoid α-error accumulation, the p-value was corrected in accordance with Bonferroni.

Differences between the clusters with regard to socio-demographic characteristics were determined using chi-square tests, whereby standardized residuals were intended to reveal precise patterns of deviation.

3. Results

3.1. Survey Results

3.1.1. Socio-Demographics

The survey (n = 963) consisted of 50.5% women and 48.7% men (0.1% other; 0.7% preferred not to answer). The age ranged from 18 to 76, with the average being 39.9 years old. Household size varied, with 10.5% living alone and 26% living in a two-person household. Larger households also account for the number of households with children (Table 1). Education levels were high, with over half holding a university degree. Income is varied but generally low, with two-thirds of respondents having a monthly net income of PLN 2000 or below. The socio-demographic overview can be found in Table 1.

Table 1. Socio-demographic overview of the sample (n = 963).

Socio-Demographic		% of Respondents
Gender	Female	50.5
	Male	48.7
	Other	0.1
	Prefer not to say	0.7
Age	<25	10.4
	25–34	24.7
	35–50	31.4
	51–65	16.4
	66+	2.8
	Prefer not to say	14.3
Household size	1 person	10.5
	2	26.0
	3	31.7
	4	21.8
	5+ people	10.1
Children	Yes	49.1
	No	50.9

Table 1. *Cont.*

Socio-Demographic		% of Respondents
Education	Primary	0.9
	Secondary	18.8
	Vocational training	27.4
	University	52.2
	Prefer not to say	0.6
Income (monthly household net)	PLN < 500	10.7
	PLN 500–1000	31.9
	PLN 1001–2000	29.4
	PLN 2001–3000	11.8
	PLN 3001–4000	5.5
	PLN > 4000	3.2
	Prefer not to say	7.5

3.1.2. Housing Conditions

The housing conditions are shaped by city size, area of the city, type of housing, and its height along with the year the buildings were constructed (Table 2). Generally speaking, the respondents resided in city centers and central, urban districts in predominantly closed blocks or clusters, with nearly a quarter living in high-rises and a fifth in houses. Therefore, unsurprisingly, most buildings tended to be three stories or more, and the majority were built after 1970.

Table 2. General living conditions.

Living Conditions		% of Respondents
City size	20,000–50,000	11.6
	50,001–100,000	16.3
	100,001–250,000	19.8
	250,001–500,000	18.5
	500,001–1.5 million	20.4
	>1.5 million	13.4
Area	City center	48.8
	Urban districts	46.2
	Suburbs	5.0
Type of housing	House	18.3
	Duplex or row house	7.1
	High-rise	23.5
	Closed block or cluster	51.2
Building height (stories)	1	6.7
	2	15.7
	3–4	48.6
	5–9	17.3
	10+	11.6
Year built	<1914	1.9
	1914–1939	4.3
	1940–1969	11.7
	1970–1989	37.6
	1990–2009	25.2
	>2010	16.3

Respondents were additionally asked about public space in their neighborhoods. Paved courts (51.8% many or dominating) and traffic spaces (74.5% many or dominating) dominated most surroundings, but there was also a significant proportion of private balconies (79.8%) and public parks (63.4%). Private gardens (35.5%) and private green spaces (30.9%) were less common.

Distances to green areas were measured in walking distance (minutes): 71.9% indicated that street greening was right at their doorstep (0–5 min walking), and 57.1% indicated playgrounds within the same distance. Parks, communal gardens, and green corridors tended to be further away with walking distances often being up to 10 or even 15 min away. Urban forests were not common (56.3%, either more than 16 min walk or not in their neighborhood at all). The proximity of green areas is also reflected in times spent outdoors: 38.5% of respondents spend two to four hours outside in a week, 22.5% at an hour. Only 15% spend less than 30 min in green areas per week.

The subjective rating of neighborhood green areas supply was good (56.9%) with 15.3% saying their surroundings had excellent provision of green areas, and 27.4% saying the green supply was poor. Only 0.4% indicated their neighborhoods had no green areas.

3.1.3. Climate Change Perceptions and Urban Strategies

Respondents were asked to answer several questions concerning their attitudes towards and possible experiences with climate change. In total, 18.1% states that climate change will occur, but the impacts will be felt later. The majority (69.7%) states that climate change is occurring and is already visible. Meanwhile, 10.1% were uncertain and 1.7% did not believe in climate change at all; 0.5% held other opinions. A total of 49% of respondents state that they already see and feel climate change impacts in their own neighborhood and of these, the vast majority indicated that they experience heat waves because of climate change.

High temperatures impact residents in different ways. Of the respondents who stated that they experience climate change impacts, 5.9% said they appreciate high temperatures, while 19% do not mind heat waves. But 61% indicate that heat waves have negative impacts on their wellbeing, and a further 13.5% state that they have experienced negative impacts on their health.

When asked if climate change was important for their local urban strategies, respondents tended to agree that these were very important (43%) or at least important (48.5%). The most important aspects for urban strategies were deemed to be the improvement in air quality, storm water management, and biodiversity conservation. Tackling heat waves and temperature reduction were given less importance.

3.2. *Cluster Analysis Results*

3.2.1. Description of Clusters

The cluster analysis used objective criteria (are of the city, type of housing, height, amount of green, year built, distance to shops for daily needs, and distance to green areas) as well as subjective criteria (rating surrounding green areas and heat stress) and resulted in four distinct clusters.

Cluster 1 (n = 288, 29.9%); block or cluster housing, poor supply of green areas, high heat stress: Respondents from this cluster are from the city center (50.7%) or urban districts (46.5%), predominantly living in closed blocks (67.7%) or high-rises (25%). Building age is on par with the overall sample (Table 2). While 62.2% have green areas in their neighborhood, only about a quarter are within proximity. Paved traffic and parking spaces dominate the surroundings (80.9%), and shops for daily needs are easily reached (73.3%). Roughly two-thirds (65.3%) find the number of green areas good, but a disproportionately large portion (34.7.%) rate the provision of green spaces as poor. All the respondents have experienced at least one heat wave and nearly 90% declare they have negative experiences with heat.

Cluster 2 (n = 176, 18.3%); block or cluster housing, good supply of green areas, less heat stress: Similarly to cluster 1, the people in this group reside in the city center or urban districts. They live in closed blocks (63.1%) and high-rises (29.5%), and nearly half of the respondents state that their building has at least five stories. Most of the buildings were built after 1970. The provision of green areas is considered very good, as 97.7% state green areas in the neighborhood, but there is still a significant number of paved areas for parking and traffic. Parks and playgrounds tend to be easily reached on foot and 12.2% are within walking distance of urban forests. It comes as no surprise that 92.6% rate the quantity of green areas as good. Only 10.8% have experienced a heat wave, and none of the respondents have had negative impacts on their health or wellbeing through high temperatures.

Cluster 3 (n = 248, 25.8%); high-rise with poor supply of green areas, less heat stress: 60.1% live in the city center and 32.3% in high-rises. A high percentage (46.8%) indicates that their housing was built in the 1970s and 80s. There are few private green areas, but infrastructure tends to be average in accessibility. Larger green areas such as parks or urban forests are not accessible on foot. Objectively, this cluster is underprivileged regarding housing and built environment. A total of 40.7%, the highest proportion of any cluster, rates the supply of green areas as poor. But nearly nobody in this group has experiences with heat waves or felt negative impacts from them.

Cluster 4 (n = 251, 26.1%); houses with high supply of green areas, low heat stress: In the final cluster, most live in urban districts (53%) with a comparatively high number of suburban residents (12.7%). The vast majority live in houses, duplex of row houses (79.3%); only few in closed blocks (8.8%) or high-rises (53%). Housing tends to be newer and built after 1990. Unsurprisingly, these neighborhoods are dominated by private gardens (93.6%) and report lower traffic spaces. Access to shops, parks, and playgrounds is low in comparison with the other clusters, but they are the closest to urban forests (16.3%). In total, 78.5% rate the quantity of green areas in their neighborhood positively. Additionally, 77.3% have not had any negative experiences with heat waves or high temperatures.

3.2.2. Comparisons Between Clusters

In further analysis, the clusters based on living conditions were compared in regard to the importance given to improving quality of life in their neighborhood. The survey asked respondents to rate the importance of air quality improvement, temperature reduction, heat wave mitigation, storm water management, biodiversity, and urban climate improvements for urban development strategies. Responses ranged from 1 (unimportant) to 4 (important). Those who responded "don't know" were excluded. Table 3 shows the number, mean, and standard deviation per cluster as well as the results of the Kruskal–Wallis Test.

The results indicate significant differences between the clusters ($p < 0.001$) on the importance of the various environmental goals for urban development. Additional Bonferroni-corrections in pair comparisons showed that cluster 1 was significantly different from the remaining three clusters. As the group which has the most experience with heat waves, all six environmental goals were more important to them than to the other clusters. Especially temperature reduction and heat wave mitigation stood out against the other clusters with medium effect size ranging from 0.24 to 0.30.

Table 3. Number (n), mean (M), and standard deviation (SD) on environmental goals to improve quality of life in neighborhood (1 = important, 4 = unimportant), divided by cluster; Kruskal–Wallis test ($p < 0.01$ **)).

		Cl. 1	Cl. 2	Cl. 3	Cl. 4	Total	KW-Test (df = 3)
Air quality improvement	n	287	171	240	247	945	KW-H = 25.596
	M	3.84	3.60	3.65	3.60	3.68	$p < 0.001$ **
	SD	0.46	0.72	0.64	0.74	0.64	
Temperature reduction	n	277	161	223	235	896	KW-H = 63.018
	M	3.44	2.86	2.87	2.98	3.07	$p < 0.001$ **
	SD	0.76	0.92	1.01	0.96	0.94	
Heat wave mitigation	n	282	166	231	239	918	KW-H = 69.212
	M	3.59	3.02	3.04	3.09	3.22	$p < 0.001$ **
	SD	0.64	0.97	0.98	0.91	0.90	
Storm water management	n	284	171	233	246	934	KW-H = 38.792
	M	3.77	3.50	3.39	3.49	3.55	$p < 0.001$ **
	SD	0.48	0.71	0.80	0.75	0.70	
Biodiversity	n	285	165	230	241	921	KW-H = 40.627
	M	3.72	3.45	3.39	3.35	3.49	$p < 0.001$ **
	SD	0.56	0.74	0.79	0.82	0.74	
Urban climate improvements	n	281	167	227	243	918	KW-H = 30.411
	M	3.76	3.43	3.48	3.49	3.56	$p < 0.001$ **
	SD	0.46	0.82	0.72	0.73	0.69	

The clusters were also compared in regard to socio-demographics (gender, age, children, household size; Table 4) and socio-economic factors (education, income; Table 5). Although there were significant differences across clusters on gender, the standardized residuals indicate only general tendencies. Cluster 2 (cluster blocks, many green areas, little heat impact) has less women than the overall sample average. Cluster 1 (closed blocks, poor supply of green areas, high heat impacts) has slightly more women. No significant differences were identified in the other two clusters. There were no significant differences identified concerning age distribution.

There were significant differences in households with children. Cluster 4 had an especially high number of children (62.5%). They live predominantly in green areas with a high portion of individual or row houses.

Household size also showed significant differences across the clusters ($p < 0.001$). Respondents of cluster 1 had an above average percentage of two-person households (32.3% compared with the sample average of 26%). As this group also had fewer children, it is an indication that there are more couples without children in cluster 1. Cluster 3 is characterized by a higher proportion of single households (15.3% compared with the sample average of 10.5%). Single and two-person household were underrepresented in cluster 4, where household size was larger (40.6% had four or more people compared with the sample average of 31.9%).

As can be seen in Table 5, there are no significant differences in income distribution. Education on the other hand did demonstrate statistically significant differences in education levels. Respondents from cluster 4 had significantly lower proportions of secondary education and a greater but not significant degree of university education. Although the standardized residuals do not indicate significant deviations, cluster 1 and 2 also deviated slightly from the overall averages. Cluster 1 had higher proportions of secondary schooling.

Cluster 2 had a lower rate of university education and a higher proportion of secondary school diploma. Cluster 3 was in line with the overall sample.

Table 4. Socio-demographic variables (gender, age, children household size) by cluster; household size: p. = person(s); χ^2-Test ($p < 0.05$ *, $p < 0.01$ **), SR = standardized residuals. (Respondents who selected "diverse" or "prefer not to answer" were not included).

		Cl. 1	**Cl. 2**	**Cl. 3**	**Cl. 4**	**Total**	**χ^2-Test**
Gender	female	55.6%	42.6%	52.9%	49.4%	50.9%	$\chi^2 = 7.936$, df = 3 $p = 0.047$ *
	SR (f)	1.1	−1.5	0.4	−0.3		
	male	44.4%	57.4%	47.1%	50.6%	49.1%	
	SR (m)	−1.1	1.6	−0.4	0.3		
Age	<25	12.9%	12.3%	10.6%	12.6%	12.1%	$\chi^2 = 7.510$, df = 12 $p = 0.822$
	25–34	29.3%	27.1%	31.7%	26.7%	28.8%	
	35–50	34.8%	38.7%	33.7%	40.3%	36.6%	
	51–65	18.8%	17.4%	21.6%	18.4%	19.2%	
	>66	4.3%	4.5%	2.4%	1.9%	3.3%	
Children (<18 years)	yes	42.4%	51.1%	41.9%	62.5%	49.1%	$\chi^2 = 28.786$, df = 3 $p < 0.001$ **
	SR (y)	−1.6	0.4	−1.6	3.0		
	no	57.6%	48.9%	58.1%	37.5%	50.9%	
	SR (*n*)	1.6	−0.4	1.6	−3.0		
Household size	1 p.	13.5%	7.4%	15.3%	4.4%	10.5%	$\chi^2 = 50.706$, df = 9 $p < 0.001$ **
	SR (1 p.)	1.6	−1.3	2.4	−3.0		
	2 p.	32.3%	24.4%	29.0%	16.7%	26.0%	
	SR (2 p.)	2.1	−0.4	0.9	−2.9		
	3 p.	26.7%	31.8%	30.6%	38.2%	31.7%	
	SR (3 p.)	−1.5	0.0	−0.3	1.9		
	>3 p.	27.4%	36.4%	25.0%	40.6%	31.9%	
	SR (>3 p.)	−1.3	1.1	−1.9	2.5		

Table 5. Socio-economic variables (education, income) by cluster; χ^2-Test ($p < 0.05$ *), SR = standardized residuals.

		Cl. 1	**Cl. 2**	**Cl. 3**	**Cl. 4**	**Total**	**χ^2-Test**
Education	primary	24.5%	30.3%	30.4%	30.1%	28.5%	$\chi^2 = 14.407$, df = 6 $p = 0.025$ *
	SR (p)	−1.3	0.4	0.5	0.5		
	secondary	22.4%	22.9%	19.0%	12.0%	18.9%	
	SR (s)	1.3	1.2	0.0	−2.5		
	university	53.1%	46.9%	50.6%	57.8%	52.6%	
	SR (u)	0.1	−1.0	−0.4	1.1		
Income	PLN < 500	14.3%	10.8%	12.3%	8.2%	11.6%	$\chi^2 = 10.118$ df = 12 $p = 0.606$
	PLN 500–1000	37.0%	33.5%	35.7%	31.0%	34.5%	
	PLN 1001–2000	29.1%	31.7%	31.3%	35.3%	31.8%	
	PLN 2001–3000	11.3%	13.2%	11.5%	15.5%	12.8%	
	PLN > 3000	8.3%	10.8%	9.3%	9.9%	9.4%	

4. Discussion

4.1. Real Living Conditions Versus Perceived Heat Experience

The cluster analysis demonstrated that objective, real living conditions and subjectively perceived conditions of green areas complimented each other, for example, in neighborhoods where residents were living in central areas characterized by closed housing blocks or high-rises with the quantity of surrounding green areas deemed as poor (cluster 1, 3) or in suburban areas with plenty of private gardens in which the amount was deemed very good (cluster 4).

This was not the case for the experience of negative impacts on health and wellbeing caused by high temperatures. Cluster 1 lives in dense housing areas with limited access to green areas and has experienced heat waves. Cluster 2 has more green areas in their surroundings and is less susceptible to suffering under high temperatures. Seebaß [37] recognized that heat stress is often related to the access to green areas, which is confirmed though cluster 1, 2, and 4's experiences with heat waves and access to green areas. Cluster 3 is an outlier in which densely built areas meet poor green supply, yet the residents can tolerate high temperatures well.

Cluster 3 has a high number of single households, and the building stock is older. They spend the least amount of time outdoors as green areas are further away. The results indicate that cluster 3's tolerance for heat is based on subjective assessments rather than objectively measurable factors. Considering the higher proportion of single households, it may be suggested that the low vulnerability to heat is due to the absence of other family or household members to consider or care for. Furthermore, this cluster could also include those who live in the city for job-related purposes and primarily only use their residence as a place to sleep—to determine this, further research would be required.

In contrast to studies from East Asia or Africa, where air conditioning plays a major role, in Poland this adaptation strategy is still uncommon. Especially for clusters such as cluster 3, who live in an older building stock and do not have greater financial means for technical aids.

H1, that residents of Polish cities experience heat stress independently from their objective living situations, is therefore confirmed. The objective built environment must be put into relation with subjective perception of neighborhood residents to understand the perceived quality of life. Both must be incorporated into NbS planning in order to implement solutions that account for both the environmental and social needs of residents.

4.2. Desires for NbS Achievements

NbSs serve many functions in urban environments, and planners must understand which functions are important not only from a functional perspective alone but also in accordance with residents' priorities and desires. When asked to rate the importance of certain goals for urban development strategies, cluster 1 clearly stood out, as presented in Section 3.2.2. As the only group who has experienced heat waves and felt their negative impact, they are experiencing climate change firsthand and therefore temperature reduction and heat wave mitigation are more pronounced goals for them than for the other clusters.

H2, that clusters with different living conditions differ in their desired goals for NbSs, is confirmed. The results have also shown that the desired effectiveness of NbS is not only linked to the built environment, but the subjective perception of environmental impacts on oneself. As cluster 1 has experienced high temperatures, these goals are important to them. For other segments of the population that do not suffer from heat, air quality improvement and storm water management are prescribed greater meaning in urban development strategies.

With vulnerabilities and an unjust distribution of green spaces in Polish cities [30,32,38], it becomes even more central to planning processes to incorporate locals' needs and incorporate NbSs that address their desires and not only tackle objectively identified issues. To increase environmental justice in Polish cities, planners must consider fair distribution of green areas alongside their functionality. Priority should thus be given to neighborhoods that currently only have limited green spaces reachable by foot. Street greening has been deemed a desirable NbS among many different urban segments [18] and would also be relatively easy to implement compared with larger-scale or more technically complex NbSs. While there are no blanket solutions, planners must ensure NbSs are accessible and meet neighborhood's individual needs to ensure site-specific desires and requirements.

4.3. Subjective Perceptions, Housing Conditions and Socio-Economic Factors

Climate change impacts affect urban residents differently and addressing these impacts in mitigation and adaptation measures are a matter of environmental justice [39]. Often socio-economically disadvantaged households are more exposed to climate change effects. Seebaß [37] recognized an impact of age on subjective heat stress and also an interaction with their individual health status. People over the age of 70 with health problems are especially vulnerable to heat. Our results did not reflect age as an influencing factor, as clusters susceptible to heat and those who were more tolerant had similar age structures. It is possible that the state of health could explain the difference in heat stress between clusters 1 and 3, both of which live in an environment with a poor supply of greenery. However, this characteristic was not surveyed in the present study and should be considered a limitation. Further studies on the topic of health variables appear necessary.

Considering the population of Poland and the entire EU is aging [40], it should be assumed that vulnerability to climate change in Polish cities will carry a growing health burden with it.

A slight gender influence could be recognized in clusters 1 (more women, less resistant to high temperatures) and cluster 2 (more men, more heat tolerant), which is in line with studies that recognize that women report more negative impacts on their health and wellbeing through heat waves and high temperatures [41].

The built environment along with traffic congestion, poor air quality, and low quality and few green areas intensify heat and cause urban heat islands [42], contributing to a higher number of negative evaluations.

Income was not a significant influence on heat perception. A connection could be made between education and perception in that those with higher levels of education and more knowledge about climate change and its impact on health and wellbeing were slightly more sensitive to heat waves. The connection between higher education levels and a greater perception of climate change impacts has also been highlighted in other publications [10,18], reflecting broader patterns in European contexts.

H3, that differences in heat perception could be explained by living conditions and socio-demographics, is rejected as neither age, income, nor the living conditions could explain heat perception. The most heat sensitive group (cluster 1) does not live in ideal conditions, and the housing conditions may be able to explain their perceptions; however, the more heat-tolerant groups (clusters 2, 3, and 4) live in very diverse housing conditions, and thus, a connection cannot be inferred across the groups. Well-rated green areas in cluster 2 and 4 may explain their tolerance levels, but socio-economic factors do not seem to influence their perceptions on heat.

5. Conclusions

Subjective factors include residents' individual needs and expectations for their neighborhood, rating green areas, heat wave experiences with objective conditions of housing and access to green areas. Cluster 1 lives in compact conditions with poor provision of green areas, while cluster 2 although also compact and central has better access to green areas and does not feel heat stress to the same degree. The heat sensitive cluster 1 also views temperature reduction as a more important aspect of urban strategies than the other clusters. Cluster 3's living conditions would lead to the presumption that they feel a strong negative impact through heat, but this is not the case. Cluster 4, in privileged peripheral areas with a large amount of private green is not facing negative impacts to the same degree as the other clusters. Although previous studies discussed socio-economic injustices in Polish cities as influencing factors, these were not evident in the clusters of this study. To increase the social benefits of NbSs, planners and practitioners must recognize that it is not merely the size of green areas but moreover the distribution that is important.

Urban green infrastructure are vital components of urban quality of life and climate change adaptation and mitigation strategies put forward by cities and desired by residents, especially considering expected temperature rises and the implication of the European Union's nature restoration law, which calls for rapid transformation and the use of NbSs such as street greening and multi-functional public green spaces. The cluster analysis made clear that access to green areas differs significantly among residents of Polish cities. This depends on both the individual living conditions in one's neighborhood and the distribution of green areas and NbSs throughout the city. Improving the accessibility to NbSs will increase residents' quality of life and help tackle climate change impacts currently affecting health and wellbeing. To achieve this, however, NbSs must be planned to account for local needs and neighborhood characteristics to assist in achieving both social and environmental justice throughout the city.

Author Contributions: Conceptualization, A.W. (Anita Walter); methodology, A.W. (Anita Walter); formal analysis, A.W. (Anita Walter); data curation, A.W. (Alice Wanner); writing—original draft preparation, A.W. (Anita Walter) and A.W. (Alice Wanner); writing—review and editing, A.W. (Alice Wanner) and U.P.-H.; visualization, A.W. (Anita Walter); supervision, A.W. (Alice Wanner) and U.P.-H.; project administration, A.W. (Alice Wanner); funding acquisition, U.P.-H. All authors have read and agreed to the published version of the manuscript.

Funding: This research was funded by UPSURGE, which received funding from the European Union's Horizon 2020 research and innovation program under grant agreement no. 101003818.

Data Availability Statement: Data available upon request.

Acknowledgments: The cluster analysis of this study was part of first author Anita Walter's master's thesis submitted to BOKU University in June 2025.

Conflicts of Interest: The authors declare no conflicts of interest.

Abbreviations

The following abbreviations are used in this manuscript:

EU	European Union
NbS	Nature-based solution

References

1. Berland, A.; Shiflett, S.A.; Shuster, W.D.; Garmestani, A.S.; Goddard, H.C.; Herrmann, D.L.; Hopton, M.E. The role of trees in urban stormwater management. *Landsc. Urban Plan.* **2017**, *162*, 167–177. [CrossRef]
2. Eisenman, T.S.; Churkina, G.; Jariwals, S.P.; Kumar, P.; Lovasi, G.S.; Pataki, D.E.; Weinberger, K.R.; Whitlow, T.H. Urban trees, air quality, and asthma: An interdisciplinary review. *Landsc. Urban Plan.* **2019**, *187*, 47–49. [CrossRef]
3. Kavehei, E.; Jenkins, G.A.; Adame, M.F.; Lemckert, C. Carbon sequestration potential for mitigating the carbon footprint of green stormwater infrastructure. *Renew. Sustain. Energy Rev.* **2018**, *94*, 47–59. [CrossRef]
4. Meo, S.A.; Abukhalaf, A.A.; Alomar, A.A.; Alessa, O.M.; Sami, W.; Klonoff, S.C. Effect of environmental pollutants PM-2.5, carbon monoxide, and ozone on the incidence and mortality of SARS-COV-2 infection in ten wildfire affected counties in California. *Sci. Total Environ.* **2021**, *757*, 143948. [CrossRef] [PubMed]
5. Taylor, L.; Hochuli, D.F. Defining greenspace: Multiple uses across multiple disciplines. *Landsc. Urban Plan.* **2017**, *158*, 25–38. [CrossRef]

6. Wanner, A.; Pröbstl-Haider, U.; Feilhammer, M. New green infrastructure european cities: Multiple ways for improvement and climate change adaptation. In *The Sustianable City XVI*; WIT Press: Southampton, UK, 2022; pp. 121–133.
7. Wilson, E.O. *Biophilia*; Harvard University Press: Cambridge, MA, USA, 1986.
8. Onur, M.; Acar, C. Circular Cities Solution with Biophilic Design and Nature-Based Solutions. In *Nature-Based Solutions for Circular Management of Urban Water*; Springer International Publishing: Cham, Switzerland, 2024; pp. 1–11.
9. Beatley, T. *Handbook of Biophilic City Planning & Design*; Island Press: Washington, DC, USA, 2016.
10. Wanner, A.; Pröbstl-Haider, U.; Jungnickel, M. Leveraging nature for just cities: Planning nature-based solutions in Greece. *Urban For. Urban Green.* **2025**, *106*, 128731. [CrossRef]
11. Seebauer, S.; Firesenecker, M.; Thaler, T.; Schneider, A.; Schwarzinger, S. Feeling hot is being hot? Comparing the mapping and the surveying paradigm for urban heat vulnerability in Vienna. *Sci. Total Environ.* **2024**, *945*, 173952. [CrossRef]
12. Bauer, W. Reframing urban nature-based solutions through perspectives of environmental justice and privilege. *Urban Plan.* **2023**, *8*, 334–345. [CrossRef]
13. Kabisch, N.; Frantzeskaki, N.; Pauleit, S.; Naumann, S.; Davis, M.; Artmann, M.; Haase, G.; Knapp, S.; Korn, H.; Stadler, J.; et al. Nature-based solutions to climate change mitigation and adaptation in urban areas: Perspectives on indicators, knowledge gaps, barriers, and opportunities for action. *Ecol. Soc.* **2016**, *21*, 39. [CrossRef]
14. Dorst, H.; Jagt, A.v.d.; Toxopeus, H.; Tozer, L.; Raven, R.; Runhaar, H. What's behind the barriers? Uncovering structural conditions working against urban nature-based solutions. *Landsc. Urban Plan.* **2022**, *220*, 104335. [CrossRef]
15. Frantzeskaki, N.; Borgström, S.; Gorissen, L.; Engermann, M.; Ehnert, F. Nature-Based Solutions Accelerating Urban Sustainability Transition in Cities: Lessonst from Dresden, Genk and Stockholm Cities. In *Ature-Based Solutions to Climate Change Adaptation in Urban Areas. Theory and Practice of Urban Sustainability Transitions*; Kabisch, N., Korn, H., Stadler, J., Bonn, A., Eds.; Springer: Cham, Switzerland, 2017.
16. Böltken, F. Wohnquartiersbeschreibung durch Befragte—Eine Anwendung. In *Regionale Standards, GESIS-Schriftenreihe*; GESIS Leibniz-Institut für Sozialwissenschaften, A.R.S.: Mannheim, Germany, 2013; Volume 12, pp. 177–181.
17. Bai, C.; Kemper, R.; Landwehr, M.; Liembd, U.; Roggo, N. *Wohnumfeldqualität: Kriterien und Handlungsansätze für die Planung*; Hochschule für Technik Rapperswil (HSR)—Kompetenzzentrum Wohnumfeld, Shaker: Aachen, Germany, 2018.
18. Pröbstl-Haider, U.; Wanner, A.; Feilhammer, M.; Mostegl, N.; Dabrowska, K. The right fit: Acceptance of nature-based solutions across European Cities. *Landsc. Urban Plan.* **2024**, *252*, 105189. [CrossRef]
19. Pröbstl-Haider, U.; Feilhammer, M.; Wanner, A. The influence of payments and time payments: Preconditions to implement nature-based solutions in Austrian cities. *Cities* **2025**, *162*, 105990. [CrossRef]
20. Grütter, J.K. *Wohnraum Planen: Architektur—Psychologie—Sozial—Gesellschaft—Kultur*; Springer: Berlin/Heidelberg, Germany, 2021.
21. Rizwan, A.M.; Dennis, L.Y.C.; Liu, C. A review on the generation, determination and mitigation of Urban Heat Island. *J. Environ. Sci.* **2008**, *20*, 120–128. [CrossRef]
22. Jay, O.; Capon, A.; Berry, P.; Broderick, C.; Dear, R.d.; Havenith, G.; Honda, Y.; Kovats, R.; Ma, W.; Malik, A.; et al. Reducing the health effects of hot weather and heat extremes: From personal cooling strategies to green cities. *Lancet* **2021**, *398*, 709–724. [CrossRef] [PubMed]
23. Georgie, J.N.; Dimitriou, D. The contribution of urban green spaces to the improvement of environment in cities: Case study of Chania, Greece. *Build. Environ.* **2010**, *45*, 1401–1414. [CrossRef]
24. Kabisch, N.; Frantzeskaki, N.; Hansen, R. Principles for urban nature-based solutions. *Ambio* **2022**, *51*, 1388–1401. [CrossRef] [PubMed]
25. Kato-Huerta, J.; Geneletti, D. Environmental justice implications of nature-based solutions in urban areas: A systematic review of approaches, indicators, and outcomes. *Environ. Sci. Policy* **2022**, *138*, 122–133. [CrossRef]
26. Olsson, J.A.; Brunner, J.; Nordin, A.; Hanson, H.I. A just urban ecosystem service governance at the neighbourhood level—Perspectives from Sofielund, Malmö, Sweden. *Environ. Sci. Policy* **2020**, *112*, 305–313. [CrossRef]
27. Cousins, J. Justice in nature-based solutions: Research and pathways. *Ecol. Econ.* **2021**, *180*, 106874. [CrossRef]
28. Kopycka, K. *Die Soziodemographische Struktur der Bevölkerung in Polen*; Budenszentrale für politische Bildung: Bonn, Germany, 2023; Available online: https://www.bpb.de/themen/europa/polen/541516/die-soziodemografische-struktur-der-bevoelkerung-in-polen/ (accessed on 11 September 2025).
29. Brinkhoff, T. Polen: Größere Städte. 2024. Available online: https://www.citypopulation.de/de/poland/cities (accessed on 11 September 2025).
30. Szczepański, M.; Śliz, A. Analyse: Wohnraum und Wohnungspolitik in Polen. *Pol. Anal.* **2023**, *304*, 2–5.
31. Wysmułek, J.; Hełdak, M.; Kucher, A. The analysis of green areas' accessibility in comparison with statistical data in Poland. *Int. J. Environ. Res. Public Health* **2020**, *17*, 4492. [CrossRef] [PubMed]
32. Suligowski, R.; Ciupa, T.; Cudny, W. Quantity assessment of urban green, blue, and grey spaces in Poland. *Urban For. Urban Green.* **2021**, *64*, 127276. [CrossRef]

33. Statistical Office of the European Union. *Eurostat Housing in Europe—2023 Edition*; Statistical Office of the European Union: Luxembourg, 2025.
34. UPSURGE. The Project. Available online: https://www.upsurge-project.eu/the-project/ (accessed on 24 July 2025).
35. Bortz, J. *Statistik für Sozialwissenschaftler*, 4th ed.; Springer: Berlin/Heidelberg, Germany, 1993. [CrossRef]
36. Rudolf, M.; Müller, J. *Multivariate Verfahren: Eine praxisorientierte Einführung mit Anwendungsbeispielen in SPSS*; Hogrefe: Göttingen, Germany, 2003.
37. Seebaß, K. Who is feeling the heat? Vulnerabilities and exposures to heat stress—Individual, social and housing explanations. *Nat. Cult.* **2017**, *12*, 137–161. [CrossRef]
38. Szafrańska, E.; de Lille, L.C.; Kazimierczak, J. Urban shrinkage and housing in a post-socialist city: Relationship between the demographic evolution and housing development in Łódź, Poland. *J. Hous. Built Environ.* **2019**, *34*, 441–464. [CrossRef]
39. Bolte, G.; Dandolo, L.; Gepp, S.; Hornberg, C.; Lumbi, S.L. Klimawandel und gesundheitliche Chancengereichtigkeit: Eine Public-Health-Perspective auf Klimagerechtigkeit. *J. Health Monit.* **2023**, *8*, 3–38. [CrossRef]
40. Abramowska-Kmon, A. Demographische Entwicklungen in Polen: Konsequenzen und Herausforderungen für die Sozialpolitik. *Pol. Anal.* **2023**, *308*, 2–6. [CrossRef]
41. Berg, U.; Iblher, T. Frauen gesundheitlich vom Klimawandel stärker betroffen als Männer. ärztin-1/2022. Des Deutscher Ärztinnenbundes e.V. 2022. Available online: https://www.aerztinnenbund.de/Frauen_gesundheitlich_vom_Klimawandel.3715.0.2.html (accessed on 11 September 2025).
42. Kunkel, A. *Maßnahmen gegen den Urban Heat Island Effect in Städten als spürbare Folge des Klimawandels—Empirische Betrachtung anhand der Stadt Aschaffenburg*; IREC Practice Paper: Mainz, Germany, 2023.

MDPI AG
Grosspeteranlage 5
4052 Basel
Switzerland
Tel.: +41 61 683 77 34

Land Editorial Office
E-mail: land@mdpi.com
www.mdpi.com/journal/land

www.ingramcontent.com/pod-product-compliance
Lightning Source LLC
Chambersburg PA
CBHW040747250926
51850CB00007BB/1428
* 9 7 8 3 7 2 5 8 7 2 8 8 6 *